This book belongs to

ADVANCES IN LIPID METHODOLOGY - TWO

edited by

William W. Christie

The Scottish Crop Research Institute, Invergowrie, Dundee (DD2 5DA), Scotland.

THE OILY PRESS

DUNDEE

ISBN 0 9514171 3 4

British Library Cataloguing-in-Publication Data. A catalogue record for this book is available from the British Library.

This is - Volume 4 in the Oily Press Lipid Library

(Volume 1 - "Gas Chromatography and Lipids" by William W. Christie; Volume 2 - "Advances in Lipid Methodology - One" edited by William W. Christie; Volume 3 - "A Lipid Glossary" by Frank D. Gunstone and Bengt G. Herslöf.)

Printed in Great Britain by Bell and Bain Ltd., Glasgow

PREFACE

This is the second volume of an occasional series of review volumes dealing with aspects of lipid methodology to be published by the Oily Press. As with the first volume, topics have been selected that have been developing rapidly in recent years. For example, Frank Gunstone has provided what must be the definitive essay on ^{13}C NMR spectroscopy of lipids, an enormously powerful technique for structural analysis of fatty acids and now for regiospecific analysis of glycerolipids. I have provided two chapters. The first deals critically with the simple but much misunderstood methods for preparation of ester (and amide) derivatives for chromatographic analysis. It is to be hoped that it will stem the needless flow of publications of "new" methods. My second topic is the humdrum but vitally important one of methods for extraction of lipids from tissue matrices. Carmen Dobarganes and her colleague discuss size exclusion high-performance liquid chromatography of lipids, including both lipid polymers with organic solvents and lipoprotein complexes with aqueous mobile phases. No substantial review on this subject has hitherto appeared in the literature. The same is true of Jean Louis Sebedio's article on the use of mercury adducts for chromatographic separation of lipids. Although this methodology is not new, it is much neglected and deserves wider recognition.

Professor Gerd Schmitz and his colleagues are certainly at the forefront with an exciting new technique for lipoprotein separation, *i.e.* capillary isotachophoresis, and this review must awaken interest in its enormous potential in all readers. Jean-Luc Le Quéré here provides an introduction to tandem mass spectrometry and its application to structural analysis of lipids. This has certainly opened my eyes to the possibilities now available both for structural analysis of fatty acid derivatives and for intact glycerolipids. This technique is proving invaluable for the analysis of acylcarnitines in biological samples, and Malcolm Rose and his colleagues have provided a further definitive review of this and the chromatographic methodologies for what was until recently an extremely difficult analytical problem.

As an appendix, I have prepared literature searches on lipid methodology for the years 1991 and 1992, continuing a feature established in the first volume.

The objective of the Oily Press is to provide compact readable texts on all aspects of lipid chemistry and biochemistry, and many more books are in the pipe-line for The Oily Press Lipid Library. If you have suggestions or comments, please let us know. By a careful choice of authors and topics, I trust that this volume will again prove to have met all our aims. This year we have moved to Dundee (in harmony with my own move to the Scottish Crop Research Institute in Invergowrie, Dundee, Scotland). This has had no effect on the service Norma and I provide to our customers or our authors, and we hope to maintain our good relations with all on our new site.

My own contributions to the book are published as part of a programme funded by the Scottish Office Agriculture and Fisheries Dept.

William W. Christie

CONTENTS

Appendix

Chapter 1

HIGH RESOLUTION ^{13}C NMR SPECTROSCOPY OF LIPIDS

Frank D. Gunstone

Chemistry Department, The University, St Andrews, Fife, (KY16 9ST), Scotland

A. INTRODUCTION

Until recently nuclear magnetic resonance (nmr) spectroscopy found only limited use in the study of lipids. This situation probably arose from a lack of basic information and from the cost of high-resolution instruments. The only exception has been the use of low-resolution ^{1}H nmr spectroscopy to measure solid/liquid ratios with cheaper instruments. This is now the standard method of choice for making this important measurement but is not included in this review [46]. Low resolution pulsed nmr is also used for the determination of oil and moisture in oil seeds [44].

High-resolution ^{1}H nmr spectroscopy has been used only in a limited manner because of the small number of signals in a ^{1}H spectrum which limits the amount of information which can be derived from it.

High-resolution ^{13}C nmr spectroscopy is becoming more significant and it is hoped that this review will assist and further this development. Instruments are now commonly 300 MHz or better and, despite their cost, are becoming more widely available.

This review will not cover imaging techniques or the underlying theory of nmr spectroscopy but further information on these is available in an article by Simoneau *et al.* [110] and in books devoted to nmr spectroscopy.

A ^{13}C nmr spectrum provides two kinds of information: (i) the chemical shift of each signal (usually 50-100 in a natural mixture) and (ii) their relative intensities. The former is of qualitative value and permits identification of important structural features, the latter, with approved safeguards, provides quantitative information of analytical value. Chemical shifts may vary slightly with concentration of the solution under study and (rather more) with the solvent. Most measurements are made with solutions of about 1M concentration and $CDCl_3$ is the solvent most commonly employed. If a solvent is not indicated in the following discussion it is probably $CDCl_3$. Other solvents include $CD_3OD/CDCl_3$ mixtures, $(CD_3)_2SO$, D_2O and C_6D_6. Figures 1.1 and 1.2 show typical spectra of vegetable oils. These are discussed in Section C2.

The chemical shift of a carbon atom depends on its total environment to a distance of six or more atomic centres. For example, in glycerol trioleate the signals for the olefinic carbon atoms (C_9 and C_{10}) differ from one another and to a further small extent depending on whether the oleate is an α or β chain (attached to primary or secondary glycerol hydroxyl group). Also the C1 signal is different for saturated and $\Delta 9$ chains. In these examples the difference is produced by structural changes up to 11 atomic centres away. This makes the spectrum more complex but also more informative when all the chemical shifts have been assigned. As another example, consider the methyl groups at the end of each acyl chain in glycerol tripalmitate. These give one signal at about 14.1 ppm which is

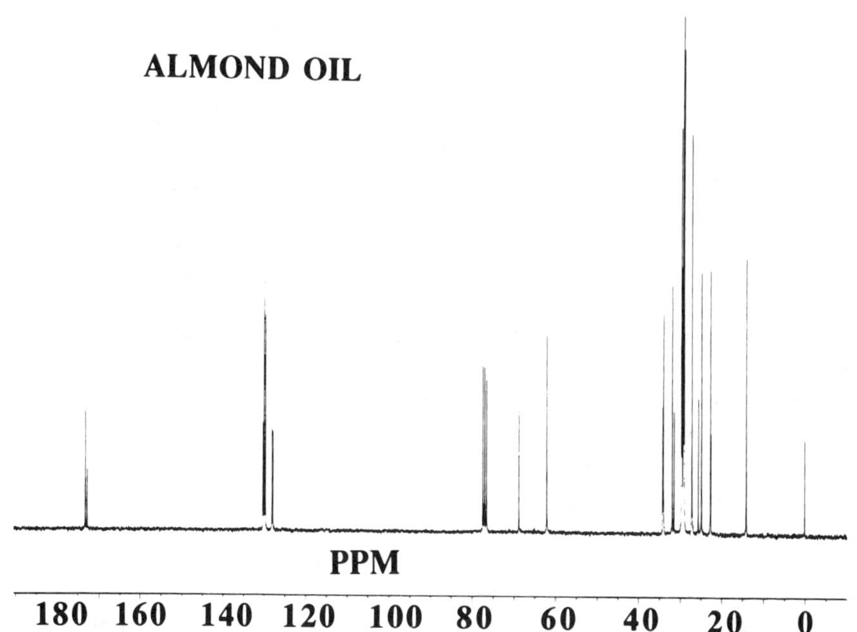

Fig 1.1. High resolution ^{13}C nmr spectrum of almond oil. Signals at approx. 173 (C1), 130 (olefinic), 77 (solvent), 69 and 62 (glycerol), 34 (C2), 32 (ω3), 29-30 (methylene envelope), 27 and 25.6 (allylic), 25 (C3), 23 (ω2) and 14 ppm (ω1).

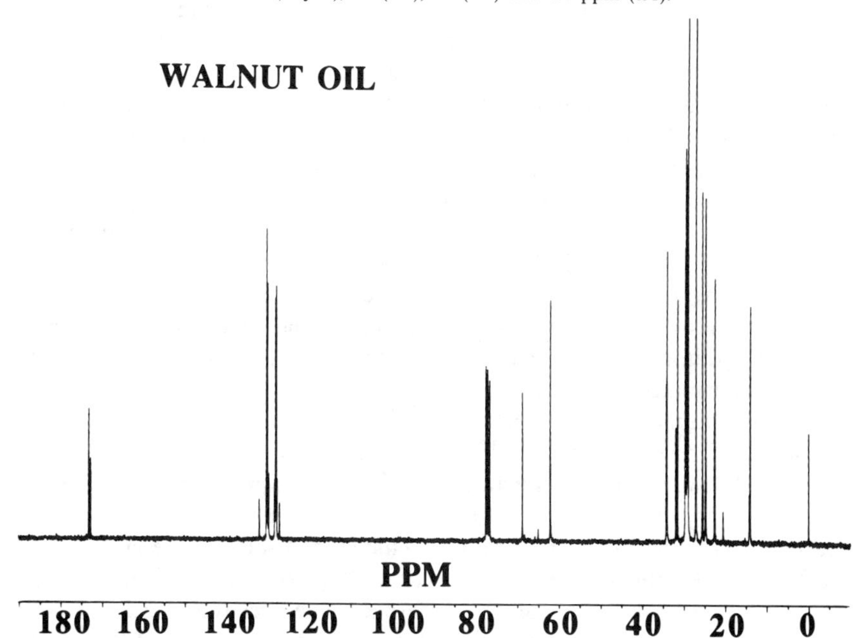

Fig. 1.2. High resolution ^{13}C nmr spectrum of walnut oil. Signals at approx. 173 (C1), 130 (olefinic), 77 (solvent), 69 and 62 (glycerol), 34 (C2), 32 (ω3), 29-30 (methylene envelope), 27 and 25.6 (allylic), 25 (C3), 23 (ω2) and 14 ppm (ω1).

well separated from other signals and hence easily recognised. The difference between α and β chains in this molecule is too small to be observed but in a vegetable oil, containing saturated and unsaturated chains, the 14.1 resonance may appear as two or more signals (a cluster). Each signal is indicative of a different environment and may result from *n*-3 or *n*-6 acids where the double bonds affect the methyl signal.

To obtain quantitative data attention has to be given to the protocol for obtaining the spectrum. In particular, the problem of relaxation has to be overcome either by adding a relaxation agent such as $Cr(acac)_3$ (chromium acetonylacetonoate) and/or by including a delay time between successive scans of the spectrum. This will add to the time required to collect the spectrum. Spectrometers now available operate at a frequency for ^{13}C of 68MHz or more and spectra are generally obtained using an NOE suppressed, inverse-gated, proton-decoupled technique. Exciting pulses have a 45-90° pulse angle and acquisition times (including delay times) are 1-20 sec per scan. The number of scans is usually 1000 or more. The sample size for a routine ^{13}C nmr spectrum is 50-100 mg but with a suitable investment in a large number of pulses high quality spectra can be obtained with as little as 1mg.

In using ^{13}C nmr data (chemical shifts and intensities) the first step is to assign as many of the chemical shifts as possible, and to assist in this Section B of this review is devoted to the spectra of individual molecular species. If the substance under study is a mixture then many individual signals will be replaced by a cluster. This makes interpretation more difficult but eventually provides additional information. It is wise to start with the signals of greatest intensity but to ignore those in the methylene envelope which result from mid-chain carbon atoms in the region 29.4-29.9 ppm and are not greatly influenced by nearby functional groups. Instead, examine the shifts associated with the following carbon atoms: ω1 (14.1 ppm), ω2 (22.8), ω3 (32.1), C1 (174.1), C2 (34.2), C3 (25.1), glycerol (68.9 and 62.1), olefinic (127-132) and allylic (27.3 and 25.6).

In collecting together new and existing chemical shift data here it has been possible to correct erroneous assignments and to make assignments where the authors merely listed chemical shifts. Following the description of individual molecular species in Section B, Section C is devoted to the spectra of synthetic and natural mixtures.

Some earlier reviews on the ^{13}C nmr spectra of fatty acids and lipids are available [35,97] but much of this work is fairly new and is not covered so completely in previous reviews. This article contains many chemical shifts and most of these are included in Tables at the end of the review to make the text easier to follow. Readers are recommended to make a copy of the tables for easy reference when reading the text. It is hoped that the collection of all this information in one place will be useful for those using nmr spectroscopy. Nevertheless not all the known data can be presented and some readers will have to go to original papers for the information they seek. Some unpublished results from the writer's studies have also been included.

Assignments of chemical shift are now often made on the basis of available knowledge such as that collated in this review. Where the necessary information is not available some advanced spectroscopic procedures can be employed. Existing information has been built up slowly over the past 20 years assisted by the study of ^2H-containing compounds and the use of chemical shift reagents.

A long-chain compound containing a CD_2 group in place of a CH_2 group will give a ^{13}C nmr spectrum in which there is no signal for the CD_2 group and with adjacent signals 2-3 times broader than usual and with a slight change of shift. Tulloch [124] has suggested that groups α and β to CD_2 change their shift values by -0.20 to -0.25 and -0.04 to -0.07 respectively. Examples of the use of this technique include the work of Tulloch *et al.* [126], Rakoff [99], Bengsch *et al.* [15] and Baenziger *et al.* [11].

Two examples of the use of chemical shift reagents such as $Yb(fod)_3$ (tris (6,6,7,7,8,8,8-heptafluoro-2,2-dimethyl-3,5-octanedionato)ytterbium) are reported by Barton *et al.* [12] and by Lunazzi *et al.* [83].

B. STUDY OF INDIVIDUAL MOLECULAR SPECIES

1. Alkanoic acids

The ^{13}C nmr spectra of long-chain saturated acids show six easily-recognised signals at 180.6, 34.2, 24.8, 32.1, 22.8 and 14.1 ppm for the C1-3 and ω3-1 carbon atoms respectively. The remaining signals, usually close together in the region 29.3-29.8 ppm, are described as the *methylene envelope*. Some of these can be identified in single compounds as can be seen, for example, in Table 1.1.

In an interesting paper concerned with medium and long-chain saturated acids Bengsch *et al.* [15] placed the chemical shifts for each chain length in decreasing order. For the C_{14}-C_{18} acids detailed in Table 1.1 they gave the sequence shown below. This sequence is often useful in making chemical shift assignments in more complex acids.

18:0	2	ω3	12	ω6	11	10	9	ω5	8	7	6	ω4	5	4	3	ω2	ω1
16:0	2	ω3	–	–	11	10	9	ω5	8	7	6	ω4	5	4	3	ω2	ω1
14:0	2	ω3	–	–	–	–	9	ω5	8	7	6	ω4	5	4	3	ω2	ω1

Acids and methyl esters give slightly different shifts for carbon atoms close to the acyl group, especially for C1 and C3 with smaller differences for C2 and C4-6 (see section B12). Glycerol esters give two signals for each of the first 6 or 7 carbon atoms but only those for C1 and C2 are observed in most spectra. These have relative intensities of 2:1 and correspond to the α and β chains respectively. Carbon atoms in the α-chain have the higher shift for C1 and C4 and the lower shift for other carbon atoms. Although the chemical shifts of the atoms in the α and β chains vary a little from spectrum to spectrum their differences are fairly constant at values around 0.41, 0.17 and 0.04 ppm for the C1-3 atoms. In saturated chains the ω1-3 signals do not change much between acids and esters or between the α and β chains. They may be different in unsaturated acids (see Sections B3 and B4) and in short- and medium- chain acids (see Tables 1.30-1.32). The characteristic shifts for short and medium-chains are significant in the spectra of butter fat and of lauric oils (see Section C1).

Several authors have given values for the effect that the end groups have on the chemical shifts of nearby carbon atoms:

	Gunstone *et al.* [56]					Bus *et al.* [27]				
	α	β	γ	δ	ε	α	β	γ	δ	ε
COOH	+4.43	-5.00	-0.58	-0.42	-0.24	-	-	-	-	-
$COOCH_3$	+4.43	-4.73	-0.48	-0.39	-0.20	+4.40	-4.75	-0.50	-0.45	-0.20
CH_3	-7.01	+2.26	-0.30	-	-	-7.00	+2.15	-0.35	-0.10	

	Johns *et al.* [60]				
	α	β	γ	δ	ε
COOH	+4.4	-5.0	-0.6	-0.45	-0.25
CH_3	-7.0	+2.2	-0.35	0	+0.1

(Gunstone: 29.80 acids, 29.84 esters; Bus 29.75 for methyl esters; Johns: 29.8 for acids)

2. *Branched chain acids*

Klein [62] reported the chemical shifts (ppm, $CDCl_3$ solution, based on CH_3O signal at 51.42) for the dimethyl ester of 11,12-dimethyleicosanedioic acid which contains the following structural unit:

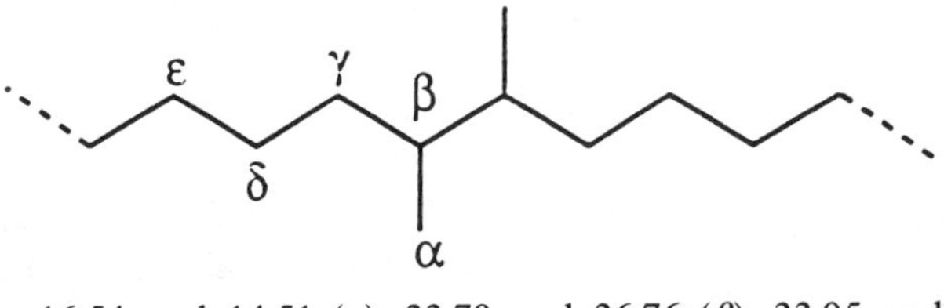

These were 16.54 and 14.51 (α), 33.70 and 36.76 (β), 33.05 and 35.06 (γ), 27.86 and 27.86 (δ), and 30.08 and 30.08 (ε) for the meso and racemic esters respectively.

In a fuller paper Gunstone [52] listed chemical shifts for 28 synthetic branched-chain acids/esters. These ranged from C_6 to C_{18} in chain length with branches of C_1 (mainly), C_2-C_6 and C_{16}. On the basis of these results the spectra of wool grease fatty acids (Table 1.2) and technical isostearic acid were interpreted.

The chemical shifts for a mid-chain methyl group and for *iso* and *anteiso* acids/esters are as follows:

19.7-19.8
30.1 37.2 CH_3
—CH_2 CH_2 CH_2 CH—
27.1 32.9

mid-chain

22.7
22.7 CH_3 39.1
CH_3 CH CH_2 CH_2—
28.0 27.4

iso

19.3
11.4 CH_3 36.7 30.1
CH_3 CH_2 CH CH_2 CH_2 CH_2—
34.5 27.2

anteiso

3. Monoenoic acids

Bus *et al.* [27,28] and Gunstone *et al.* [57] have provided useful information on the chemical shifts of *cis* and *trans* monoene acids and their methyl esters. Lie Ken Jie. [65] has now added the glycerol triesters of the series of *cis* and *trans* acids $CH_3(CH_2)_7CH{=}CH(CH_2)_nCOOH$ with n = 0-9, *i.e.* a series of $\Delta 2$ (C_{11}) to $\Delta 11$ (C_{20}) esters. Table 1.3 contains a list of monoene acids/esters reported in these three major studies. Other papers with results on one or two monoene acids/esters have not been included though some are covered in other parts of this review.

The major points of significance for the chemical shifts are: (i) The effect of end groups (acyl and methyl) on olefinic and allylic signals with respect to configuration, position in the acyl chain, and presence in α or β chains in glycerol esters. (ii) The effect of *cis* and *trans* double bonds in various positions on the easily recognisable C1-3 and ω1-3 signals.

These observations become significant in other topics discussed later in this review such as acids/esters with double bonds in less common positions (*e.g.* $\Delta 6$ or *n*-7) and in partially-hydrogenated oils containing many different configurational and positional 18:1 isomers.

Monoene acids with unsaturation which is $\Delta 2$ - $\Delta 11$ or $\omega 2$ - $\omega 4$ have different signals for the two olefinic carbon atoms. These two chemical shifts and, more significantly, the numerical difference between them provide a valuable insight into the position of the double bond. Differences in chemical shift reported by three research groups [27,28,57,65] are summarised in Table 1.4. These values do not differ much between *cis* and *trans* isomers but give information about double bond position. The difference is large for Δ2 alkenes and falls off as the double bond moves from the acyl group. By the Δ12 position it is not always possible to see two signals and there is then only one olefinic signal until the unsaturated carbon atoms come under the influence of the end methyl group (ω4). This means that in the C_{18} chain only the Δ12 and Δ13 isomers give a single olefinic signal (at about 129.9 (*cis*) and 130.4 (*trans*)).

With glycerol esters the olefinic signals may be different in the α and β chains. The difference values are set out in Table 1.5 and are to be interpreted as in the following figures for glycerol trioleate and glycerol trielaidate from Lie Ken Jie [65]:

	α9	β9	(diff)	α10	β10	(diff)
oleic glyceride	129.719	129.693	(+0.026)	130.014	130.029	(-0.015)
elaidic glyceride	130.196	130.173	(+0.023)	130.502	130.517	(-0.015)

These figures show the small differences in the chemical shift in α and β chains. The differences become harder to observe, especially in complex mixtures, as the value gets smaller. Thus for oleate and elaidate it is easier to get two signals for C9 than for C10. Unexpectedly the differences do not change regularly as the double bond moves from the acyl group (Table 1.5) and this explains an earlier observation that it was possible to see double signals with oleic glycerides (Δ9, difference 0.026 and 0.015) but not with petroselinic glycerides (Δ6, difference 0.000 and 0.009).

The selection of values cited in Tables 1.4 and 1.5 do not show the differences in chemical shift between *cis* and *trans* isomers. These are significant and useful and are discussed later (Section C5 on partially-hydrogenated oils).

Sometimes allylic shifts can also be used to determine double bond position:

$$\underset{}{CH_3(CH_2)_n}\overset{\alpha'}{CH_2}\,CH{=}CH\overset{\alpha}{CH_2}(CH_2)_mCOOR$$

The olefinic acids or esters show one or two allylic signals (generally α′ > α) and these are different for *cis* (27.2-27.3) and *trans* isomers (32.6-32.7). From the data collected in Table 1.6 it is apparent that there are distinct allylic signals for the Δ3-Δ7 (α) isomers and for the ω2-ω5 (α′) isomers in both the *cis* and *trans* series.

The effect of *cis* and *trans* olefinic centres on nearby carbon atoms is given by the following numbers:

	α	β	γ	δ	ε
cis isomers					
Bus *et al.* [27]	-2.45	+0.05	-0.35	-0.10	-
Gunstone *et al.* [57]	-2.50	-0.05	-0.45	-0.20	-0.10
Johns *et al.* [60]	-2.50	0.00	-0.40	-0.20	0.00
trans isomers					
Bus *et al.* [27]	+2.90	-0.05	-0.50	-0.10	-
Gunstone *et al.* [57]	+2.85	-0.10	-0.55	-0.20	-0.10
Johns *et al.* [60]	+2.80	-0.10	-0.50	-0.20	0.00

These figures show the marked effect of the double bond on allylic (α) carbon atoms and the difference between *cis* and *trans* systems. There is also a minor but observable effect on the γ-carbon atoms and between the *cis* and *trans* isomers at this carbon atom. These observations are exploited in the study of hydrogenated fats.

Changes in the α and γ carbon atoms are easily seen in the C1-C3 and ω1-ω3 carbon atoms and it is clear from Table 1.7 that it should be possible to observe one or both of the α and γ effects in the Δ4-Δ6 and ω3-ω6 (*i.e.* Δ15-Δ12) C_{18} monoenes.

4. Polyenoic acids

Most natural polyene esters have methylene-interupted *cis* olefinic unsaturation and are grouped in families depending on the position of the first double bond with respect to the methyl end group. The best known are the *n*-3 (18:3, 18:4, 20:4, 20:5, 22:5 and 22:6) and *n*-6 families (18:2, 18:3, 20:3, 20:4, 22:4 and 22:5). Chemical shifts (from various laboratories) are summarised in Tables 1.8 (*n*-3 acids/esters), 1.9 (*n*-6 acids/esters), 1.10 (Δ6 esters) and 1.11 (*cis*/*trans* 18:2 stereoisomers). Unfortunately it has not been possible to list only acids or only methyl esters. Information on glycerol esters is limited to studies of natural mixtures (see later sections) and to the glycerol esters of docosahexaenoic and arachidonic acid [5] and of linoleic and/or linolenic acids [9].

The Tables are laid out to show that in the *n*-3 compounds the carbon atoms ω1-5 have similar chemical shifts independent of total unsaturation and of chain length (in the range 18-24). The same holds for the shifts of carbon atoms ω1-8 in *n*-6 compounds and of carbon atoms Δ1-8 in Δ6 esters. This suggests that in polyene acids/esters with methylene-interrupted *cis* double bonds it is possible to identify the position of the outer double bonds with respect to the appropriate end groups. (Tables 1.8-1.10)

Mono- and diallylic carbon atoms usually have values around 27.3 and 25.65 ppm but these change if one double bond is close to an end group (see previous section on monoene acids). It is thus possible to distinguish *n*-3 and Δ4, 5 and 6 acyl chains: *n*-3 from ω2 and ω5 signals, Δ, 5 and 6 from C3 and so forth.

Bus *et al.* [27,28] and Gunstone *et al.* [57] have described ways of calculating the olefinic chemical shifts in all-*cis* methylene-interrupted polyenes and calculated values have been used in making some of the assignments listed in the Tables. Most of the olefinic chemical shifts are between 128 and 129 ppm with only a few signals lying outwith this range. The unpublished results of Sacchi agree reasonably well with those of Aursand *et al.* [5] for EPA (Table 1.8), but agreement is poorer for the DHA signals and further studies of this are clearly needed. In all the acids/esters listed in Tables 1.8 and 1.9 the following pattern is observed.

	n-6	*n*-3
lowest shift	$\omega 7$	$\omega 4$
highest shift	$\omega 6$	$\omega 3$
second highest shift	olefinic carbon closest to the acyl function	

The chemical shifts observed in all the *cis/trans* isomers of methyl linoleate are listed in Table 1.11. As already indicated in the discussion on monoene compounds a change from *cis* to *trans* configuration has a considerable effect on the shifts of olefinic and allylic carbon atoms.

Rakoff [99,100] has reported the chemical shifts of some deuterated polyenes (18:3 and 18:4). Moine *et al.* [89] have also made a thorough study of 18:4, and the spectra of EPA isomers containing partially conjugated systems have been reported (Burgess *et al.* [25], 5*c*8*c*10*t*12*t*14*c* and Lopez *et al.* [82], 5*c* (5*t*)7*t*9*t*14*c*17*c*)

5. *Conjugated unsaturated acids*

Information is available on acids or esters containing conjugated diene, triene and tetraene systems. The trienes are most common and have therefore been most extensively studied. These are generally C_{18} compounds with unsaturation at Δ9,11,13 or Δ8,10,12 and with several *cis, trans* isomers. Chemical shifts for the olefinic carbon atoms and for some of the other carbon atoms are given in Table 1.12. It is easy to decide whether the outer double bonds in a polyene system are *cis* or *trans* from the chemical shift of the allylic carbon atoms. The 9,11,13-trienes are also easily distinguished from isomeric 8,10,12-trienes by the influence of the triene system on the $\omega 1$-$\omega 4$ carbon atoms [16,45,123,125].

The analysis of seed oils containing these conjugated acids by gas chromatography presents some difficulties and attempts have been made to use the intensities of olefinic carbon atoms in ^{13}C nmr spectra to derive this information [45,125].

Some 9,11- and 10,12-C_{18} dienes have been reported [27,28] and data are given in Table 1.13.

Among less common conjugated systems are those in bosseopentaenoic acid (20:5 5*c*8*c*10*t*12*t*14*t*, [25]) and a pair of 20:5 acids with conjugated trienes (5*c*7*t*9*t*14*c*17*c* and its 5*t* isomer, [82]).

The keto ester which also contains a 9,11,13- triene system (licanate) is discussed along with other keto esters (Table 1.18).

6. Acetylenic and allenic acids

Significant reports on the ^{13}C nmr spectra of acetylenic acids and esters have been made by Gunstone *et al.* [56], (acids), Bus *et al.* [27, 28] (methyl esters) and more recently by Lie Ken Jie. [65] (glycerol esters). Each of these papers covers a wide range of acetylenic acids/esters and it is only possible to give some typical examples (Table 1.14) and to discuss some generalisations arising from these results.

The effects of a triple bond on the chemical shifts of nearby carbon atoms have been derived by Gunstone *et al.* [56] and by Bus *et al.* [27]. The values are in close agreement. They are larger than those observed with olefinic groups.

	α	β	γ	δ	ε	ζ
Gunstone *et al.* [56]	-10.96	-0.56	-0.84	-0.53	-0.16	-0.11
Bus *et al.* [27]	-11.00	-0.65	-0.80	-0.45	-0.10	-

The shifts for acetylenic (80.19 ppm) and propargylic carbon atoms (18.85) are quite distinctive and are easily distinguished from olefinic and allylic carbon atoms.

The acyl group influences the chemical shift of acetylenic carbon atoms so that up to about Δ11 the two unsaturated carbon atoms have different chemical shifts. As with olefinic acids/esters the difference between the two shifts is characteristic of the distance of the unsaturated centre from the acyl carbon. The results for acids, methyl esters and glycerol esters are given in Table 1.15. The difference values observed with these three groups are close enough to indicate clearly where the unsaturated group is situated in the chain.

The earlier papers [27,28,56] contain information on diacetylenic compounds and along with more recent studies [69,119] cover compounds with the structural unit -C≡C$(CH_2)_n$C≡C- when n is 0-6 and also some with acetylenic and olefinic unsaturation. The 9*a*11*a*, 9*a*12*a* and 9*a*13*a* acids/esters are included in Table 1.14.

Lie Ken Jie *et al.* [77] have examined a series of allenic acids. The allenic carbon atoms have chemical shifts of approximately 91, 204 and 91 ppm but the unit has only a small effect on the α-carbon atoms (29.06 ppm). This compares with values of 27.2 for a *cis* double bond, 32.6 for a *trans* double bond and 18.8 for an acetylenic bond. Two examples of allenic esters are included in Table 1.14.

7. Cyclic acids

Cyclopropene acids occur in seed oils of the *Sterculiaceae*, in kapok seed oil (about 12%), and at a much lower level (~1%) in cottonseed oil.

These acids have attracted attention because they inhibit the desaturation process by which monoene acids are produced biosynthetically. Spectroscopic information is available for sterculic acid and its 6,7-isomer [96] and for 2-hydroxysterculic acid [117]. In addition, data on malvalic and sterculic esters have been obtained from the spectrum of kapok seed oil (Table 1.16A). Buist *et al*, [24] have given information on (*cis*)-dihydrosterculate. The results from all these sources are summarised in Table 1.16B. This shows the effect of the cyclopropene system on adjacent carbon atoms. The difference between the chemical shifts for the two olefinic carbon atoms can be used to determine the position of the cyclopropene ring in the chain. Pollard has given shielding parameters for the cyclopropene ring of -3.73(α), -2.38(β), -0.3(γ) and -0.25(δ) though he admits that the α and β parameters might be inverted. The results of Arsequell *et al.* [4] for C_{16} acids are somewhat different from the remainder.

Other cyclic acids which have been examined include those set out below. Details are available from the original papers.

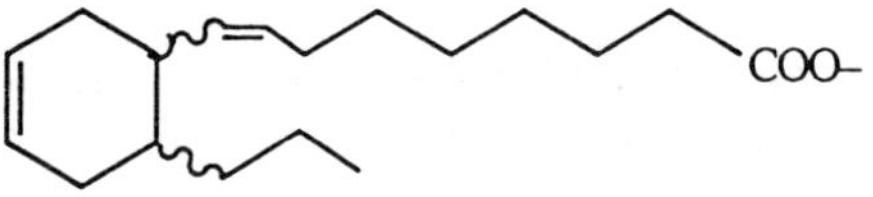

COO–

$(CH_2)_{11-n}COOCH_3$

$(CH_2)_nCH_3$

Awl *et al.* [10]

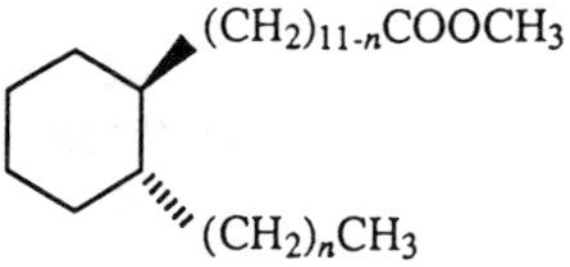

n = 3,5,7 Vatèla *et al.* [128]

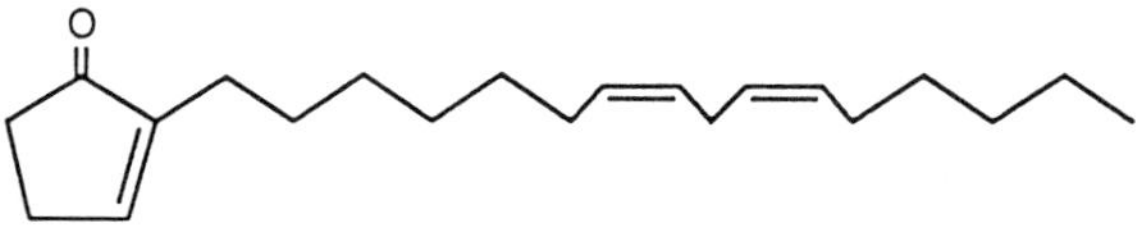

O

Frega *et al.* [43]

8. Fatty acids containing oxygenated functions

(a) *Saturated hydroxy and acetoxy compounds*

Tulloch et al. [127] have reported the ^{13}C nmr spectra of most of the regioisomeric hydroxystearates and acetoxystearates. The authors conclude that for mid-chain isomers the effect of these substituents on neighbouring carbon atoms is given by:

X	CHX	α	β	γ	δ	ε	ζ
OH	+42.2	+7.8	-4.0	+0.06	-0.06	-0.09	-0.05
OAc	+44.7	+4.4	-4.4	-0.20	-0.20	-0.10	-0.05

(relative to CH_2 groups at 29.65 ppm)

(b) *Unsaturated mono-hydroxy and acetoxy acids*

The monohydroxy acids of this type which have been examined contain the structural unit $-CH(OH)(CH_2)_nCH=CHCH_2-$ where n is 0-2 and where the double bond is *cis* or *trans*. Most of the available results are summarised in Table 1.17. Additional data not included in the Table are reported for some diacetoxy 18:4 esters [113] and for some alken-l-ols [13]. From the data in Table 1.17 Pfeffer *et al.* [95] concluded that the effect of an OH function on nearby double bonds is (the values for $n = 0$ have been calculated by the reviewer):

$n = 0$	2.1 (β) and 2.9 (γ) *cis*	1.6 (β) and 2.9 (γ) *trans*
$n = 1$	3.08 (γ) and -4.63 (δ) *cis*	4.06 (γ) and -4.18 (δ) *trans*
$n = 2$	-0.73 (δ) and 0.73 (ε) *cis*	

These values are added to or subtracted from the chemical shifts normally associated with a double bond in the appropriate position.

The values for ricinoleate (12-hydroxyoleate) are useful in the interpretation of the spectrum obtained with castor oil containing about 90% of this acid. The spectrum contains dominant signals associated with ricinoleic ester including: 27.38 (C8), 125.39 (C9), 133.00 (C10), 35.37 (C11), 71.44 (C12), 36.85 (C13), 25.73 (C14), 31.86 (C16) and 22.63 ppm (C17) [55].

(c) *Dihydroxy acids*

Data presented by Rakoff *et al.* [98] for *threo* and *erythro* saturated and unsaturated dihydroxy acids are given in Table 1.18. The *threo* and *erythro* isomers are distinguishable by their nmr spectra, particularly by the chemical shifts of methylene groups α and β to the diol unit. The chemical shifts for 7,10-dihydroxystearic acid given by Knothe *et al.* [63] have been assigned by the reviewer (Table 1.18).

(d) *Hydroperoxy and other oxidised acids*

Neff *et al.* [91] and Frankel *et al.* [41] have examined the ^{13}C nmr spectra of mono-, bis- and tris-hydroperoxy esters produced from the autoxidation of trilinolein and trilinolenin. Chemical shifts for the carbon atoms α and β to the hydroperoxy group are given below. These presumably occur in the structural unit

$$-CH_2CH(OOH)CH=CHCH=CHCH_2- \text{ (trans, cis)}.$$

	trilinolein			trilinolenin		
	mono	*bis*	*tris*	*mono*	*bis*	*tris*
C-OOH	86.4	86.4	86.6	86.7	88.0-87.0	88.0-87.0
*C*C-OOH	25.0	25.0	25.0	24.84	25.51	25.58

Typical shifts for the hydroperoxy peroxy unit have been given by Frankel *et al.* [41,42]:

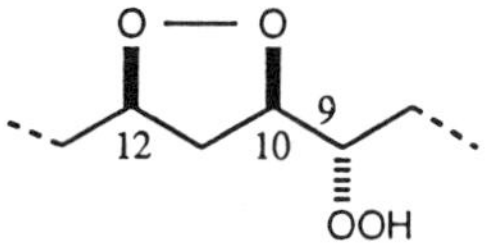

a 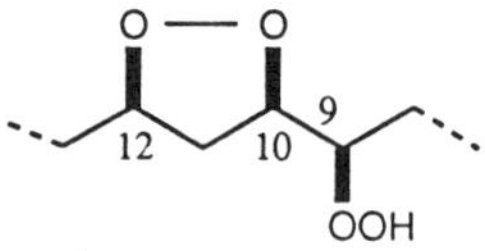 *b*

carbon atom	8	9	10	11	12	13
a	33.5	83.3	81.3	42.4	72.5	33.9
b	32.0	84.1	82.0	42.5	73.0	32.9

Oxidation of the cyclic C_{18} ester shown below gave allylic hydroperoxides which were reduced to allylic alcohols. Spectroscopic data has been reported for esters having oxygenated groups at C7, 9, 11, 12, 13, 14 and some unidentified components [7,8].

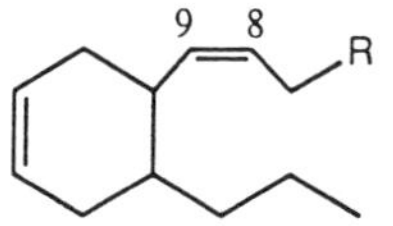

R = $(CH_2)_5CO_2Me$

In the presence of α-tocopherol, oxidised linoleate becomes attached to a tocopherone structure at C9 or C13 through a peroxy bond. Full ^{13}C assignments for eight regio- and stereoisomeric products are given in the original paper [134].

(e) *Primary alcohols and hydroperoxides*

Bascetta and Gunstone [13] examined a series of C_8-C_{18} primary alcohols (RCH_2OH) and hydroperoxides (RCH_2OOH). Chemical shifts for the C_{18} compounds are included in Table 1.18A. They concluded that the influence of the CH_2X group on the α (C2) and β (C3) carbon atoms is +3.34 and -3.69 for CH_2OH and -1.89 and -3.58 for CH_2OOH relative to the CH_2 group at 29.30 ppm.

(f) *Epoxy Acids*

The same authors [13] reported the chemical shifts for 13 isomeric *cis* and *trans* epoxyoctadecanoates, seven epoxyoctadecenoates and for some 1,2-epoxyalkanes and epoxycycloalkanes. The 9,10-epoxystearates are included in Table 1.18A. In mid-chain epoxy esters the epoxidic carbon atoms have a single resonance at 56.85 (*cis*) or 58.51 ppm (*trans*). Under the influence of the CO_2Me group (up to 7,8-) or the CH_3 group (to ω4,5) two signals are observed. The influence of a mid-chain epoxy group on adjacent carbon atoms is:

cis α-1.71 (27.60), β-2.93 (26.38), γ-0.38 (29.00)
trans α+2.57 (31.87), β-3.50 (25.80), γ-0.23 (29.02)

These values are based on a value of about 29.3 ppm for undisturbed CH_2 groups. The seven olefinic epoxy esters contain the group shown below: both functional groups have *cis* configuration and n=1-5. Methyl vernolate (12,13-epoxyoleate) is included in this group.

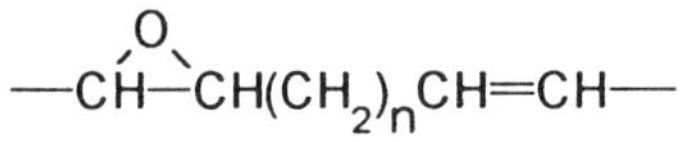

$$\text{—}\overset{\overset{\displaystyle O}{\frown}}{CH\text{—}CH}(CH_2)_nCH{=}CH\text{—}$$

A preliminary study of epoxidised soybean oil and epoxidised linseed oil gave some promising results. The C1 and C2 signals distinguish between saturated acyl chains and those having an epoxide group at position 9,10 but the ω1-ω3 signals are more informative since, together, they allow distinction between 9,10-, 12,13- and 15,16- epoxides. These presumably relate to epoxidised oleate, linoleate and linolenate respectively. Saturated esters and 9,10-epoxides are not separated in the ω1 and ω2 signals but are separated in the ω3 signals. These last do not show any 15,16-epoxide since the ω3 carbon atom is also epoxidic.

The chemical shifts are listed in Table 1.18B along with the analytical data derived from intensity values. These figures make good sense for these two epoxidised oils but the quantitative agreement is not as good as might be wished. It is known that carbon atoms close to the methyl end of an acyl chain have relatively long relaxation times and no allowance was made for this in obtaining these data. Better quantitative studies must await the more detailed investigation which is in process [55].

(g) *Oxo(keto) acids*

Tulloch [122,127] reported the ^{13}C nmr spectra of all 16 methyl oxo-octadecanoates. Values for 9-oxostearate (Table 1.18A) are typical of the mid-chain oxo esters with slightly different values being reported for esters with the oxo group close to the CO_2Me or CH_3 group. Chemical shifts on nearby carbon atoms are influenced by the oxo group thus: C=O +182.10, α +13.10, β -5.75, γ -0.40, δ -0.25, ε -0.20, ζ -0.08

Smith *et al.* [111] reported the chemical shifts of four heneicosadien-11-ones. The signals α to the keto function are: 42.83 and 41.96 (1,6*c*), 42.86 and 41.87 (1,6*t*), 42.76 and 41.96 (2*t*,6*c*) and 42.85 and 41.96 ppm (3*t*,6*c*).

The methyl ester of licanic acid (4-oxo 18:3 9*c*11*t*13*t*) shows the chemical shifts listed in Table 1.18A [118].

Other oxo acids or esters with ^{13}C spectra which have been fully or partly assigned include: 13-hydroxy-12-oxo 18:1 9*c*; 13-hydroxy-12-oxo 18:2 9*c*15*c*; and the phytodiene acid [32]; 9,12-dioxo 18:1 10*t*; 9-oxo 18:1 10*t*; 12-oxo 18:1 10*t*; 9-oxo 18:1 11*t*; and 12-oxo 18:1 9*t* [79].

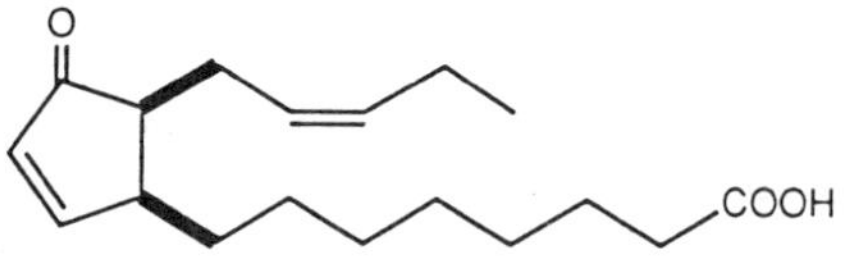

phytodiene acid

(viii) *Furanoid acids*

Furanoid acids, present in some fish oils and in some plant sources, have the general structure shown below with 0-2 methyl groups attached to the furan ring.

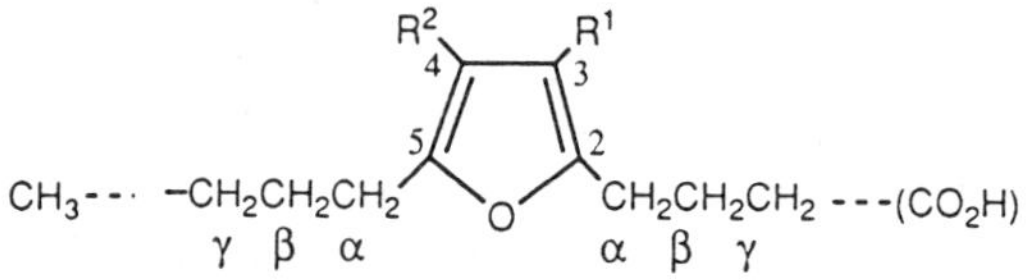

furanoid acid

Many of these have been synthesised [68] including the complete series of C_{18} acids with $R^1 = R^2 =$ H. Some chemical shifts (ppm) are given in Table 1.18C. There is not a full agreement between these literature values.

9. Other acids and esters

A number of interesting long-chain compounds which do not fit into any of the above categories are listed in this section. It is not possible to detail the spectra which must be obtained from the original reports.

epithio compounds [80]
dithiastearates [66]
sulphinyl and sulphonyl derivative of thialaurates [67]
thiol esters of 2-mercaptoethanol and 3-mercaptopropane-1,2-diol [114]
selenophene fatty acids (Se analogues of furanoid acids) [65]
aza, aziridine, azetidine fatty acids [75]
l-pyrrolino fatty acids [74]
esters of trimethylolpropane and pentaerythritol [20,112]
alkyl glucopyranosides and alkylglucoside polyesters [2,39]
gem-dichlorocyclopropanes [79]
betaine lipids [130]
cyano lipids [87]
isanic acid (9*a*11*a*17*e*-18:3) [88]
l-phenylalkyl esters [72]
fluorohexadecanoic acids [4]

10. Phospholipids

Chemical shifts and coupling constants for phosphatidylcholines (PC) and phosphatidylethanolamines (PE) are listed in Tables 1.19-1.22 (some other phospholipids are detailed in Table 1.23). Synthetic compounds (single molecular species) are listed separately from natural mixtures. Three different solvents are commonly employed: chloroform, chloroform-methanol and chloroform-methanol-water.

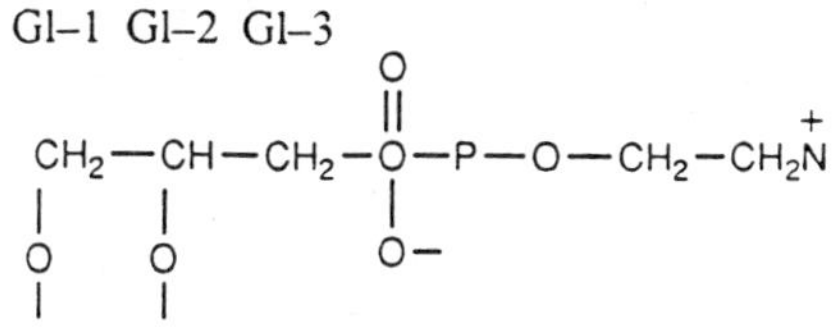

```
Gl–1  Gl–2  Gl–3
                     O
                     ||                  +
CH2—CH—CH2—O—P—O—CH2—CH2N
 |      |             |
 O      O             O–
 |      |
```

Interest is centred in the head groups (formulated above) which have six carbon atoms in PC and five carbon atoms in PE, and also in the acyl chains attached to Gl1 and Gl2 (α and β chains respectively). Questions to be considered are (i) can individual phospholipids be identified? (ii) can mixtures of phospholipids (possibly with triacylglycerols also) be characterised? and (iii) do the spectra provide information about the nature of the acyl chains and about their distribution? Since chemical shifts are influenced by changes of solvent it is important to make appropriate comparisons, and it is difficult to make useful generalisations.

Four of the signals in the head group are split by coupling with phosphorus. The carbon atoms G1-3 and OCH_2 show two-bond coupling whilst Gl-2 and CH_2N show three-bond coupling. The coupling constants do not change very much with structural details or with solvent and are sometimes used to assist in assignment. The signals for OCH_2 (3.9-5.0 Hz) and Gl-3 (4.0-5.5Hz) have lower coupling constants than those for CH_2N (4.8-7.0 Hz) and Gl-2 (7.2-8.2Hz). The following table shows the range of chemical shifts (ppm) for each carbon in the head group. These figures encompass changes of solvent but lyso compounds and some phospholipids with short chains (C_4-C_8) may fall outside these values.

	PC	PE	TAG
Gl-1	62.9-63.3	62.6-63.2	62.1-62.5
Gl-2	70.4-70.7	70.3-71.2	69.4-69.9
Gl-3	63.3-63.8	63.9-64.5	62.1-62.3
OCH_2	59.3-59.7	62.0-62.3	-
CH_2N	66.1-66.8	40.4-42.3	-
NMe_3	54.3-54.5	-	-

These three categories of glycerol esters can easily be distinguished from each other from the OCH_2, $CH_2\overset{+}{N}$, and $\overset{+}{N}Me_3$ signals. Though less obvious the glycerol carbon atoms also provide distinctive signals in the spectra of a mixture taken in one solvent. Examples will be discussed later.

For quantitative purposes it is probably best to use the Gl-1 signal which is not split by coupling. For the simple PC PE mixture discussed below this indicated a PE content of 32%. Other signals gave values between 25 and 33%. When producing this spectrum no attempt was made to get the best quantitative results. Several papers describe the use of ^{31}P spectra for this purpose. These spectra are simpler in that there is only one signal for each phospholipid class [29,26,86].

The synthetic phospholipids show two signals for C1 and C2 which relate to the presence of these carbon atoms in the α and β-chains. The higher chemical shift is associated with the α-chain for C1 and the β-chain for C2. In some of the natural mixtures there are three or four C1 signals for saturated and unsaturated chains in the α and β-chains. This is most obvious in the spectra of egg yolk phospholipids which contain acids such as arachidonic having a double bond closer to the carboxyl group.

Two interesting examples of mixtures are cited [55]. In the first a mixture of dilauro PC and PE in $CDCl_3$ solution was examined. Many signals were identical for both species but the following were different:

	Gl-1	Gl-2	Gl-3	$CH_2\overset{+}{N}$	OCH_2	$\overset{+}{N}Me_3$	Cl
PC	62.94	70.54 70.44	63.50 63.44	66.36 66.28	59.41 59.35	54.30	173.48 173.14
PE	62.85	70.44 70.32	63.78 63.72	40.93	62.08	–	173.48 173.08

Table 1.24 gives the results for a mixture of the triacylglycerols and the PC and PE esters of soybean. Each of these classes give a number of distinctive signals which can be used for identification or quantification.

11. Glycerol esters and other surface active compounds

The qualitative recognition and quantitative analysis of mixtures of mono-, di and triacylglycerols have been successfully investigated by ^{13}C nmr spectroscopy. Sacchi *et al.* [104], Gunstone [50] and Mazur *et al.* [85] have all reported chemical shifts for the glycerol carbon atoms in five categories of acylglycerols: 1- and 2-mono-, 1,2- and 1,3-di-, and triacylglycerols, and these are summarised in Table 1.25A. These figures do not agree completely and therefore when investigating such compounds it is useful to have some authentic standards available. However, it should be noted that there is good agreement in the sequence order of these chemical shifts. The relative intensities of the signals can also assist in identification since "symmetrical" glycerol esters (such as the 2-mono-, 1,3-di- and triacylglycerols) have two signals in a 1:2 ratio whilst the remaining glycerol esters are "unsymmetrical" and have three signals of equal intensity.

Gunstone [55] has shown that the chemical shifts are solvent-dependent and that chloroform-methanol (2:1) gives a better resolution of signals than chloroform alone (see Table 1.25A).

In studying the spectra of many natural oils and fats we have observed that the two signals associated with the glycerol carbon atoms in triacylglycerols (68.93 and 62.10 ppm) are sometimes accompanied by smaller signals at around 68.35 and 65.05 ppm (ratio 1:2) resulting from the presence of 1,3-diacylglycerols. For example, a sample of palm oil gave signals at 68.93 (intensity 2.96), 68.35 (0.27), 65.05 (0.49) and 62.10 (5.33) corresponding to 8.4% (on a molar basis) of 1,3-diacylglycerol. There was no evidence for the presence of 1,2-diacylglycerols though special care has to be taken to detect very small signals. It is noteworthy that the 1,3-diester is not the isomer expected on the basis of either incomplete biosynthesis or subsequent lipolysis but the identified product may result from acyl migration of the 1,2-isomer to the more stable form.

Dawe *et al.* [34] preferred to study monoacylglycerols as isopropylidene derivatives and used these in a method of quantitative analysis. The signal for the quaternary carbon in the hetero ring and those for the glycerol carbon atoms are all well resolved:

```
CH2OCOR                          CH2O
|                                |    \      / Me
CHO       Me              RCOOCH       C
|    \  /                        |    /      \ Me
|     C                          CH2O
CH2O /  \ Me
```

109 and 73.84, 66.56, and 64.59 ppm   98 and 66.07 and 62.04 ppm

Ahmed [1] gave chemical shifts for the glycerol 1-monoester 3-sulphate in D_2O solution *viz* 71.6 (CH_2O acyl), 70.0 (CHOH) and 67.5 (CH_2OSO_3Na).

Commercial monoacyl glycerols (which may contain some diacylglycerols) are acetylated to give further surface active compounds. The products are mainly monoacyl glycerols with their two monoacetates and the diacetates. These have the structures and chemical shifts shown in Table 1.25B.

12. Other end groups

This review has been centred on natural and synthetic long-chain acids and their methyl and glycerol esters apart from some discussion of alcohols and acetates (section B.8). Here we focus on a wider range of end groups with both saturated and unsaturated compounds.

Table 1.26 shows the effect of end groups on carbon atoms 1-3 for C_{14}-C_{18} compounds. There is a significant difference between acyl and alkoxy compounds especially for C1 and smaller differences for changes within each category. The nitriles are quite distinct from the other compounds.

Table 1.27 shows the small changes which occur in olefinic carbon atoms (Δ9 or Δ9,12) with different end groups. A pattern is apparent in monounsaturated compounds. The difference between the two olefinic chemical shifts is about 0.25-0.35 for acyl compounds and nitriles but lower (0.13-0.20) for alkoxy compounds. These changes reflect the polarity of the end group.

An interesting example of the use of these different chemical shifts is found in a study of the enzymic reaction between carboxylic acids and alcohols. The appearance of a signal at 64.3 ppm (alkoxy C1 in ester) and the reduction in signal at 63.0 ppm (alkoxy C1 in alcohol) was used to chart the progress of this reaction [92].

The effect of Δ9 unsaturation on the alkyl and acyl chains of a wax ester is apparent from the results obtained with two mixtures (12:0/18:1 and 18:1/12:0; 12:0/18:2 and 18:2/12:0). Small differences are noticed (sometimes only after resolution enhancement) in the C1, C1′, C9 and C10 signals. (Table 1.28). It should be possible by careful measurement, and perhaps by admixture with known compounds, to distinguish between unsaturation in acyl and alkyl chains [55].

C STUDY OF MIXTURES

1. Butter fat and lauric oils

These two groups of materials are taken together because both contain short-chain acids which give distinctive nmr signals. Pfeffer *et al.* [93] suggested that the C2 signal should be used to recognise butyric acid in butter fat (Table 1.29) and Lie Ken Jie *et al.* [70] have given data for the glycerides of a range of saturated acids. Those for the C_4-C_{10} acids are listed in Table 1.30 where they are compared with corresponding values for glycerol tristearate.

Table 1.31 gives chemical shifts for C1-C3, ω1-ω3, and the olefinic and allylic carbon atoms of a butter fat [54]. Using the intensity data it was concluded that the fat examined contained (on a mole % basis): 13% of butyric acid (based on C1, C2, ω1 and ω2 signals), 6% of hexanoic acid (based on C3, and ω1-ω3 signals) and 3% of octanoic acid (based on the ω3 signal). There was no spectroscopic evidence for the presence of short-chain acids in the β position. The olefinic and allylic signals indicated the presence of oleic acid and *trans*-vaccenic acid.

Lauric oils are characterised by high levels of lauric acid (about 50%) and by the presence of C_6, C_8 and C_{10} acids at lower levels. Since these last can be recognised by their ^{13}C nmr spectra it is possible to identify oils of this kind. In a simple experiment with the methyl esters of 8:0, 10:0 and 12:0 and a mixture of these three we showed that the C_8 ester (22.66, ω2; 31.73, ω3; and 29.00, C5) and the C_{10} ester (31.95, ω3) were easily recognised in the mixture. Lie Ken Jie *et al.* [70] have reported the chemical shifts of glycerol esters of type GA_3 where A is a range of saturated acids. Some characteristic signals for the C_4-C_8 glycerides can be recognised in Table 1.30. We have also examined the spectra of several lauric oils (Table 1.32). These oils contain short, medium and long-chain acids along with oleate (*n*-9) and linoleate (*n*-6). The most informative signals are those related to the ω2 and ω3 carbon atoms as set out in Table 1.32.

2. *Vegetable oils*

The ^{13}C nmr spectrum of a vegetable oil contains between 50 and 100 nmr signals (see Figures 1.1 and 1.2). Most of these can be assigned thus providing valuable information about natural mixtures of fatty acid glycerides. With due attention to appropriate parameters it is possible to obtain quantitative data. Useful information comes from the ω1-3, C1-3, glycerol, olefinic and allylic carbon atoms leaving out only the signals of the methylene envelope. The following discussion is based largely on a study of over 40 different vegetable oils [55].

(i) *Glycerol carbon atoms.* Vegetable oils have two glycerol signals in their spectra at around 68.9 (β) and 62.1 ppm (α) sometimes accompanied by less intense signals at 65.05 and 68.35-68.40 ppm. (This last signal, close to the larger signal at 68.9, is not always observed even when the peak at 65.05 is apparent). These smaller signals are expected for 1,3-diacylglycerols (see section B.11). This is not the diacyl ester expected as a product of either biosynthesis or metabolism but must be produced from 1,2/2,3-diacylglycerols by acyl migration to the more stable 1,3-isomer. These additional signals are apparent in several oils at low levels but are most noticeable in palm oil and its fractions (see Table 1.33). This kind of information has also been used by Sacchi *et al.* [104] in their study of virgin and refined olive oils.

(ii) *ω1-ω3 signals.* In an individual triacylglycerol such as tripalmitin there are distinct signals from the ω1 (14.1), ω2 (22.6) and ω3 (31.95 ppm) carbon atoms. In a natural mixture, however, these generally appear as clusters of two or more signals because the chemical shift is affected by nearby double bonds and by the acyl function if the chain is short enough (see sections B1 and C1). In this way *n*-3 and *n*-6 unsaturated acyl chains are distinguished from one another and from *n*-9/saturated chains. This last pair are not normally distinguishable though they sometimes show two signals for the $\omega 3$ carbon after resolution enhancement. Some average chemical shifts are listed in Table 1.34. Although chemical shifts in different spectra vary by ± 0.02ppm differences between the signals within one spectrum are more constant. The difference between *n*-9/sat and *n*-6 signals are 0.04, 0.11 and 0.40 ppm for the $\omega 1$, $\omega 2$ and $\omega 3$ carbon atoms. The $\omega 2$ and $\omega 3$ signals give the more reliable results because of the greater difference.

A signal at 31.83 ppm which sometimes appears in the $\omega 3$ cluster is indicative of *n*-7 acids. In most vegetable oils this will be 16:1(9) and/or 18:1(11). This signal is very large in the spectrum of macadamia oil which contains about 20% of hexadecenoic acid and was also observed in the spectra of oils from almond, grape, olive, passionflower, wheatgerm, soya, corn, caraway, carrot, parsnip, borage and evening primrose.

(iii) *C1, C2 and C3 signals.* These signals can be used to recognise acids with double bonds close to the acyl group and to determine the distribution of some fatty acids between the α and β positions.

The C1 and C2 signals (and sometimes the C3 signal after resolution enhancement) appear as doublets because the chemical shifts differ slightly in the α and β positions. These differ by 0.41 (C1, α larger), 0.16 (C2, β larger) and 0.025 (C3, β larger). Some significant values are collected in Table 1.35.

The differences between saturated and $\Delta 9$ acids (oleic, linoleic, α-linolenic) are small and are observed only after resolution enhancement or when spectroscopic data is collected over extended time-periods. This has been exploited by Wollenberg [131] in his study of groundnut, corn and high-oleic sunflower oils (Table 1.36).

When the double bond is closer to the carboxyl group than $\Delta 9$ the chemical shifts for C1, C2 and C3 are easily distinguishable from those for saturated and $\Delta 9$ acids and it is possible to determine both the total content of such acids and their distribution between α and β positions. This has been done for oils containing DHA (first double bond $\Delta 4$) and EPA (first double bond $\Delta 5$) in studies of fish oils (Section C4) and for vegetable oils containing petroselinic acid [49,84], γ-linolenic acid [47] and columbinic acid [49]. Typical information is given in Table 1.37.

(iv) *Allylic signals.* Oils containing oleic and linoleic acids have allylic signals at 27.2 (adjacent to one double bond) and 25.65 ppm (between two adjacent double bonds). Additional or alternative allylic signals are present in the spectra of oils containing α-linolenic acid, γ-linolenic acid, petroselinic acid and columbinic acid [47,49]. Chemical shifts are listed in Table 1.38.

(v) *Olefinic signals.* Of over forty oils examined one (Illipé) showed only two olefinic signals corresponding to oleic acid. The remainder generally had five signals associated with oleic and linoleic acid (with overlap of L9 and O10) or ten signals associated with oleic, linoleic and α-linolenic acids (overlap of Ln9 and L13 and of L9 with O10). The chemical shifts reported by Wollenberg [131] for these unsaturated acids as α and β chains are listed in Table 1.39 along with values for Δ11 (20:1) and Δ13(22:1) acids. Oils containing petroselinic, γ-linolenic and columbinic acid give the olefinic signals listed in Table 1.40. Many of the oils examined contain a signal between 129.79 and 129.84. This probably represents 11*c* acids (see data in Table 1.4 and 1.41) and was observed as a minor signal in the spectra of the following oils: almond, avocado, corn, macadamia, mango, olive, palm, rape, soya and wheatgerm.

It is known that in a series of monoene acids the two olefinic signals get closer together as unsaturation moves from the acyl group. This is illustrated in Table 1.41 for glycerol esters containing petroselinic, oleic, *cis*-vaccenic and erucic acids.

(vi) *Other contributions.* In his study of palm oil Soon Ng used ^{13}C nmr spectra to determine the distribution of saturated, oleic and linoleic acids between the α and β positions [115] and to demonstrate the presence of vaccenic acid [116]. Sacchi *et al.* [105B] also reported the distribution of saturated, oleic and linoleic acids between the α and β positions in olive oil and used the results to detect adulteration of virgin olive oil.

Mallet *et al.* [84] measured the ratio of oleic to petroselinic acid in oils containing the latter by comparing peak heights of the olefinic signals at 129.69 (O9) and 128.94 (P6).

Two Indian groups [58 and 129] used the intensity of signals at 24.8 (C3, all acyl groups), 25.7 (double allylic) and 27.3 ppm (single allylic) to analyse vegetable oils for the content of saturated, monoene and diene acids. Some of their results are given in Table 1.42.

3. *Animal fats*

Much less attention has been given to the ^{13}C nmr spectra of animal fats (excluding the fish oils which are discussed in section C4). Bonnet *et al.* [22,23] use the signals at 24.8 (C3), 25.7 (double allylic) and 27.3 ppm (single allylic) to determine the proportions of saturated, monoene and diene acids; the C1 signal to study the distribution of saturated and unsaturated acids between the α and β positions; and olefinic signals around 130 (O9, O10, L9 and L13) and around 128 ppm (L10 and L12) to obtain a monoene-diene ratio. Some of their results are assembled in Table 1.43.

4. *Fish oils*

Fish oils are generally more complex than vegetable oils and most other animal fats. Quantitative analysis by gas chromatography presents a number of difficulties based partly on the range of chain-length (at least C_{14}-C_{22}) but mainly on the high unsaturation of some of the acids (4-6 double bonds). The fact that DHA (22:6(*n*-3)) and EPA (20:5(*n*-3)) have unsaturation starting at Δ4 and Δ5 respectively also means that the usual lipolytic procedures for studying fatty acid distribution between the α and β glycerol carbon atoms cannot be satisfactorily employed. However, this has not stopped the publication of results obtained in this way.

Having double bonds close to the carboxyl group is helpful in the study of fish oils by ^{13}C nmr spectroscopy. Preliminary data on the *n*-3 polyene acids especially by Gunstone [48] and by Aursand *et al.* [5] have already been reviewed (section B4). The same authors have then used their results to study fish oils [6,51]. In their later paper Aursand *et al.*[6] examined the white muscle of Atlantic Salmon (*Salmo salar*) both as intact muscle and as extracted lipid. The spectroscopic protocol was such as to give quantitative results and chromium acetylacetonate was added as a relaxation agent when appropriate.

Table 1.44 shows the assignment of some of the observed chemical shifts used in the subsequent calculations, the results of which are set out in Table 1.45. Table 1.46 indicates that for a mixture of linoleate with EPA and DHA esters the composition (expressed on a % mole basis) obtained by nmr agrees satisfactorily with those obtained by weighing. Finally, using appropriate signals, it was possible to determine the distribution of EPA and DHA between the α and β chains (Table 1.47). These figures show that DHA is more heavily concentrated in the β-position. Since the salmon extract is not very rich in DHA (7%) or EPA (4-5%) it will be interesting to apply this study to oils having higher levels of these two important polyene acids.

5. *Partially-hydrogenated vegetable oils*

Vegetable oils rich in unsaturated C_{18} acids and fish oils, containing saturated and monoene through hexaene acids, are subjected to partial hydrogenation using a heterogeneous nickel catalyst on a large scale for use in spreading fats and in baking and frying fats. During the reaction some unsaturated centres are hydrogenated but others undergo double-bond migration and/or change of configuration from *cis* to *trans*. A partially-hydrogenated fat is thus a very complex mixture of acids with *cis* and *trans* unsaturation at many positions along the carbon chain. This is in contrast to natural fats where, for the most part, *cis* double bonds occupy only a limited number of positions.

The full analysis of hydrogenated fat is a difficult challenge. Although the proportion of *cis* and *trans* isomers can be measured by gas chromatography or by infrared spectroscopy there is no simple way of defining double bond positions. Even the most sophisticated gas chromatography will not separate all the isomers. Hydrogenated fish oils are even more complex than hydrogenated vegetable oils and this discussion is confined to the latter group.

Some time ago Pfeffer *et al.* [94] described the use of ^{13}C nmr spectroscopy to determine the *cis/trans* composition of catalytically hydrogenated lipid mixtures. This was based on the signals for allylic carbon atoms which have very different shifts for *cis* and *trans* isomers (see section B3). They concluded that, with the equipment then available, the olefinic carbon atoms gave broad ill-defined signals which were not useful. However Gunstone [53] has now shown that ^{13}C nmr spectroscopy can be successfully exploited to study partially hydrogenated fats. The application to partially-hydrogenated fish oils remains for the future.

A preliminary study with synthetic glycerides containing oleic and elaidic acid alone and as mixtures demonstrated that these simple mixtures give useful diagnostic signals, especially for allylic and olefinic carbon atoms (Table 1.48).

The successful interpretation of the complex spectra obtained with partially-hydrogenated vegetable oils depends largely on the earlier studies of Gunstone *et al.* [57] and Bus *et al.* [27,28]. These have already been discussed (Section B3) and are now supported by more recent information on glycerol esters (Lie Ken Jie [65]). Valuable insight into the nature of hydrogenated fat is available from the following:

(i) *Signals* α to the double bond (allylic). These are very different for *cis* (about 27.2) and *trans* (about 32.6) isomers and it is possible to obtain a *cis/trans* ratio by comparing intensities of all the *cis* allylic signals with those of all the *trans* allylic signals. It is obviously important to identify all the signals in each of these two classes. As previously reported (section B3) it is possible to distinguish some allylic signals as coming from a single 18:1 isomer: *viz* the Δ5-7 and Δ13-15 isomers of both the *cis* and *trans* types.

(ii) *Signals* γ to the double bond. It was also indicated in Section B3 that double bonds provide a minor perturbation on the chemical shifts of γ-carbon atoms and these are slightly different for *cis* and *trans* isomers. When this shift is superimposed on other easily observed signals then it is possible to recognise the *cis* and *trans* isomers of Δ5, 6 and 12-14 18:1 isomers (see Table 1.49 for some examples).

(iii) The most valuable information comes from the olefinic signals. There may be up to 25 of these and when the spectrum is collected over a longer period of time this number is almost doubled. The signals can be assigned on the basis of (*a*) the observed chemical shift, (*b*) differences between chemical shifts for the two olefinic signals of each isomer, and (*c*) differences in chemical shifts between adjacent isomers. Some signals are split depending on whether they are α or β chains. Some typical chemical shifts collected from the study of 12 hydrogenated fats are given in Table 1.50 and the kind of quantitative data which can be obtained is illustrated in Table 1.51. The figures here are merely indicative since they come from a spectrum not optimised to give the best quantitative data.

6. *Spreading and baking fats*

The ^{13}C spectra of the fats present in spreading and baking fats provide valuable information about their composition. In particular it is possible to determine the presence or absence of butter fat, lauric oils, partially-hydrogenated vegetable oil, and of oleic, linoleic and α-linolenic acids.

Butterfat contains several signals which distinguish it from other fats and which can be used to recognise its presence. It is easy to identify signals for butanoate, hexanoate, octanoate and olefinic signals which show the presence of oleate and *trans*-vaccenate. In butterfat C4>C6>C8 but in the lauric oils C8>C6. (see section C1)

The appearance of signals for *trans* acids (olefinic, allylic and ω3 (*n*-6, *t*)) is related to the presence of partially-hydrogenated oils.

α-Linolenic acid, shown by the presence of appropriate olefinic, allylic, and ω1 and ω2 signals, is usually indicative of the presence of rapeseed or soybean oil. Oleic, linoleic and α-linolenic are distinguished through their olefinic signals.

Four examples, taken from a larger set of results [54], are collected in Table 1.52 and refer to three spreading fats (A-C) and one baking fat (D). Careful perusal of these data show: (i) Samples A and B contain butterfat (approximately 15% and 30% by comparison with the spectrum of butterfat). (ii) Samples B and C contain α-linolenic acid but A and D do not. (iii) The ratio of oleic: linoleic: α-linolenic is 16:1:0, 10:3:1, 11:4:1 and 1:2.5:0 in fats A-D: this suggests that A contains a high oleic oil, B and C contain rapeseed oil, and D contains sunflower seed oil. (iv) All the fats contain some hydrogenated fat but this is quite low in D.

Spectra of the fat from samples such as these contain about 50 signals plus those associated with glycerol carbon atoms and in the methylene envelope. At first sight the information appears complex but it is more easily understood and interpreted when the chemical shifts are appropriately grouped as in Table 1.52.

7. *Acyl glycerols and other surface active compounds*

The chemical shifts observed with glycerol and its esters and with acetylated monoacylglycerols were discussed in section B11. The present section extends that by describing some practical applications of the assignments already reported. Quantitative results in this section are only approximate since the spectroscopic data collection procedure was not optimised. Nevertheless the results available are sufficient to indicate the potential of the technique as the basis of a quantitative method of analysis. The nmr data furnish results on a % mol basis. These are different from % wt, especially when the components of the mixture have very different molecular weights. It is usually possible to determine or to assume values which allow the interconversion of % mol and % wt.

A commercial emulsifier containing mono-, di- and triesters of glycerol was examined on three separate occasions and analysed on the basis of signal intensities for the glycerol carbon atoms and for C1 and C2. The results show reasonable reproducibility (Table 1.53). A second commercial emulsifier was analysed on the basis of the glycerol and C1 signals and compared with the specification provided with the sample (Table 1.54). This was probably obtained by gas chromatography. The glycerol signals seem to give the superior result [50]. Sacchi *et al.* [104] used this technique to determine the level of 1,2- and 1,3-diacylglycerols in olive oil.

A product obtained by the glycerolysis of butter oil and described as butter monoacylglycerol was shown to be a mixture of the α-monoacyl ester and residual glycerol [55].

An acetylated monoglyceride examined by ^{13}C nmr spectroscopy gave the results shown in Table 1.55 and these are compared with the specification of the sample.

Propylene glycol (propane-1,2-diol, $CH_3CH(OH)CH_2OH$) can furnish two monoesters (one primary and one secondary) and one diester. Appropriate chemical shifts are listed in Table 1.56 along with semiquantitative data for two rather different samples of propylene glycol esters. These have been obtained from several signals and show good agreement. Before being used as a routine analytical procedure the method of collecting the spectra would have to be optimised [55].

D. OTHER STUDIES

1. Individual seeds

Attempts have been made to determine the fatty acid composition of individual seeds by ^{13}C nmr spectroscopy. Information of this kind, obtained by a non-destructive method, would be very useful in seed breeding programmes. Attempts made to achieve this using the conditions normally employed for solutions gave spectra with some resolution of signals but no quantitative data was offered (*e.g.* Schaefer *et al.* [108,109]).

Greater progress was made with the techniques appropriate for solids (CP-MAS) and Rutar *et al.* [102,103] described spectra obtained from single fir seeds. They recognised fatty acids with 5*c*9*c* and 5*c*9*c*12*c* patterns of unsaturation. More recently Wollenberg [132] has reported the fatty acid composition (or at least the monoene-diene ratio) for single sunflower seeds of both oleic-rich and linoleic-rich types (Table 1.57). Rape seeds, containing about 2 mg of oil per seed, are still too small for individual examination but have been studied in groups of four (Table 1.58).

2. *Study of physical properties*

The physical nature of solid fatty acids and their glycerides has attracted much study over many decades. In the solid state most of these compounds are polymorphic (*i.e.* exist in more than one structural arrangement) and display different properties (such as melting point) for each polymorph. Polymorphic nature and properties have been extensively studied by X-ray crystallography, by infrared spectroscopy, and by several thermal procedures. With recent developments in nmr spectroscopy of solids it has become possible to apply this technique to the nature of solid lipids.

Using CP-MAS (cross polarization, magic angle spinning) Bociek *et al.* [21] compared liquid (melted) tripalmitin and tristearin with the solid α, β and β' polymorphic forms. They found that β forms show three different signals for the glycerol and acyl carbon atoms indicating the non-equivalence of the two α-chains in this polymorph. The liquid and the α and β' polymorphs showed only two signals for each of these carbon atom types. These were in a 2:1 ratio and indicated the equivalence of the two α chains in these forms (Table 1.59).

This work has been extended by Culot *et al.* [33] who studied the polymorphs of 12 triglycerides containing stearic acid and/or oleic acid and/or elaidic acid and reported chemical shifts for the glycerol, acyl (C1) and methyl (ω1) carbon atoms.

ACKNOWLEDGEMENTS

I am pleased to acknowledge the financial assistance of Karlshamns (LipidTeknik) for my work on high resolution nmr including the preparation of this review and I thank those who kindly made information available to me prior to publication (Dr Lie Ken Jie, Ms. M. Aursand, Ms C. Culot and Dr R. Sacchi).

REFERENCES

1. Ahmed,F.U., *J. Am. Oil. Chem. Soc.*, **67**, 8-14 (1990).
2. Akoh,C.C. and Swanson,B.G., *J. Am. Oil. Chem. Soc.*, **66**, 1295-1301 (1989).
3. Ando,Y., Ofa,T. and Takagi,T., *J. Am. Oil Chem. Soc.*, **66**, 1323-1325 (1989).
4. Arsequell,G., Fabrias,G., Gosalbo,L. and Camps,F., *Chem. Phys. Lipids.*, **63**, 149-158 (1992).
5. Aursand,M. and Grasdalen,H., *Chem. Phys. Lipids*, **62**, 239-251 (1992).
6. Aursand,M., Rainuzzo,J.R. and Grasdalen,H., *J. Am. Oil. Chem. Soc.*, in the press (1993).
7. Awl,R.A., Frankel,E.N., Brooks,D.D. and Weisleder,D., *Chem. Phys. Lipids*, **41**, 65-80 (1986).
8. Awl,R.A., Frankel,E.N. and Weisleder,D., *Lipids*, **22**, 721-730 (1987).
9. Awl,R.A., Frankel,E.N. and Weisleder,D., *Lipids*, **24**, 866-872 (1989).
10. Awl,R.A., Neff,W.E., Frankel,E.N., Plattner,R.D. and Weisleder,D., *Chem. Phys. Lipids*, **39**, 1-17 (1986).
11. Baenziger,J.E., Smith,I.C.P. and Hill,R.J., *Chem. Phys. Lipids*, **54**, 17-23 (1990).

12. Barton,F.E. (II), Himmelsbach,D.S. and Walters,D.B., *J. Am. Oil Chem. Soc.*, **55**, 574-576 (1978).
13. Bascetta,E. and Gunstone,F.D., *Chem. Phys. Lipids*, 36, 253-261 (1985),
14. Basti,M.M. and La Planche,L.A., *Chem. Phys. Lipids.*, **54**, 99-113 (1990).
15. Bengsch,E., Perly,B., Deleuze,C. and Valero,A., *J. Mag. Resonance.*, **68**, 1-13 (1986).
16. Bergter,L. and Seidl,P.R., *Lipids*, **19**, 44-47 (1984).
17. Bilyk,A., Bistline,R.G., Piazza,G.J., Feairheller,S.H. and Haas,M.J., *J. Am. Oil Chem. Soc.*, **69**, 488-491 (1992).
18. Birdsall,N.J.M., Ellar,D.J., Lee,A.G., Metcalfe,J.C. and Warren,G.B., *Biochim. Biophys. Acta*, **380**, 344-354 (1975).
19. Birdsall,N.J.M., Feeney,J., Lee,A.G., Levine,Y.K. and Metcalfe,J.C., *J. Chem. Soc. Perkin II.*, 1441-1445 (1972).
20. Black,K.D. and Gunstone,F.D., *Chem. Phys. Lipids*, **56**, 169-173 (1990).
21. Bociek,S.M., Ablett,K S. and Norton,I.T., *J. Am. Oil. Chem. Soc.*, **62**, 1261-1266 (1985).
22. Bonnet,M., Denoyer,C. and Renou,J.P., *Internat. J. Food. Sci. Technol.*, **25**, 399-000 (1990).
23. Bonnet,M. and Renou,J.P., *Analusis*, **18**, 123-000 (1991).
24. Buist,P.H.and MacLean,D.B., *Canad. J. Chem.*, **59**, 828-838 (1981).
25. Burgess,J.R., de la Rosa,R.I., Jacobs,R.S. and Butler,A., *Lipids*, **26**, 162-165 (1991).
26. Burns,R.A. and Roberts,M.F., *Biochemistry*, **19**, 3100-3106 (1980).
27. Bus,J., Sies,I. and Lie Ken Jie,M.S.F., *Chem. Phys. Lipids*, **17**, 501-518 (1976).
28. Bus,J., Sies,I. and Lie Ken Jie,M.S.F., *Chem. Phys. Lipids*, **18**, 130-144 (1977).
29. Capuani,G., Aureli,T., Miccheli,A., Di Cocco,M.E., Ramacci,M.T. and Delfini,M., *Lipids*, **27**, 389-391 (1992).
30. Carminati,G., Cavalli,L. and Buosi,F., *J. Am. Oil. Chem. Soc.*, **65**, 669-677 (1988).
31. Convert,O., Michel,E., Heymans,F. and Godfroid,J.J., *Biochim. Biophys. Acta*, **794**, 320-325 (1984).
32. Crombie,L. and Morgan,D.O., *J. Chem. Soc. Perkin. Trans I*, 577-580 and 581-587 (1991).
33. Culot,C., Gibon,V. and Durant,D., Poster presented at the First International Conference on Applications of Magnetic Resonance in Food Science, Guildford, UK (1992).
34. Dawe,R.D. and Wright,J.L.C., *Lipids*, **23**, 355-358 (1988).
35. Eads,T.M. and Croasman,W.R., *J. Amer Oil Chem. Soc.*, **65**, 78-83 (1988).
36. Eisele,J.L., Neumann,J.M. and Chachaty,C., *Chem. Phys. Lipids*, **55**, 351-354 (1990).
37. El-Emary,M. and Morgan,L.O., *J. Am. Oil. Chem. Soc.*, **55**, 593-599 (1978).
38. Fairchild,E.H., *J. Am. Oil. Chem. Soc.*, **59**, 305-308 (1982).
39. Focher,B., Savelli,G. and Torri,G., *Chem. Phys. Lipids*, **53**, 141-155 (1990).
40. Frankel,E.N., Garwood,R.F., Khambay,B.P.S., Moss,G.P. and Weedon,B.C.L., *J. Chem. Soc. Perkin Trans I*, 2233-2240 (1984).
41. Frankel,E.N., Neff,W.E. and Miyashita,K., *Lipids*, **25**, 40-47 (1990).
42. Frankel,E.N., Weisleder,D. and Neff,W.E., *Chem. Comm.*, 766-767 (1981).
43. Frega,N., Bocchi,F., Capozzi,F., Luchinat,C., Capella,P. and Lercker,C., *Chem. Phys. Lipids*, **60**, 133-142 (1991).
44. Gambhir,P.N., *Trends Food Sci. and Technol.*, **3**, 191-196 (1992).
45. Gaydou,E.M., Miralles,J. and Rasaozanakolona,V., *J. Am. Oil Chem. Soc.*, **64**, 997-1000 (1987).
46. Gribneau,M.C.M., *Trends Food Sci. and Technol.*, **3**, 186-190 (1992).
47. Gunstone,F.D., *Chem. Phys. Lipids*, **56**, 201-207 (1990).
48. Gunstone,F.D., *Chem. Phys. Lipids.*, **56**, 227-229 (1990).
49. Gunstone,F.D., *Chem. Phys. Lipids*, **58**, 159-167 (1991).
50. Gunstone,F.D., *Chem. Phys. Lipids.*, **58**, 219-224 (1991).
51. Gunstone,F.D., *Chem. Phys. Lipids*, **59**, 83-89 (1991).
52. Gunstone,F.D., *Chem. Phys. Lipids*, **63**, 155-163 (1993).
53. Gunstone,F.D., *J. Am. Oil. Chem. Soc.* in the press (1993).
54. Gunstone,F.D., *J. Am. Oil. Chem. Soc.*, **70**, 361-366 (1993).
55. Gunstone, F.D., Unpublished information.
56. Gunstone,F.D., Pollard,M.R., Scrimgeour,C.M., Gilman,N.W. and Holland,B.C., *Chem. Phys. Lipids*, **17**, 1-13 (1976).

57. Gunstone,F.D., Pollard,M.R., Scrimgeour,C.M. and Vedanayagam,H.S., *Chem. Phys. Lipids*, **18**, 115-129 (1977).
58. Husain,S., Sastry,G.S.R., Kifayatullah,Md., Sarma,G.V.R. and Nageswara Rao,R , *J. Oil Technol. Assoc. India,* **19**, 11-13 (1987).
59. Jiang,R.T., Shyy,Y.T. and Tsai,M.D., *Biochemistry*, **23**, 1661-1667 (1984).
60. Johns,S.R., Leslie,D.R., Willing,R.J. and Bishop,D.G., *Austral. J. Chem.*, **30**, 813-822 (1977).
61. Kalinoski,H.T. and Jensen,A., *J. Am. Oil. Chem. Soc.*, **66**, 1171-1175 (1989).
62. Klein,R.A., *Chem. Phys. Lipids,* **26**, 173-185 (1980).
63. Knothe,G., Bagby,M.O., Peterson,R.E. and Hou,C.T., *J. Am. Oil. Chem. Soc.*, **69**, 367-371 (1992).
64. Koritala,S. and Bagby,M.O., *J. Am. Oil. Chem. Soc.*, **69**, 575-578 (1992).
65. Lie Ken Jie,M.S.F. *et al.* Unpublished information.
66. Lie Ken Jie,M.S.F. and Bakare,O., *Chem. Phys. Lipids*, **53**, 203-209 (1990).
67. Lie Ken Jie,M.S.F. and Bakare,O., *Chem. Phys. Lipids*, **61**, 139-147 (1992).
68. Lie Ken Jie,M.S.F., Bus,J., Groenewegen,A. and Sies,I., *J. Chem. Soc.*, Perkin II, 1275-1279 (1986).
69. Lie Ken Jie,M.S.F., Cheung,Y.K., Chau,S.H. and Yan,B.F.Y., *Chem. Phys. Lipids*, **60**, 179-188 (1991).
70. Lie Ken Jie,M.S.F., Lam,C.C. and Yan,B.F.Y., *J. Chem. Research, S,* page 12-13, *M*, page 250-272 (1992).
71. Lie Ken Jie,M.S.F., Lam,C.C. and Yan,B.F.Y., *J. Chem. Research*, in the press (1993).
72. Lie Ken Jie,M.S.F. and Leung,D.W.Y., *Chem. Phys. Lipids*, **50**, 155-161 (1989).
73. Lie Ken Jie,M.S.F., Sinha,S. and Ahmad,F., *J. Am. Oil. Chem. Soc.*, **60**, 1777-1782 (1983).
74. Lie Ken Jie,M.S.F. and Syed-Rahmatullah,M.S.K., *Lipids*, **26**, 842-846 (1991).
75. Lie Ken Jie,M.S.F. and Syed-Rahmatullah,M.S.K., *J. Am. Oil. Chem. Soc.*, **69**, 359-362 (1992).
76. Lie Ken Jie,M.S.F. and Wong,C.F., *Lipids*, **27**, 59-64 (1992).
77. Lie Ken Jie,M.S.F. and Wong,C.F., *Chem. Phys. Lipids*, **61**, 243-254 (1992).
78. Lie Ken Jie,M.S.F. and Wong,K.P., *Lipids*, **26**, 837-841 (1991).
79. Lie Ken Jie,M.S.F. and Wong,K.P., *Chem. Phys. Lipids*, **62**, 177-183 (1992).
80. Lie Ken Jie,M.S.F. and Zheng,Y.F., *Chem. Phys. Lipids*, **49**, 167-178 (1988).
81. Lingh,I. and Stawinski,J., *J. Org. Chem.*, **54**, 1338-1342 (1989).
82. Lopez,A. and Gerwick, W.H., *Lipids*, **22**, 190-194 (1987).
83. Lunazzi,L., Plaucucci,G., Gross,L. and Strocchi,A., *Chem. Phys. Lipids,* **30**, 347-352 (1982).
84. Mallet,J.F., Gaydou,E.M. and Archavlis,A., *J. Am. Oil Chem. Soc.,* **67**, 607-610 (1990).
85. Mazur,A.W., Hiler,G.D., Lee,S.S.C., Armstrong,M.P. and Wendel,J.D., *Chem. Phys. Lipids.*, **60**, 189-199 (1991).
86. Meneses,P. and Glonek,T., *J. Lipid. Res.*, **29**, 679-689 (1988).
87. Mikolajczak,K.L. and Weisleder,D., *Lipids*, **13**, 514-516 (1978).
88. Miller,R.W., Weisleder,D., Plattner,R.D. and Smith,C.R. Jr., *Lipids*, **12**, 669-675 (1977).
89. Moine,G., Forzy,L. and Oesterhelt,G., *Chem. Phys. Lipids*, **60**, 273-280 (1992).
90. Murari,R., El-Rahman,M.A., Wedmid,Y., Parthasarathy,S. and Baumann,W.J., *J. Org. Chem.*, **47**, 2158-2163 (1982).
91. Neff,W.E., Frankel,E.N. and Miyashita,K., *Lipids*, **25**, 33-39 (1990).
92. O'Connor,C.J., Petricevic,S.F., Coddington,J.M. and Stanley,R.A., *J. Amer Oil Chem. Soc.*, **69**, 295-300 (1992).
93. Pfeffer,P.E., Luddy,F.E., Unruh,J. and Shoolery,J.N., *J. Am. Oil. Chem. Soc.,* **54**, 380-386 (1977).
94. Pfeffer,P.E., Sampugna,J., Schwartz,D.P. and Shoolery,J.N., *Lipids*, **12**, 869-871 (1977).
95. Pfeffer,P.E., Sonnet,P.E., Schwartz,D.P., Osman,S.F. and Weisleder,D., *Lipids*, **27**, 285-288 (1992).
96. Pollard,M.R., PhD Thesis, University of St Andrews (1977).
97. Pollard,M.R., *Analysis of Oils and Fats*, pp. 401-434 (1986) (ed. R.J. Hamilton and J.B. Rossell, Elsevier Applied Sciences, London).

98. Rahn,C.H., Sand,D.M., Wedmid,Y., Schlenk,H., Krick,T.P. and Glass,R.L., *J. Org. Chem.*, **44**, 3420-3424 (1979).
99. Rakoff,H., *Chem. Phys. Lipids*, **35**, 117-125 (1984).
100. Rakoff,H., *Lipids*, **25**, 130-134 (1990).
101. Rakoff,H., Weisleder,D. and Emken,E.A., *Lipids*, **14**, 81-83 (1979).
102. Rutar,V., Kovac,M. and Lahajnar,G., *J. Mag. Resonance*, **80**, 133-138 (1988).
103. Rutar,V., Kovac,M. and Lahajnar,G., *J. Am. Oil Chem. Soc.*, **66**, 961-965 (1989).
104. Sacchi,R., Addeo,F., Giudicianni,I. and Paolillo,L., *Riv. Ital. Sostanze Grasse*, **67**, 245-252 (1990).
105A. Sacchi,R., Addeo,F., Giudicianni,I. and Paolillo,L., *Ital. J. Food Sci.*, **4**, 253-262 (1991).
105B. Sacchi,R., Addeo,F., Giudicianni,I. and Paolillo,L., *Ital. J. Food. Sci.*, **4**, 117-123 (1992).
106. Santaren,J.F., Rico,M., Guilleme,J. and Ribera,A., *Org. Mag. Reson.*, **18**, 98-103 (1991).
107. Santaren,J.F., Rico,M. and Ribera,A., *Org. Mag. Reson.*, **21**, 238-242 (1983).
108. Schaefer,J. and Stejskal,E.O., *J. Am. Oil Chem. Soc.*, **51**, 210-213 (1974).
109. Schaefer,J. and Stejskal,E.O., *J. Am. Oil Chem. Soc.*, **52**, 366-369 (1975).
110. Simoneau,C., McCarthy,M.J., Reid,D.S. and German,J.B., *Trends Food Sci. Technol.*, **3**, 208-211 (1992).
111. Smith,L.M., Smith,R.G., Loehr,T.M., Daves,G.D., Daterman,G.E. and Wohleb,R.H., *J. Org. Chem.*, **43**, 2361-2366 (1978).
112. Smith,P.W.R., *J. Am. Oil. Chem. Soc.*, **69**, 352-354 (1992).
113. Solem,M.L., Jiang Zhi,D. and Gerwick,W.H., *Lipids*, **24**, 256-260 (1989).
114. Sonnet,P.E. and Moore,G.G., *Lipids*, **24**, 743-745 (1989).
115. Soon Ng, *Lipids*, **20**, 778-782 (1985).
116. Soon Ng and Heng Fui Koh, *Lipids* **23**, 140-143 (1988).
117. Spitzer,V., *J. Am. Oil Chem. Soc.*, **68**, 963-969 (1991).
118. Spitzer,V., Marx,F., Maia,J.G.S. and Pfeilsticker,K., *J. Am. Oil Chem Soc.*, **68**, 440-442 (1991).
119. Spitzer,V., Marx,F., Maia,J.G.S. and Pfeilsticker,K., *Fat Sci. Technol.*, **93**, 169-174 (1991).
120. Stewart,L.C., Kates,M. and Smith,I.P.P., *Chem. Phys. Lipids.*, **48**, 177-188 (1988).
121. Takagi,T., Kaneniwa,M. and Itabashi,Y., *Lipids*, **21**, 430-433 (1986).
122. Tulloch,A.P., *Canad. J. Chem.*, **55**, 1135-1142 (1977).
123. Tulloch,A.P., *Lipids*, **17**, 544-550 (1982).
124. Tulloch,A.P., *Prog. Lipid Res.*, **22**, 235-256 (1983).
125. Tulloch,A.P. and Bergter,L., *Lipids*, **14**, 996-1002 (1979).
126. Tulloch,A.P. and Bergter L., *Chem. Phys. Lipids*, **28**, 347-355 (1981).
127. Tulloch, A.P. and Mazurek, M., *Lipids*, **11**, 228-234 (1976).
128. Vatéla,J.M., Sébèdio,J.L. and Le Quéré,J.L., *Chem. Phys. Lipids*, **48**, 119-128 (1988).
129. Vedanayagam,M.S. and Gangiah,K., *J. Oil Technol. Assoc. India*, **19**, 6-11 (1987).
130. Vogel,G., Woznica,M., Gfeller,H., Müller,C., Stämpfli,A.A., Jenny,T.A. and Eichenberger,W., *Chem. Phys. Lipids*, **52**, 99-109 (1990).
131. Wollenberg,K.F., *J. Am. Oil Chem. Soc.*, **67**, 487-494 (1990).
132. Wollenberg,K., *J. Am. Oil Chem.*, **68**, 391-400 (1991).
133. Yamauchi,R., Matsui,T., Kato,K. and Ueno,Y., *Lipids*, **25**, 152-158 (1990).

Table 1.1
Chemical shifts (ppm) of C_{14}, C_{16}, and C_{18} saturated acids and esters.

	acids[a]			methyl esters[a]			glycerol esters (α and β chains)[b]					
	14:0	16:0	18:0	14:0	16:0	18:0	14:0		16:0		18:0	
1	180.68	180.60	180.58	174.04	174.08	174.06	173.26	172.84	173.27	172.86	137.27	172.88
2	34.25	34.26	34.24	34.15	34.18	34.18	34.07	34.24	34.07	34.24	34.07	34.24
3	24.80	24.84	24.81	25.12	25.12	25.13	24.90	24.94	24.89	24.94	24.91	24.94
4	29.22	29.26	29.23	29.34	29.37	29.37	29.16	29.12	29.15	29.12	29.16	29.12
5	29.39	29.41	29.38	29.44	29.46	29.47	29.31	29.35	29.31	29.34	29.32	29.34
6	29.56	29.60	29.56	29.64	29.65	29.66	29.52	29.54	29.52	29.54	29.52	29.54
(max value)	(29.78)	(29.84)	(29.82)	(29.83)	(29.85)	(29.88)	(29.73)		(29.74)		(29.76)	
ω5	-	-	-	-	-	-	29.70		29.70		29.72	
ω4	29.49	29.53	29.51	29.51	29.55	29.56	29.41		29.40		29.42	
ω3	32.06	32.11	32.07	32.11	32.11	32.12	31.97		31.98		31.98	
ω2	22.80	22.82	22.79	22.84	22.84	22.84	22.73		22.72		22.73	
ω1	14.12	14.13	14.12	14.13	14.14	14.14	14.13		14.12		14.11	

a Reference: [56]
b Reference: [70]

Table 1.2

Chemical shifts (ppm), intensities, and assignments of wool grease fatty acids.

chemical shift (ppm)	intensity	assignment
180.09	0.84	C1, all acids
70.49	0.24	CHOH
39.08	1.19	ω3, *iso*
36.66	1.15	ω4, *anteiso*
34.41	1.35	ω3, *anteiso*
34.10	1.76	C2, all acids
31.95	1.42	ω3, unbranched acids
30.06	1.50	ω6, *anteiso*
(28.8 - 30.0)	(40.96)	methylene envelope
27.98	1.26	ω2, *iso*
27.45	1.31	ω4, *iso*
27.14	1.22	ω5, *anteiso*
25.06	0.26	CH_2 β to CHOH
24.70	2.31	C3, all acids
22.68*	3.12	[ω1 and branched methyl, *iso* [ω2, unbranched acids
19.23	1.18	branched methyl, *anteiso*
14.14	1.35	ω1, unbranched acids
11.42	1.06	ω1, *anteiso*

Reference: [52]

* on resolution enhancement this gives peaks at 22.72 (smaller) and 22.68 ppm (larger).

ω1 - ω6 refer to carbon atoms with respect to the end methyl group; *iso* acids have a methyl group on ω2 and *anteiso* acids have a methyl group on ω3.

On the basis of the above figures the sample contains unbranched acids (~30%), *iso* acids (~32%), *anteiso* acids (~32%), and others (including hydroxy acids ~6%).

Table 1.3

Monoene acids, esters, and glycerides for which chemical shifts have been reported.

Chain length	double bond position and configuration (author)
8	3*t*(B), 7*e*(G)
9	7*t*(G)
10	4*c*(B), 7*t*(G), 8*c*(G), 8*t*(G)
11	2*c*(B,L), 2*t*(L), 3*c*(B), 4*c*(B,G), 5*c*(B,G), 6*c*(B,G), 7*c*(B), 8*c*(B,G), 9*c*(B), 10*e*(B,G)
12	2*t*(B), 3*c*(G,L), 3*t*(L), 4*c*(G), 5*c*(G), 6*c*(B), 7*c*(G), 8*t*(G), 10*c*(G)
13	3*c*(G), 4*c*(L), 4*t*(L), 5*c*(G), 6*c*(G), 7*c*(G), 9*c*(G), 10*c*(G)
14	3*c*(G), 4*c*(G), 5*c*(B,G,L), 5*t*(L), 6*c*(G), 7*c*(G), 8*c*(B,G)
15	6*c*(L), 6*t*(L)
16	7*c*(B,L), 7*t*(G,L), 8*c*(G), 8*t*(G), 9*c*(A), 10*c*(B)
17	8*c*(L), 8*t*(L), 11*c*(B), 12*c*(G)
18	2*c*(G), 3*c*(G), 4*c*(G), 5*c*(B,G), 5*t*(B,G), 6*c*(B,G), 6*t*(B,G), 7*c*(B,G), 8*c*(B,G), 8*t*(B,G), 9*c*(A,B,G,L), 9*t*(B,G,L), 10*c*(B,G), 10*t*(B,G), 11*c*(B,G), 11*t*(B,G), 12*c*(B,G), 12*t*(G), 13*c*(G), 13*t*(G), 14*c*(G), 14*t*(G), 15*c*(G), 15*t*(G), 17*e*(G)
19	10*c*(L), 10*t*(L)
20	11*c*(A,B,G,L), 11*t*(L)
22	13*c*(A)

References: A [5], B [27,18], G [57], and L [65]

Table 1.4
Difference in chemical shift (ppm) between the two olefinic carbon atoms in monoene acids and esters.

double bond position	acids		methyl esters				glycerol triesters	
	c^a	t^a	c^a	c^b	t^a	t^b	c^c	t^c
2	-	-	31.28	31.80	-	28.85	33.19* 33.23	30.36* 30.53
3	14.12	-	12.40	12.90	-	13.50	13.79 13.78	14.42 14.39
4	4.89	-	4.01	4.30	-	4.30	4.79	4.49 4.51
5	3.23	3.26	2.77	2.95	-	2.95	3.13	3.20 3.19
6	1.60	1.64	1.30	1.50	-	1.50	1.60	1.63 1.64
7	0.85	0.72	0.73	0.90	0.73	0.90	0.87 0.91	0.88 0.90
8	0.50	0.54	0.45	0.50	0.43	0.50	0.54	0.53 0.57
9	0.26	0.31	0.24	0.35	-	0.35	0.30 0.34	0.31 0.34
10	0.17	0.16	-	0.20	-	0.20	0.17 0.21	0.18 0.21
11	0.07	0.08	-	0.10	-	0.10	0.11 0.12	0.11 0.13
ω4	0.48	0.50	-	-	0.43	-	-	-
ω3	2.16	2.46	-	-	2.29	-	-	-
ω2	7.34	-	-	-	7.31	-	- -	

References: *a* [57], *b* [27, 28], *c* [65]
* double values refer to α and β chains respectively, single values apply to α *and* β chains.

Table 1.5
Differences in chemical shift (ppm x10^{-3}) of olefinic carbon atoms in α and β chains.

double bond position	*cis*		*trans*	
	$n+1$	n	$n+1$	n
2	-168	-126	237	-68
3	0	-8	41	7
4	27	27	7	31
5	7	0	7	0
6	-6	-11	0	9
7	-16	23	0	19
8	-20	-27	-16	26
9	-15	26	-15	23
10	-12	23	-9	20
11	8	16	-7	16

Reference: [65]
interpretation: in the glycerol esters of oleic acid, C9 will show two signals differing by 26x10-^{3} ppm (α larger) and C_{10} will (perhaps) show the two signals differing by 15x10^{-3} ppm (β larger).

Table 1.6
Effect of double bond position on the chemical shift (ppm) of allylic carbon atoms in 18:1 acids.

position of double bond (allylic carbons)		*cis* isomers		*trans* isomers	
2	(-,4)	-	-	-	-
3	(2,5)	33.12	27.49	-	-
4	(3,6)	22.66	27.33	-	-
5	(4,7)	26.58	27.36	31.91	32.65
6	(5,8)	26.88	27.36	32.21	32.66
7	(6,9)	27.13	27.39	32.40	32.67
8	(7,10)	27.28	27.28	32.58	32.68
9	(8,11)	27.31	27.31	32.64	32.66
10	(9,12)	27.30	27.30	32.68	32.68
11	(10,13)	27.29	27.29	32.69	32.69
12	(11,14)	27.29	27.29	32.65	32.65
13	(12,15)	27.27	26.99	32.69	32.36
14	(13,16)	27.33	29.69 (ω3)	32.69	34.79 (ω3)
15	(14,17)	27.21	20.60 (ω2)	32.65	25.65 (ω2)
16	(15,18)	-	-	-	-
17	(16,-)	33.91 (ω3)	-	-	-

References: [57,65]

Table 1.7
Effect of olefinic groups on the chemical shifts of the C1-C3 and ω1-ω3 signals in 18:1 acids.

double bond position	C1	C2	C3	ω3	ω2	ω1
cis-isomers						
4	180.04†	34.27	22.66*	32.06	22.75	14.15
5	180.45	33.58†	24.74	32.06	22.79	14.13
6	180.51	34.13	24.39†	32.04	22.78	14.11
7	180.30	34.22	24.73	32.10	22.82	14.13
8	180.54	34.22	24.78	32.06	22.80	14.12
9	180.55	34.24	24.80	32.06	22.80	14.13
10	180.46	34.22	24.81	32.01	22.76	14.12
11	180.55	34.22	24.76	31.87	22.72	14.10
12(ω6)	180.53	34.31	24.82	31.64†	22.68	14.09
13(ω5)	180.54	34.18	24.74	32.06	22.40†	14.00
14(ω4)	180.58	34.25	24.78	29.69*	22.96	13.80†
15(ω3)	180.61	34.23	24.79	-	20.60*	14.41
trans-isomers						
4						
5	180.58	33.47†	24.57	32.04	22.78	14.12
6	180.51	34.06	24.22†	32.04	22.76	14.12
7	180.53	34.20	24.63	32.05	22.78	14.11
8	180.43	34.20	24.74	32.01	22.76	14.11
9	180.61	34.19	24.75	32.01	22.76	14.11
10	180.62	34.21	24.76	32.00	22.77	14.11
11	180.57	34.22	24.77	31.88	22.73	14.10
12(ω6)	180.57	34.20	24.77	31.49†	22.62	14.07
13(ω5)	180.59	34.21	24.77	31.96	22.27†	13.96
14(ω4)	180.59	34.21	24.78	34.79*	22.82	13.64†
15(ω3)	180.57	34.21	24.78	-	25.65*	14.04

Reference: [57]
* α to a double bond
† γ to a double bond

Table 1.8

Chemical shifts (ppm) of *n*-3-polyene acids and methyl esters.

Δ†	4	5	6	6	7	8	9	11
carbon atom	22:6(A,Au)*°	20:5(A,Au)*°	24:6(E,T)*	18:4(A,Au)*	22:5(E,G)*	20:4(E,G)*	18:3(A,Au)*	20:3(E,G)*
1	179.38	179.87	174.01	180.07	174.14	174.22	180.56	173.84
2	34.02	33.42	34.02	34.01	34.05	34.08	34.14	34.11
3	22.54	24.54	24.65	24.35	24.88	24.93	24.64	25.10
4	129.64[a] (129.52)	26.51	29.15	29.07	28.81	29.07	29.06	29.35
5	127.58[a] (127.49)	129.10 (128.96)	26.92	26.91	29.27	28.92	29.10	29.42
6	25.63	128.81 (128.65)	129.69	129.58	27.06	29.44	29.18	29.42
7	128.01 (128.24)	25.68	128.45	128.60	130.04	27.18	29.60	29.62
8	128.35 (127.92)	128.22 (128.08)	25.68	25.60	128.22	*d*	27.23	29.35
9	25.68	128.23 (128.06)	128.07	128.32[b]	-	*d*	130.24	29.82
10	123.28 (128.18)	25.68	128.61	128.18	128.46	25.65	127.80	27.37
11	128.31 (128.03)	128.14 (127.99)	25.68	25.71	128.12	*d*	25.65	130.28
12	25.68	128.32 (128.17)	128.23	127.98	25.66	*d*	128.30[b]	127.87
13	128.13 (128.02)	25.68	128.23	128.35[b]	127.99	25.65	128.27[b]	25.75
14	128.13 (128.20)	127.94 (127.79)	25.68	-	127.99	*d*	-	128.33
15	25.57(?)	128.63 (128.46)	128.23	-	25.66	*d*	-	128.33
16	127.92 (127.82)	-	128.23	-	127.89	-	-	
17	128.60 (128.50)	-	25.68	-	128.57	-	-	
			c					
ω5	25.57	25.59	25.68(?)	25.69(?)	25.56	25.56	25.56	25.71(?)
ω4	127.08 (126.98)	127.08 (126.96)	127.09	127.12	127.04	127.04	127.16	127.31
ω3	132.03 (131.93)	132.10 (131.89)	132.02	132.05	132.02	132.00	131.95	131.86
ω2	20.59	20.61	20.59	20.61	20.57	20.58	20.58	20.67
ω1	14.27	14.31	14.25	14.29	14.27	14.29	14.28	14.30

† position of first double bond with respect to the acyl function

* A = acid, E = methyl ester, Au = Aursand [5], T = Takagi [121], G = Gunstone [57]

° figures in parenthesis provided by Sacchi (unpublished)

? assignments considered doubtful by reviewer

a values interchanged by reviewer, *b* these values are interchangeable (author), *c* also 128.23, 128.23, *d* chemical shifts reported but not fully assigned.

Table 1.9
Chemical shifts (ppm) of *n*-6 polyene acids and methyl esters.

Δ† carbon atom	5 20:4(A,Au)*	6 18:3(E,G)*	7 19:3(E,G)*	7 22:4(A,Au)*	8 20:3(E,G)	9 18:2(A,Au)
1	180.36	-	174.03	180.14	174.19	180.54
2	33.48	33.96	34.07	34.01	34.12	34.15
3	24.51	24.63	24.92	24.56	24.98	24.70
4	26.48	29.16	28.85	28.69	28.95	29.08
5	129.09	26.92	29.34	29.31	29.12	29.12
6	128.77	129.57	27.09	27.00	29.45	29.40
7	25.63	128.28	129.96	129.94	27.24	29.63
8	128.27[b]	25.68	128.06	128.05	130.19	27.22
9	128.12[b]	128.07	25.71	25.62	127.91	130.02
10	25.63	128.39	128.19	128.34	25.70	128.12
11	127.88	-	128.39	127.94	128.27	-
12	128.60	-	-	25.62	128.39	-
				c		
ω8	25.63	25.68	25.71	25.62	25.70	25.67
ω7	127.57	127.65	127.71	127.56	127.71	127.95
ω6	130.52	130.39	130.40	130.45	130.44	130.21
ω5	27.26	27.27	27.27	27.21	27.24	27.25
ω4	29.37	29.41	29.34	29.22	29.45	29.19
ω3	31.57	31.59	31.58	31.51	31.59	31.58
ω2	22.63	22.65	22.62	22.56	22.58	22.62
ω1	14.12	14.10	14.05	14.04	14.03	14.09

symbols: as in previous table
c also 127.91, 128.51

Table 1.10
Chemical shifts (ppm) of Δ6 polyene methyl esters.

carbon atom	*n*-1 16:4 (E,An)*	*n*-3 24:6 (E,T)*	*n*-3 18:4 (E,G)*	*n*-4 16:3 (E,An)*	*n*-6 18:3 (E,G)*	*n*-9 18:2 (,B)*
1	174.06	174.01	174.03	174.06	-	-
2	34.02	34.02	33.97	33.97	33.96	33.8
3	24.65	24.65	24.60	24.60	24.63	24.3
4	29.15	29.15	29.12	29.09	29.16	29.0
5	26.92	26.92	26.88	26.87	26.92	26.8
6	129.69	129.69	129.64	129.59	129.57	129.3
7	128.18	128.45	128.30	128.23	128.28	128.7
8	25.62	25.68	25.65	25.62	25.68	25.7

*A = acid, E = methyl ester, An [3], B [11], G [48], T [121]

Table 1.11
Chemical shifts (ppm) of *cis/trans* stereoisomers of 18:2 (methyl esters)

carbon atom	8	9	10	11	12	13	14
Gunstone *et al.* [57]							
9*c*12*c*	27.32	130.01	128.24	25.77	128.08	130.20	27.32
9*c*12*t*	27.18	130.31	127.94	30.55	128.40	130.84	32.64
9*t*12*c*	32.67	130.64	128.59	30.56	127.83	130.41	27.22
9*t*12*t*	32.59	130.92	128.79	35.68	128.65	131.08	32.59
Bus *et al.* [27,28]							
9*c*12*c*	-	130.05	128.25	-	128.10	130.20	-
9*c*12*t*	-	130.40	127.95	-	128.40	130.90	-
9*t*12*c*	-	130.80	128.60	-	127.85	130.55	-
9*t*12*t*	-	131.00	128.85	-	128.70	131.20	-

Table 1.12
Chemical shifts (ppm) of 18:3 conjugated triene acids.

	9,11,13					8,10,12	
	c t c	*c t t*	*t t c*	*t t t*		*c t c*	*t t c*
	a	*b*	*c*	*d*		*e*	*f*
olefinic carbon atoms (shifts listed in increasing order)							
	127.87 (11)*	125.98 (11)*	126.12	130.60 (14)*		127.81	126.21
	128.02 (12)	128.95 (9)	128.79	130.73 (9)		128.07	128.71
	128.86 (14)	130.73 (14)	130.82	130.81 (10)		128.86	130.85
	128.96 (9)	131.52 (13)	131.75	130.95 (13)		129.07	132.04
	132.38 (10)	132.91 (10)	132.77	133.98 (12)		132.09	132.71
	132.62 (13)	134.90 (12)	134.70	134.19 (11)		132.57	134.66
other carbon atoms							
					5	28.89	28.85
8	27.89	27.83	32.84	32.80	7	27.81	32.78
15	27.65	32.53	27.58	32.53	14	27.89	27.89
16	31.94	31.58	31.95	31.58	16	31.57	31.56
17	22.39	22.28	22.38	22.28	17	22.62	22.62
18	14.02	13.95	14.00	13.95	18	14.10	14.11

References: [16,45,123,125]
trivial names: *a* punicic, *b* α-eleostearic, *c* catalpic, *d* β-eleostearic, *e* jacaric, f calendic
* this assignment of olefinic signals to particular carbon atoms is not accepted by all authors (see original papers)
also α-parinaric acid (18:4 9*c*11*t*13*t*15*c*): olefinic signals at 128.12, 128.15, 128.25, 128.91, 132.69, 132.87, 132.87 and 134.38: and signals at 14.30 (C18), 21.30 (C17) and 27.94 (C8)

Table 1.13
Chemical shifts (ppm) of 18:2 conjugated diene esters.

	9*c*11*t*	10*c*12*t*	10*t*12*c*	10*t*12*t*
1	174.35	174.30	174.30	
	51.45	51.35	51.35	
2	34.20	34.20	34.20	
3	25.00	25.05	25.05	
4	29.15	29.25	29.25	
5	29.20	29.25	29.25	
6	29.20	29.50	29.50	
7	29.75	29.40	29.25	
8	27.70	29.80	29.50	
9	130.00	27.75	32.90	
10	125.75	130.00	134.55	132.30
11	128.90	125.75	128.80	130.60
12	134.80	128.80	125.90	130.55
13	32.95	134.70	130.10	132.45
14	29.45	32.90	27.75	
15	29.00	29.25	29.40	
16	31.80	31.55	31.60	
17	22.70	22.60	22.60	
18	14.10	14.10	14.10	

References: [27,28]

Table 1.14
Chemical shifts (ppm) for selected acetylenic and allenic acids and esters.

acyl compound unsaturation reference	acid 9*a* *a*	glycerol esters α 9*a* *b*	 β 9*a*	acid 9*a* 12*a* *a*	methyl ester 9*a* 13*a* *c*	methyl ester 9*a* 11*a* *d*	methyl ester 9*e* 10*e** *e*	methyl 12*e* 13*e** *e*
1	180.41	173.16	172.76	180.42	174.30	174.09	174.25	174.33
					51.45	51.36	51.44	51.43
2	34.19	34.02	34.18	34.14	34.10	34.08	34.24	34.14
3	24.76	24.88	24.85	24.72	24.95	24.97	25.14	24.98
4	29.12	29.06	29.01	28.78	29.10†	29.06		
5	28.92	28.85	28.86	28.78	28.80	28.79		
6	28.73	28.69	28.70	28.78	28.65	28.71		
7	29.28	29.14		28.78	29.00†	28.39		
8	18.85	18.76		18.76	18.75	19.26	29.06	
9	80.04	80.02	79.99	80.24	81.10	77.25	90.93	
10	80.34	80.31	80.32	74.80	78.90	65.50	204.10	
11	18.87	18.80		9.73	19.60	65.47	91.09	28.74
12	29.28	29.19		74.63	19.60	77.47	29.06	90.87
13	29.02	28.93		80.39	78.85	19.26		203.85
14	29.34	29.23		18.76	81.15	28.47		90.90
15	29.34	29.29		28.78	18.50	28.60		29.03
16	32.00	31.91		31.17	31.20	31.39	32.07	31.42
17	22.77	22.72		22.31	21.95	22.59	22.83	22.19
18	14.11	14.13		14.00	13.65	14.03	14.22	13.95

References: *a* [56], *b* [65], *c* [27], *d* [69], *e* [77]
* allenic esters
† could be interchanged

Table 1.15
Effect of acyl group on acetylenic carbon atoms with position in the chain.

position of unsaturation		2	3	4	5	6	7	8	9	10	11
Gunstone *et al.* [56], acids (80.19)											
	n	−7.44	−9.50	−2.44	−1.57	−0.83	−0.45	−0.28	−0.15	−0.05	-
	n+1	+12.46	+4.40	+1.28	+1.34	+0.63	+0.35	+0.21	+0.15	+0.12	-
	diff	19.90	13.90	3.72	2.91	1.46	0.80	0.49	0.30	0.17	-
Bus *et al.* [28], methyl esters (80.20)											
α	n	−7.25	−8.90	−2.15	−1.45	−0.80	−0.40	−0.20	−0.10	−0.05	−0.05
	n+1	+9.80	+3.70	+1.00	+1.20	+0.60	+0.30	+0.25	+0.15	+0.05	0
	diff	17.05	12.60	3.15	2.65	1.40	0.70	0.45	0.25	0.10	0.05
Lie Ken Jie *et al.* [65], glycerol esters (80.22)											
α	*n*	−7.75	−9.50	−2.48	−1.68	−0.95	−0.53	−0.33	−0.20	-	-
	n+1	+11.08	+4.07	+1.15	+1.22	+0.59	+0.30	+0.18	+0.09	-	-
	diff	18.83	13.58	3.63	2.90	1.55	0.83	0.51	0.29	0.17	0.10
β	*n*	−7.73	−9.53	−2.53	−1.67	−0.96	−0.55	−0.35	−0.23	-	-
	n+1	+11.45	+4.11	+1.15	+1.17	+0.59	+0.30	+0.19	+0.10	-	-
	diff	19.18	13.64	3.69	2.84	1.55	0.84	0.54	0.33	0.20	0.12

Interpretation (using figures of Gunstone *et al.* [56]) : a long-chain acid with acetylenic unsaturation at Δ6 will have chemical shifts of 80.19 - 0.83 (79.36 ppm) for C6 and 80.19 + 0.63 (80.82 ppm) for C7 and these two shifts differ by 1.46.

Table 1.16A
Chemical shifts (ppm) and intensities for kapok seed oil.

C1	-	178.63	0.28	olefinic		
	α	173.16	3.30	L13	130.14	7.58
		173.14	3.41	O10	129.97	4.56
	β	172.75	3.15	L9	129.92	6.11
				O9	129.67	2.61
C2	β	34.19	5.34	L10	128.09	6.39
	α	34.04	7.48	L12	127.91	6.97
		33.92	1.16	M9	109.56	0.67
				S10	109.47	0.27
C3		24.90	12.06	S9	109.20	0.24
				M8	109.11	0.80
methylene envelope		28.5-30.3 (15 signals)		allylic and others influenced by cyclopropene		
ω1	-	15.81	0.34	M6,11	27.43	3.52
	-	15.77	0.26	S7,12	27.37	1.40
	n-6	14.14	12.04	O,L	27.23	18.07
	n-9, sat	14.11	8.32	M10, S11	26.07	2.76
	-	10.97	0.22	S8	26.01	1.30
	S,M CH_2	7.42	2.63	M7	25.96	2.13
ω2	*n*-9, sat	22.74	11.06	-	25.83	0.19
	n-6	22.63	7.99	L11	25.66	7.71
ω3	*n*-9, sat	31.98	7.98			
	-	31.85	0.59			
	n-6	31.57	8.38			
Gl	β	68.94	5.10			
	(DG)	65.02	0.18			
	α	62.10	8.23			

Reference: [55]
O = oleic, L = linoleic, S = sterculic, M = malvalic, DG = diacylglycerol (1,3)
some small signals have not been assigned.

Table 1.16B
Chemical shifts (ppm) for some cyclopropene acids.

	position of unsaturated ring	$\leftarrow CO_2H$ β	α	C	(CH_2)	C	α	$CH_3\rightarrow$ β
sterculic [96, 97]	9,10	27.40	26.06	109.55	7.41	109.55	26.06	27.40
sterculic [55]	9,10	27.37	26.01	109.20	7.42	109.47	26.07	27.37
2-OH sterculic [117]	9,10	27.33	26.02	109.23	7.43	109.52	26.10	27.45
6,7-isomer [96, 97]	6,7	27.02	25.75	108.88	7.46	110.20	26.12	27.48
malvalic [55]	8,9	27.43	25.96	109.11	7.42	109.56	26.07	27.43
C_{16} acid [4]	12,13	-	-	109.6	7.3	109.7	24.7	26.0
C_{16} acid [4]	11,12	-	24.5	109.6	7.3	109.7	24.5	26.9
C_{16} acid [4]	10,11	26.7	24.6	109.7	7.3	109.8	24.6	26.9
cis-dihydrosterculic [24]	9,10	30.6	29.2	16.2	11.3	16.2	29.2	30.6

Table 1.17

Chemical shifts (ppm) for unsaturated hydroxy acids.

	CH_2	CH_2	CH(OH)	CH_2	CH_2	CH =	CH	CH_2-	ref
cis isomers									
$n=0$									
9-OH 10*c*		37.6	67.8	-	-	132.1	132.8	27.6	*b*
12-OH 10*c*		37.6	67.5	-	-	131.9	132.8	27.7	*b*
$n=1$									
12-OH 9*c*	25.8 .	37.0	71.6	35.5	-	133.0	125.5	27.4	*c*
10-OH 12*c*		36.6	71.2	35.2	-	125.1	132.9	27.2	*d*
10-OH 12*c* 15*c*		36.7	71.2	35.2	-	131.9	125.4	24.7[a]	*d*
$n=2$									
9-OH 12*c*	25.50	37.36	71.71	37.46	23.60	129.19	130.69	27.23	*e*
10-OH 6*c*	25.64	37.22	71.43	37.54	23.53	129.65	129.85	26.75	*e*
9-OH 5*c*	25.70	37.24	71.51	37.75	23.55	129.09	130.55	26.56	*e*
trans isomers									
9-OH 10*t*		37.4	73.2	-	-	132.1	133.2	32.1	*b*
12-OH 10*t*		37.4	73.1	-	-	131.9	133.3	32.2	*b*
12-OH 9*t*	25.7	36.9	71.2	40.9	-	134.4	126.2	32.7	*c*

a 126.7 (C15) 130.9 (C16); References: *b* [40], *c* [101], *d* [64], and *e* [95]

* These assignments are listed so that CHOH and CH=CH chemical shifts lie in the same vertical column. The CH_2 groups between these are absent when denoted by -. Other carbon atoms can be identified from these fixed points. Some of the values may be also affected by $COOH(CH_3)$ or CH_3 end groups. Some of these assignments have been made or adjusted by the author of this review.

Table 1.18A
Chemical shifts (^{13}C nmr) for selected oxygenated acids.

	1	2	3	4	5	6	7	8	9	10	11
1					177.6	62.81	76.89	173.69	173.84	173.93	173.30
2					34.9	32.71	27.32	33.63	33.82	34.01	27.27
3					26.1	25.64	25.67	24.54	24.68	24.88	37.03
4					30.3	28.71	29.41	28.85	28.98	28.93	208.91
5					26.8	29.55	29.41	29.18	28.98	29.01	42.59
6					38.3	29.55	29.41	28.85	28.98	29.05	23.40
7	26.0	25.6			72.3	29.55	29.41	26.25	25.80	23.77	29.21
8	31.3	33.7	27.4	32.7	34.4	29.55	29.41	27.48	31.89	42.66	27.59
9	74.8	74.6	125.1	126.1	34.4	29.55	29.41	56.67	58.54	211.03	129.11
10	74.8	74.6	133.0	133.9	72.4	29.55	29.41	56.67	58.54	42.78	133.12
11	31.3	33.7	31.8	37.2	38.5	29.55	29.41	27.48	31.89	23.89	125.78
12	26.0	25.6	74.0	74.0	26.5	29.55	29.41	26.25	25.80	29.21	135.41
13			73.9	73.8	30.9	29.55	29.41	28.85	28.98	29.40	131.05
14			33.7	33.7	30.8	29.55	29.41	29.18	29.26	29.40	130.54
15			25.4	25.5	30.4	29.55	29.41	29.18	29.26	29.25	32.52
16					33.1	31.80	31.67	31.51	31.63	31.80	31.43
17					23.7	22.55	22.40	22.29	22.42	22.62	22.25
18					14.5	13.92	13.77	13.66	13.81	14.05	13.96
(OCH_3)										51.31	51.78

	Compound	Reference	Solvent	Standard
1	methyl *erythro* -9,10-dihydroxystearate	Rakoff *et al.* [98]	$CDCl_3$	TMS
2	methyl *threo*-9,10-dihydroxystearate	Rakoff *et al.* [98]	$CDCl_3$	TMS
3	methyl *threo*-12,13-dihydroxy-*cis*-9-octadecenoate	Rakoff *et al.* [98]	$CDCl_3$	TMS
4	methyl *threo*-12,13-dihydroxy-*trans*-9-octadecanoate	Rakoff *et al.* [98]	$CDCl_3$	TMS
5	7,10-dihydroxystearic acid	Knothe *et al.* [63]	CD_3OD	-
6	octadecane-1-ol	Bascetta *et al.* [13]	$CDCl_3$	TMS
7	octadecane-1-hydroperoxide	Bascetta *et al.* [13]	$CDCl_3$	TMS
8	methyl *cis*-9,10-epoxystearate	Bascetta *et al.* [13]	$CDCl_3$	TMS
9	methyl *trans*-9,10-epoxystearate	Bascetta *et al.* [13]	$CDCl_3$	TMS
10	methyl 9-oxostearate	Tulloch [122]	$CDCl_3$	TMS
11	methyl licanate (4-oxo 18:3 9*c*11*t*13*t*)	Spitzer *et al.* [118]	-	-

Table 1.18B
Chemical shifts (ppm) for ω1-ω3 signals in epoxidised soybean oil and epoxidised linseed oil.

		ppm	% mol			
			soya		linseed	
ω1	sat + 9,10- ep	14.11	35		27	
	12,13- ep	13.99	58		21	
	15,16- ep	10.88 ⎤ 10.61 ⎥ 10.49 ⎦	7		52	
ω2	sat + 9,10- ep	22.67	33		26	
	12,13- ep	22.57	64		23	
	15,16- ep	21.24 ⎤ 21.17 ⎦	3		51	
				(*a*)		(*b*)
ω3	sat	31.93	21	20	26	12
	9,10- ep	31.86	22	26	41	20
	12,13- ep	31.68 ⎤ 34.47 ⎦	57	49	33	16

Reference: [55]
a The ω3 signal does not include 15,16- epoxides and therefore the quantitative results have been adjusted to give totals of 95% for soybean and 48% for linseed, these values being derived from the ω1 and ω2 signals.

Table 1.18C
Chemical shifts (ppm) of some furanoid acids.

R^1	R^2	2	3	4	5	α	β	γ	br Me	ref
H	H	156.5	104.9	104.9	154.4	28.1	28.1	29.2	-	*a*
H	H	154.5	105.1	104.9	154.8	[28.0	28.0	29.9*	-	*b*
						[28.2	28.1	29.3†		
Me/H	H/Me	[153.3	107.9	113.9	149.3	-	-	-	9.9	*d*
		[153.5	113.9	107.9	149.5					
Me	H	149.4	113.8	107.7	153.5	[25.9	28.7	29.2*	9.9	*a*
						[28.0	27.9	31.5†		
Me	H	153.5	107.7	113.9	149.5	31.5	-	26.0	9.9	*c*
Me	Me	148.4	114.4	114.4	148.4	31.5	-	26.1	8.3	*c*

References: *a* [73], *b* [68], *c* [98], *d* [78]
* CO_2H side, † CH_3 side

Table 1.19A
Chemical shifts (ppm) of phosphatidylcholines (synthetic compounds).

sample	1	2	3	4	5	6	7[c]
solvent[a]	C	C		C	C	C	C
$\overset{+}{N}Me_3$	54.27	54.25	-	54.26	54.51	54.49	54.42
$\overset{+}{N}CH_2$	66.23(5.5)	66.22(5.9)	66.8(6)	66.23(5.6)	66.42	66.39	66.45
OCH_2	59.34(4.3)	59.37(4.0)	59.7(5)	59.37(4.1)	59.31	59.34(4.9)	59.47
Gl-3	63.30(4.7)	63.38(4.6)	64.1(5)	63.34(4.6)	63.48	63.39(4.4)	63.59
Gl-2	70.59(7.5)	70.52(7.3)	71.1(8)	70.58(7.4)	70.60	70.60(7.2)	70.61
Gl-1	63.01	63.03	63.0	63.04	63.00	63.01	63.09
C1[b]	173.51	173.52	173.7	173.49	173.50	173.54	173.53
	173.15	173.17	173.4	173.14	173.15	173.16	173.19
C2[b]	34.14	34.16	34.2	34.16	34.12	34.16	34.22
	34.34	34.35	34.3	34.36	34.32	34.35	34.37
C3[b]	24.89	24.93	25.1	24.93	24.76	24.92	24.99
	24.98	25.02		25.02	24.89	24.98	
ω3	31.88	31.96	32.9	31.96	31.53	31.93	31.99
ω2	22.67	22.71	23.1	22.71	22.58	22.70	22.70
ω1	14.10	14.12	14.0	14.12	14.08	14.12	(14.10)

[a] C = $CDCl_3$; CM = $CDCl_3$, CD_3OD (2:1); CMW = $CDCl_3$, CD_3OD, D_2O (50:50:15); [b] α and β-chains respectively; [c] the authors cite values related to the ω1 signal = 0.00: the reviewer has raised all the values by 14.10.

1 dilauro PC Gunstone [55]
2 dimyristo PC Gunstone [55]
3 dipalmito PC Birdsall *et al.* [18]
4 distearo PC Gunstone [55]
5 dilinoleo PC Gunstone [55]
6 1-16:0, 2-18:1 PC Gunstone [55]
7 1-16:0, 2-18:1 PC Santarén *et al.* [107] (also in D_2O)

Table 1.19B
Chemical shifts (ppm) of phosphatidylcholines (synthetic compounds).

sample	8[c]	9[c]	10	11	12	13	14
solvent[a]	C	C	C	C	CM	CMW	CMW
$\overset{+}{N}Me_3$	54.42	54.46	54.04	54.04	54.00	54.52	54.44
$\overset{+}{N}CH_2$	66.50	66.81	66.08	66.65	66.52	66.85(7.2)	66.71(7.6)
OCH_2	59.42	59.41	59.55	60.21	58.96(5.2)	59.71(3.9)	59.62(4.8)
Gl-3	63.36	63.32	63.96	64.68	63.54(5.3)	64.08(4.5)	64.00(4.8)
Gl-2	70.62	71.05	70.79	70.99	70.40(7.9)	70.97(7.7)	70.92(7.9)
Gl-1	63.09	63.00	62.60	62.79	62.60	63.30	63.23
C1[b]	173.57	173.52	176.46	174.2	173.87		
	173.19	172.94	176.00	173.9	173.50		
C2[b]	34.20	33.78	35.68	34.14	33.96		
	34.21	34.15	35.85	34.34	34.11		
C3[b]	24.99	24.92	-	24.95	24.75		
	-	-	-	-	24.78		
ω3	31.68	31.94	-	31.67	31.56		
	31.57	31.44	-	-	-		
ω2	22.73	22.69					
	22.59	22.60	18.01	22.65	22.44		
ω1	(14.10)	(14.10)	12.85	13.52	13.68		

footnotes: see previous page

8 1-18:0, 2-18:2 PC Santarén *et al.* [107] (also in D_2O)
9 1-18:0, 2-20:4 PC Santarén *et al.* [107] (also in D_2O)
10 $(4:0)_2$ PC Burns *et al.* [26] (also in micelles)
11 $(6:0)_2$ PC Burns *et al.* [26] (also in micelles)
12 $(8:0)_2$ PC Basti *et al.* [14] (also thio PC)
13 $(16:0)_2$ PC Murari *et al.* [90] (and several related compounds)
14 $(18:1)_2$ PC Murari *et al.* [90] (and several related compounds)

Table 1.20
Chemical shifts (ppm) of phosphatidylcholines (natural mixtures).

sample	1	2	3	4[c]	5	6
solvent[a]	C	CM	CM	CM	CMW	CMW
$\overset{+}{N}Me_3$	54.29	54.30	54.34	54.32	54.45	54.47
$\overset{+}{N}CH_2$	66.26(5.9)	66.59	66.67	66.62(2.7)	66.79(7.0)	66.67
OCH_2	59.34(4.3)	59.28(4.7)	59.33(4.8)	59.32(4.7)	59.65(4.8)	59.71(4.8)
Gl-3	63.33(4.8)	63.76(5.1)	63.81(5.0)	63.78(5.2)	64.02(4.8)	67.72(5.6)
Gl-2	70.51(7.7)	70.59(7.7)	70.68(7.4)	70.63(7.9)	70.94(7.9)	69.08(7.2)
Gl-1	63.01	62.91	62.96	62.94	63.26	63.54
C1[b]	173.52	174.07	174.05	174.11		
	173.49	174.04	-	-		
	173.13	173.69	173.69	173.71		
C2[b]	34.12	34.25	34.29	34.29		
	34.31	34.40	34.44	34.43		
C3[b]	24.89	25.09	25.09	25.09		
	24.97	-	25.13	25.11		
ω3 *n*-9, sat.	31.93	32.12	32.13	32.14		
	-	-	32.01	32.00		
n-6	31.52	31.71	31.74	31.73		
ω2 *n*-9, sat.	22.70	22.87	22.87	22.88		
n-6	22.58	27.75	22.77	22.77		
n-3	20.56	20.72	20.74			
ω1 *n*-3	14.29	14.35	14.35	14.35		
n-9, sat.	14.13	14.18	14.17	14.19		
n-6	14.09	14.15	-	-		

a C = $CDCl_3$; CM = $CDCl_3$, CD_3OD (2:1); CMW = $CDCl_3$, CD_3OD, D_2O (50:50:15)
b α- and β-chains respectively *c* also 33.86, 25.31, 24.87,

1 soybean PC — Gunstone [55]
2 soybean PC — Gunstone [55]
3 rape PC — Gunstone [55]
4 egg yolk PC — Gunstone [55]
5 egg yolk PC — Murari *et al.* [90]
6 egg yolk PC (2-lyso) — Murari *et al.* [90]

Table 1.21
Chemical shifts (ppm) of phosphatidylethanolamines (synthetic compounds).

sample	1	2	3	4	5	6
solvent[a]	C	C	CMW	CM	-	-
$\overset{+}{N}CH_2$	40.43	40.57	40.86 (6.0)	41.34 (4.8)	41.05 (4.8)	40.94 (5.0)
OCH_2	62.22	62.27	62.13 (4.7)	62.23 (5.0)	62.04 (4.5)	62.12 (3.8)
Gl-3	63.91	64.06 (4.1)	64.11 (4.6)	64.51 (5.0)	64.26 (4.9)	64.14 (4.0)
Gl-2	70.29 (7.4)	70.47 (7.4)	70.93 (8.2)	71.24 (7.9)	70.91 (7.8)	70.84 (7.9)
Gl-1	62.58	62.69	63.24	63.29	62.97	62.97
C1[b]	173.41	173.34		174.93	174.03	174.06
	173.13	173.07		174.56	173.63	173.72
C2[b]	34.10	34.18		34.67	34.46	34.44
	34.29	34.37		34.83	34.63	34.59
C3[b]	24.90	24.97		25.47	25.30	25.34
	24.97	25.04				
ω3	31.95	31.98		32.48	32.36	32.37
ω2	22.71	22.72		23.16	22.99	23.11
ω1	14.12	14.09		14.27	14.11	14.26

a C= $CDCl_3$; CM= $CDCl_3$, CD_3OD (2:1); CMW= $CDCl_3$, CD_3OD, D_2O (50:50:15).
b α and β-chains respectively.

1 dilauro PE Gunstone [55]
2 dipalmito PE Gunstone [55]
3 dipalmito PE Murari *et al.* [90] (and several related compounds)
4 dipalmito PE Jiang *et al.* [59] (also thio PE and other solvents)
5 dipalmito PE Birdsall *et al.* [18]
6 dioleo PE Birdsall *et al.* [18]

Table 1.22
Chemical shifts (ppm) of phosphatidylethanolamines (natural mixtures).

sample	1	2	3	4
solvent[a]	CM	CM	CMW	CMW
$\overset{+}{N}CH_2$	40.69	40.59	40.74(6.2)	40.86(7.0)
OCH_2	62.08(5.0)	62.13	62.07(4.8)	62.29(4.6)
Gl-3	64.07(4.7)	64.07	64.07(4.7)	67.21(5.5)
Gl-2	70.59(8.0)	70.55[c]	70.88(7.5)	69.11(7.6)
Gl-1	62.81	62.81	63.16	65.56
	174.04	173.99		
C1[b]	174.01	-		
	173.64	173.61		
	-	173.39		
	-	172.90		
C2[b]	34.29	34.26		
	34.43	34.40		
C3[b]	25.13	25.13		
	-	25.09		
ω3 *n*-9, sat.	32.17	32.16		
n-6	31.76	31.74		
ω2 *n*-9, sat.	22.90	22.90		
n-6	22.79	22.79		
n-3	20.76	20.75		
ω1 *n*-3	14.36	14.37		
n-9, sat ⎤	14.17	14.20		
n-6 ⎦	-	-		

a CM = $CDCl_3$, CD_3OD (2:1); CMW = $CDCl_3$, CD_3OD, D_2O (50:50:15); *b* α and β-chains respectively; *c* also 71.97 and 70.68 ppm

1 soybean PE Gunstone [55]
2 egg yolk PE Gunstone [55]
3 egg yolk PE Murari *et al.* [90]
4 egg yolk lyso PE Murari *et al.* [90] (and several other compounds)

Table 1.23
Chemical shifts (ppm) and coupling constants for some other phospholipids.

sample	1	2	3	4	5	6
solvent [a]	CMW (50:50:15)	CM (9:1)	CM (2:1)	C	CMW (55:55:15)	CMW (55:55:15)
$\overset{+}{N}Me_e$	54.48	-	-	-	-	-
$\overset{+}{N}CH_2$	66.89(7.3)	-	-	-	-	-
OCH_2	59.61(4.2)	63.68	63.67	-	-	-
Gl-3	65.54(5.4)	64.83	64.18	64.7	70.2	71.7(67.5)[e]
Gl-2	78.95(8.3)	70.25	70.49(7.5)	70.8	78.4	78.7(71.4)
Gl-1	72.16	62.55	62.77	63.4	61.4	65.9(63.2)
other		54.11[b]	54.31[b]	[174.1[c]	[67.7[d]	[68.8[d]
		170.00	169.62	173.7]	68.7]	68.9]
				[35.3		
				35.2]		
				25.8		

1 bisether PC (**1**, R= R^1 = $C_{16}H_{33}$) Murari *et al.* [90]
2 PS (**2**) Lingh *et al.* [81]
3 PS (**2**) Gunstone [55]
4 PA Birdsall *et al.* [18]
5 PA (**3**, diphytanyl ether) Stewart *et al.* [120]
6 PG (**4**, diphytanyl ether, R^1 = H) Stewart *et al.* [120] also deoxy compounds
7 PGP (**4**, diphytanyl ether R^1 = PO_3H_2) Stewart *et al.* [120] also deoxy compounds
8 PAF (**1**, R = $C_{18}H_{37}$,R^1 = Ac) Convert *et al.* [31]
9 lyso PAF (**1**, R = $C_{16}H_{33}$, R^1 = H) Murari *et al.* [90]
10 PC ether (**1**, R = $C_{16}H_{33}$,R^1 = $C_{15}H_{31}CO$) Murari *et al.* [90]

a C = chloroform, M = methanol, W = water (all deuterated)
b CH and COOH in serine unit, *c* C1-C3 respectively, *d* C-1 in ether groups,
e Gl-1′ etc from second glycerol unit, *f* C-1 in ether group, *g* CH_3CO,

Table 1.23 (continued)

Chemical shifts (ppm) and coupling constants for some other phospholipids.

sample solvent[a]	7 CMW (55:55:15)	8 M	9 CMW (50:50:15)	10 CMW (50:50:15)
$\overset{+}{N}Me_3$	-	54.80	54.49(3.5)	54.50
$\overset{+}{N}CH_2$	-	67.34(6.0)	66.89(6.9)	67.03(7.0)
OCH_2	-	60.67(4.5)	59.65(5.1)	59.60(4.6)
Gl-3	71.5(66.9)[e]	65.50(4.5)	67.92(5.8)	64.63(4.8)
Gl-2	78.8(71.3)	73.19(7.4)	70.20(7.5)	72.45(9.1)
Gl-1	65.9(66.9)	70.14	72.07	69.76
other				
	69.3[d]	72.66[f]	72.30[f]	72.22[f]
	70.4			
		[172.30[g]		
		21.10]		

Footnotes on previous page.

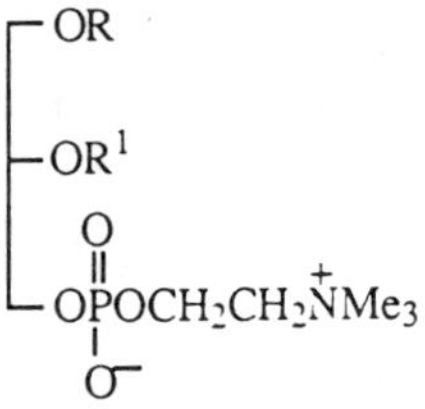

OR
OR[1]
$OP(=O)(O^-)OCH_2CH_2\overset{+}{N}Me_3$

1

$OCOC_{15}H_{31}$
$OCOC_{15}H_{31}$
$OP(=O)(O^-)OCH_2CH(^+NH_3)COOH$

2

OPO_3H_2
RO
OR

3

O P O (=O, OH)
RO — OH
RO — OR

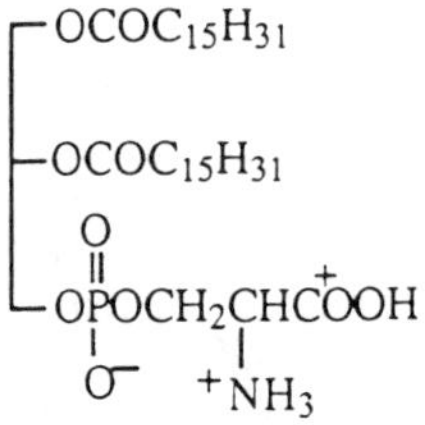

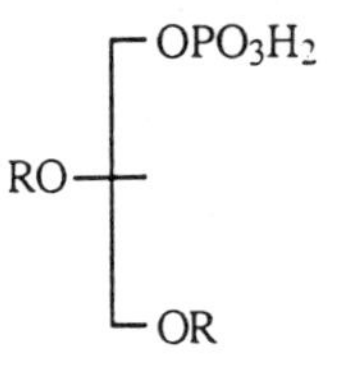

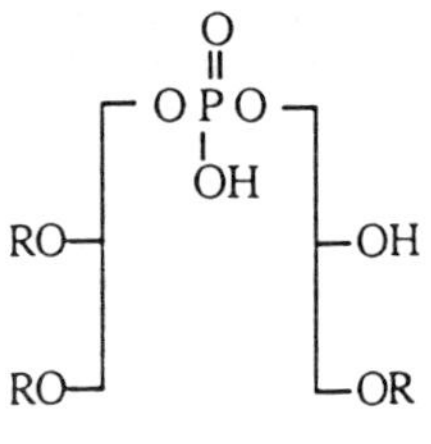

4

Table 1.24
Some chemical shifts (ppm) from the spectrum of a mixture of TAG, PE, and PC from soybean ($CDCl_3$, CD_3OD 2:1).

	TAG	PE	PC
174.14	-	-	C1α
174.11	-	C1α	-
173.91	C1α	-	-
173.79	-	-	C1β
173.74	-	C1β	-
173.49	C1β	-	-
[70.72, 70.62]	-	G1-2	G1-2
69.38	G1β	-	-
66.65	-	-	$CH_2\overset{+}{N}$
[64.06, 63.99]	-	G1-3	-
63.88	-	-	G1-3
62.86	-	G1-1	G1-1
62.46	G1α	-	-
[61.98, 61.91]	-	CH_2O	-
[59.44, 59.38]	-	-	CH_2O
54.30	-	-	N^+Me_3
40.87	-	$CH_2\overset{+}{N}$	

Reference: [55]
Other signals in this spectrum are derived from all three components of the mixture and are not listed here.

Table 1.25A
Chemical shifts (ppm) for the glycerol carbon atoms of acyl glycerols in descending order (mainly).

	Sacchi *et al*[a]*	Gunstone[b]*	Mazur *et al*[c]*	Gunstone[d]*
2-mono-	72.12	74.97	74.70	75.36
1,2-di-	71.83	72.25	72.06	72.40
1-mono-	69.90	70.27	-	70.24
tri-	68.84	68.93	68.85	69.41
1,3-di-	67.55	68.23	68.15	67.71
†				
1-mono	64.64	65.04	-	65.58
1,3-di	64.73	65.04	64.96	65.36
1-mono	63.12	63.47	-	63.46
1,2-di	62.22	62.20	62.10	62.92
tri	62.03	62.12	62.06	62.50
2-mono	61.02	62.05	60.90	61.06
1,2-di	60.74	61.58	61.37	60.82

a Sacchi *et al.* [104]: figures based on oleic glycerides.
b Gunstone [50]: figures based on palmitic glycerides.
c Mazur *et al.* [85]: figures based on glycerol esters of 22:0 and/or 10:0 and/or 8:0.
d Gunstone [55].
* Measurements in $CDCl_3$ except for the fourth colum (CD_3Cl, CD_3OD2:1).
† Figures above this sign relate to the β carbon atom, those below to α carbon atoms.
Chemical shifts for glycerol ($CDCl_3/CD_3OD$) : 72.78 and 63.66 ppm

Table 1.25B
Structural and chemical shifts (ppm) of acetylated monoacylglycerols.

	OCOR OH OH	OCOR $OCOCH_3$ OH	OCOR OH $OCOCH_3$	OCOR $OCOCH_3$ $OCOCH_3$
Gl-1	65.15	62.07	65.00	62.00
Gl-2	70.26	72.39	68.14	69.16
Gl-3	63.39	61.40	65.26	62.33
C-1	174.31	173.84	173.96	173.38
$COCH_3$ (β)	-	21.00 170.98	-	20.88 170.16
$COCH_3$ (α)	-	-	20.79 171.10	20.69 170.57

Table 1.26
Chemical shifts (ppm) for C1, C2, and C3 signals with different end groups (saturated chains).

name/structure			C1	C2	C3
carboxylic acid [a]	-COOH		180.62	34.25	24.80
methyl ester[a]	$-COOCH_3$		174.05	34.17	25.12
long-chain ester [b]	-COOR	acyl	173.94	34.48	25.06
		alkyl	64.39	28.70	25.98
glycerol ester [c]	α		173.27	34.07	24.90
	β		172.86	34.24	24.94
amide [b]	$-CONH_2$		176.37	36.02	25.57
sec. amide [d]	-CONHBu	acyl	172.9	36.9	25.8[f]
		alkyl	39.1	31.8(?)	20.0
nitrile [b]	-CN		119.82	25.43	17.14[g]
alcohol [b]	$-CH_2OH$		63.01	32.81	25.78
[e]			62.79	32.68	25.65
acetate [b]	$-CH_2OAc$		64.64	28.67	25.97[h]
hydroperoxide [e]	$-CH_2OOH$		76.89	27.38	25.70
epoxide[e]	-CH(O)CH-		47.0, 52.3	32.7	26.2

a Gunstone *et al.* [56], *b* Gunstone [55], *c* Lie Ken Jie [65]
d *N*-butylpalmitamide, *e* Bascetta *et al.* [13], *f* Bilyk *et al.* [17],
g C4 28.71, C5 28.81 *h* also $COCH_3$ 171.14, 20.97.

Table 1.27

Chemical shift (ppm) for olefinic carbon atoms in oleic, elaidic and linoleic derivatives.

	18:2 (9*c*)			18:1 (9*t*)			18:2 (9*c* 12*c*)			
structure	C9	C10	diff	C9	C10	diff	C9	C10	C12	C13
COOH [a]	129.78	130.09	0.31	130.23	130.54	0.31	130.01	128.24	128.08	130.20
$COOCH_3$ [a]	129.78	130.02	0.24	-	-	-	-	-	-	-
CO_2R [b] acyl	129.76	130.00	0.24	-	-	-	130.05	128.06	127.93	130.21
CO_2R [b] alkyl	129.78	129.97	0.19				130.06	128.03	127.93	130.19
glycerol ester [c] α	129.72	130.01	0.29	-	-	-	127.97	128.08	127.91	130.18
glycerol ester [c] β	129.69	130.04	0.35							
$CONH_2$ [b]	129.71	129.98	0.27	-	-	-	-	-	-	-
CN [b]	129.68	130.03	0.35	-	-	-	128.10	127.97	127.89	130.20
CH_2OH [b]	129.82	129.95	0.13	130.28	130.42	0.14	130.11	128.01	127.94	130.20
CH_2OAc [b]	129.78	129.98	0.20	130.25	130.45	0.20	130.05	128.07	127.96	130.17

References: *a* [57], *b* [55], *c* [65]

Table 1.28
Chemical shifts (ppm) for selected signals in mixtures of long-chain esters.

	lauryl oleate + oleyl laurate		lauryl linoleate + linoleyl laurate	
acyl chain	sat	unsat	sat	unsat
C1	173.97	173.95	173.99	173.95
C2	——34.43——		——34.42——	
C3	——25.06——		25.05	25.04
alkyl chain				
C1′	64.41	64.38	64.42	64.38
C2′	——28.69——		——28.69——	
C3′	——25.97——		——25.96——	
olefinic signals	acyl	alkyl	acyl	alkyl
C13	-		——130.20——	
C12	-		——127.93——	
C10	129.98		130.08	130.05
C9	129.76	129.79	128.04	128.05

Reference: [55]

Table 1.29
Chemical shifts (ppm) for C2 signals in butter fat and some butanoic acid glycerides.

	C_4		long-chain	
	1,3	2	1,3	2
butter	35.94	-	34.07	34.24
BBO	35.96	36.14	34.09	-
PPB	35.98	-	34.13	34.30
BPB	-	36.14	34.11	-
BBB	35.96	36.12	-	-

Reference: [93]; B=4:0, P=16:0, O=18:1.

Table 1.30
Chemical shifts (ppm) for $(4{:}0)_3$ to $(10{:}0)_3$ and $(18{:}0)_3$ - α and β chains.

	$(4{:}0)_3$	$(6{:}0)_3$	$(8.0)_3$		$(10{:}0)_3$		$(18{:}0)_3$
C1	173.14	173.31	173.29		173.31		173.27
	172.74	172.89	172.89		172.90		172.88
C2	35.94	34.03	34.08		34.08		34.07
	36.09	34.19	34.24		34.25		34.24
C3	-	24.56	24.88		24.90		24.91
		24.59	24.92				24.94
C4	-	-	29.09		29.15		29.16
			29.05		29.11		29.12
C5	-	-	28.94	(29.00)*	29.31		29.32
			28.96		29.33		29.34
C6	-	-	-		29.47		-
					29.48		
ω4	-	-	-		29.31		29.42
ω3	-	31.26	31.68	(31.74)*	31.94	(31.95)*	31.98
		31.22	31.70				
ω2	18.37	22.31	22.63	(22.66)*	22.70		22.73
	18.40						
ω1	13.63	13.90	14.07		14.12		14.13
	13.57						

Reference: [70]
*Distinctive values observed in the spectrum of mixture of the methyl esters of 8:0, 10:0, and 12:0 (see text)

Table 1.31
Chemical shifts (ppm) and intensities for butter fat.

C1α	173.26	3.04	ω1 *n*-9, sat	14.14	8.51
C1α 4:0	173.06	0.57	ω1	14.03	0.35
C1β	172.85	1.84	ω1 6:0	13.91	0.71
			ω1 4:0	13.63	1.26
C2α 4:0	35.91	1.51	ω2 *n*-9, *n*-6, sat	22.72	9.17
C2β	34.22	3.90	ω2 6:0	22.32	0.73
$C_2α$	34.05	5.73	ω2 4:0	18.36	1.50
C3 α,β	24.88	7.25			
C3 6:0	24.55	0.81	ω3 *n*-9, sat	31.96	8.34
			ω3 *n*-7	31.79	0.65
9*c*	130.00	1.69	ω3 8:0	31.69	0.41
9*c*	129.70	1.43	ω3 6:0	31.26	0.67
11*t*	130.30	0.38			
11*t*	130.29	0.40	allylic *t*	32.64	1.04
			c	27.24	2.44
			c	27.19	2.32
			cc	25.63	0.33

Reference: [54]

Table 1.32
Chemical shifts (ppm) for ω2 and ω3 carbon atoms in lauric oils.

	CNO	PKO	PKO	PKO(st)	PKO(ol)	Cuphea	Scab	MCT
ω3 signals								
?	32.26	-	-	-	-	-	32.25	-
n-9, sat	32.03	31.99	31.95	31.96	31.95	31.99	31.96	-
C_{10}	-	31.94	-	-	-	31.94	31.92	31.90
C_8	31.77	31.73	31.69	31.70	31.69	31.73	31.72	31.69
n-6	31.63	31.60	31.56	-	31.56	31.60	31.58	-
C6	31.35	-	-	-	31.27	-	-	-
ω2 signals								
?	23.00	22.95	-	-	22.94	-	-	-
n-9, sat	22.78	22.73	22.71	22.72	22.71	22.74	22.72	22.69
n-6, C_8	22.70	22.64	22.63	22.64	22.63	-	22.65	22.63
?	22.55	22.50	-	-	22.49	-	-	-
C6	22.39	-	-	-	22.32	-	-	-

Reference: [55]

Table 1.33
Triacylglycerols and 1,3-diacylglycerols in palm oil and its fractions.

chemical shift (ppm)	intensities		
	oil	olein	stearin
68.91-68.93	2.96	3.25	2.39
62.10-62.12	5.33	5.79	4.31
65.35-65.40	0.27	0.27	0.15
65.05-65.06	0.49	0.47	0.31
TAG (% mole)	91.6	92.4	93.6
DAG (% mole)	8.4	7.6	6.4

Reference: [55]

These figures were not obtained under conditions required for the best quantitative results but indicate the potential value of this method.

Table 1.34
Chemical shifts (ppm) for ω1, ω2, and ω.3 carbon atoms.

acyl chain	sat	*n*-9	*n*-7	*n*-6	*n*-3	8:0	10:0
ω1	———	14.13	———	14.09	14.29	14.09	14.12
ω2	———	22.60	———	22.71	20.57	22.63	22.69
ω3	—31.95—		31.83	31.55	-	31.69	31.90

Table 1.35
Chemical shifts (ppm) for C1, C2, and C3 carbon atoms.

	C1		C2		C3	
	α	β	β	α	β	α
saturated	173.26	172.85	34.24	34.07	24.90	
oleic	173.20	172.83	34.20	34.04	24.87	
linoleic	173.17	172.77	34.19	34.02	24.85	
erucic	173.23	172.82	34.21	34.04	24.88	
petroselinc	173.05	172.66	34.11	33.95	24.54	24.51
α-linolenic	173.04	172.64	34.07	33.92	24.51	24.49
columbinic*	173.02	172.62	33.50	33.32	24.70	24.64

*18:3 (5*t*9*c*12*c*)

Table 1.36
Distribution of 18:1 and 18:2 between α and β acyl chains on the basis of the carbonyl signal (C1).

	nmr		lipolysis	
oil	18:1	18:2	18:1	18:2
groundnut	46.3	30.5	48.3	31.5
corn	27.0	57.0	27.4	57.7
high-oleic sunflower	80.5	10.0	82.1	9.1

Reference: [131]

Table 1.37
Analysis of oils containing petroselenic acid by ^{13}C nmr spectroscopy.

	total petroselinic acid				α, β (%)	
signal	C1	C2	C3	olef	C1	C2
carrot	70.9	70.3	70.5	70.5	83,48	81,51
celery	69.3	67.9	68.1	66.7	78,54	73,59
parsnip	60.1	60.0	58.2	58.7	82,16	78,25
parsley	79.8	77.7	79.0	78.5	82,77	75,81
caraway	38.9	41.6	-	36.8	34,47	34,54
coriander	75.8	-	74.8	75.0	90,49	?,53

Reference: [49]

Table 1.38
Chemical shifts (ppm) for allylic carbon atoms.

	single	double
general value	~27.2	25.65
α-linolenic	20.57 (C17, ω2)	25.55 (C14, ω5)
petroselinic	26.84 (C5) 27.30 (C8)	-
γ-linolenic	26.86 (C5)	-
columbinic*	31.87 (C4), 32.62 (C7) 31.83	25.72 (C11)

*18:3 (5*t*9*c*12*c*)
The shifts cited above are alternative or additional to the general values.

Table 1.39
Olefinic signals (ppm) for oleic, linoleic, and α-linolenic glycerides.

α-linolenic	linoleic	oleic
131.946, 131.949 (16)		
130.223, 130.188 (9)	130.210, 130.218 (13)	
	130.002, 129.977 (9)	130.022, 130.037 (10)
		129.721, 129.695 (9)
128.311, 128.321 (13)		
128.262, 128.250 (12)		
	128.106, 128.124 (10)	
	127.939, 127.927 (12)	
127.794, 227.811 (10)		
127.157, 127.150 (15)		

also 20:1 11*c*, 129.939 and 129.948 (12) and 129.841 and 129.813 (11)
22:1 13*c*, 129.912 and 129.916 (14) and 129.876 and 129.856 (13)

Reference: [131], with corrections made with the author's consent:chemical shifts are for α and β chains and are listed in decreasing order: numbers in parentheses refer to carbon atoms.

Table 1.40
Chemical shifts (ppm) for olefinic carbon atoms in some less common glycerides.

petroselinic (18:1 6*c*)	128.95(6), 130.55(7)
γ-linolenic (18:3 6*c*9*c*12*c*)	129.45(6), 128.35(7),128.00(9), 128.42(10), 127.60(12), 130.40(13)
columbinic (18:3 5*t*9*c*12*c*)	129.27(5), 131.00(6), 129.23(9), 128.42(10), 127.81(12), 130.22(13)

Reference: [47,49]

Table 1.41
Chemical shifts (ppm) for olefinic carbon atoms in petroselinate, oleate, *cis*-vaccenate and erucate.

	chemical shifts (ppm)		diff
18:1 6*c*	130.55	128.95	1.60
18:1 9*c*	130.01	129.68	0.33
18:1 11*c*	129.92	129.84	0.08
22:1 13*c*	129.883	129.858	0.025

Table 1.42
Fatty acid composition (%) and iodine values (IV) of some edible oils.

	^{13}C nmr				gas chromatography				IV
	sat	monoene	diene	IV	sat	monoene	diene	IV	obs
safflower	12	19	68	140	11	15	73	145	142
rice bran	18	43	33	98	15	42	33	98	90
sesame	18	49	45	126	14	45	41	114	119
groundnut	22	41	36	102	21	47	32	100	98
soybean	25	31	49	117	20	28	53	121	135
cottonseed	31	18	52	110	28	21	51	111	112
palm	51	38	11	54	50	43	8	53	56
coconut	90	4	1	5	92	6	2	9	8

Reference: [58]

Table 1.43
The distribution of saturated and unsaturated acids between the α and β positions in selected animal fats.

			saturated			*unsaturated*		
animal	source	diet	α	β	total	α	β	total
duck	subcutaneous	-	28	6	34	40	26	66
duck	foie-gras	-	31	4	35	35	29	64
rabbit	perirenal	low fat	42	18	60	21	18	39
rabbit	perirenal	coconut oil	52	23	75	15	10	25

Reference: [23]

Table 1.44
^{13}C nmr signals (chemical shift, ppm) observed in the spectra of lipid extracted from Atlantic salmon†.

	n-3	*n*-6	other		
ω1	14.29	14.09	14.13		
ω2	20.56	22.71	22.63, 22.68		
ω3	-	31.80	31.92, 31.94		
	22:5	18:4	20:5*	22:6	
	Δ7	Δ6	Δ5	Δ4	other
C1 β	-	-	172.63, 172.59	172.11	172.81, 172.77(?)
C1 α	-	-	173.04, 172.98	172.51	173.23
C2 α	-	-	33.36	-	33.88, 34.03, 34.18
C2 β	-	-	33.55	-	
C3	-	24.67	-	-	24.73, 24.86
C4	28.75	-	26.49	-	
C5	-	26.84	-	-	
C6	27.00	-	-	-	
olefinic signals (selected)					
all *n*-3 esters		127.00 (ω3) and 131.99 (ω4)			
22:6		129.44 (C5) and 127.61 (C4)			
20:5		128.76 (C6) and 128.95 (C5)			

†for details of other signals see original paper [6]
*including signals from 20:4(*n*-6) which is likely to be small

Table 1.45

Composition of lipid extracted from Atlantic salmon by ^{13}C nmr spectroscopy and by gas chromatography.

	nmr		gc
18:1	12.4	(13.6)	14.7
20:1, 22:1	26.5	(25.0)	23.2
16:1, 18:1 (*n*-7)	11.5	(10.3)	12.5
18:2	3.4	(3.1)	2.8
18:3	0.9	(0.7)	1.0
18:4	1.9	(2.3)	2.0
20:4	3.1	(1.5)	1.4
20:5	3.9	(5.2)	5.2
22:5	2.0	(2.1)	1.4
22:6	7.1	(7.1)	6.7
other (sat)	27.3	(29.1)	29.1

Reference: [6]
*with added $Cr(acac)_3$

Table 1.46

Composition of a mixture of 18:2, 20:5, and 22:6 by nmr spectroscopy.

	by wt*	nmr without*	[$Cr(acac)_3$] with*
18:2	53.8	54.5	55.3
20.5	12.3	12.2	11.6
22:6	33.9	33.3	33.1

Reference: [6]
*results are % mole

Table 1.47

Distribution of 20:5 and 22:6 between α and β positions in Atlantic salmon.

	α (%)	β (%)	based on signals at (ppm)		
20:5	57.0	43.0	33.36	33.55	(C2)
22:6	18.4	81.6	172.51	172.11	(C1)

Reference: [6]

Table 1.48

Chemical shifts for C8-C11 for triolein, trielaidin, and a mixture of the two.

	OOO	EEE	mixture
C8	27.19	32.58	27.19 and 32.59
C9	129.70	130.18	129.70 and 130.17
C10	130.00	130.49	130.00 and 130.48
C11	27.24	32.63	27.24 and 32.64

Reference: [53]

Table 1.49

Information available from ω1-ω3 signals.

	major signal	additional signals		
ω1	14.12	-	13.66 (14*t*)	14.40 (15*c*)
ω2	22.72	22.37 (13*c*)	22.21 (13*t*)	20.53 (15*c*)
ω3	31.95	31.56 (12*c*)	31.43 (12*t*)	

Reference: [53]

Table 1.50

Chemical shifts and assignments of 23 olefinic signals observed in one or more of 12 hydrogenated vegetable fats.

131.85 (5t), 131.05 (6t), 130.74 (7t), 130.59 (8t, 14t), 130.49 (9t), 130.43 (10t), 130.40 and 130.30 (11-13t), 130.25 (10t), 130.18 (9t), 130.09 and 130.04 (8c, 8t, 14c, 14t), 130.02 (9c), 129.95 (10c), 129.92 and 129.83 (11-13c, 7t), 129.78 (10c), 129.70 (9c), 129.58 (8c), 129.55 (?), 129.41 (6t), 129.37 (?), and 128.64 (5t).

Reference: [53]

Table 1.51

18:1 isomer distribution (%) in a rape seed oil hydrogenated to m.p 43° (saturated acids are excluded).

	5	6	7	8	9	10	11-13	14,15	total
cis	-	-	-	3.4	7.2	7.2	2.8	-	20.6
trans	2.3	4.8	7.6	13.8	13.7	15.5	18.3	3.4	79.4

Reference: [53]

Table 1.52

Chemical shifts (ppm) and intensities for three spreading fats (A-C) and one baking fat (D).

		A		B		C		D	
C1	α	173.17	4.29	173.21	3.28	173.26	3.03	⌈173.28	2.93
								⌊173.24	4.20
	C_4	-		173.02	0.39	-		-	
	β	172.77	2.28	172.79	1.79	172.83	1.83	172.83	2.89
C2	C_4	35.92	0.29	35.92	0.70	-		-	
	β	34.21	4.79	34.21	4.03	34.21	4.13	34.20	5.56
	α	34.04	9.12	34.05	7.28	34.05	7.17	34.03	9.44
C3	αβ	24.88	11.74	24.89	9.93	24.89	9.87	24.87	11.26
	C_6	-		24.56	0.33	-		-	
ω1	*n*-3	-		14.29	0.42	14.30	0.36	-	
	n-9, sat	14.12	13.63	14.13	10.72	14.14	11.10	⌈14.14	8.73
	n-6	13.98	0.47	13.96	0.28	-		⌊14.11	9.15
	C_6	13.90	0.30	13.90	0.30	-		-	
	C_4	13.62	0.32	13.63	0.58	-		-	
ω2	*n*-6	22.72	12.46	22.72	10.93	22.71	10.42	22.71	8.15
	n-9, sat	22.61	1.34	22.61	1.69	22.60	1.67	22.60	8.19
	C_6	-		22.32	0.35	-		-	
	n-3	-		20.58	0.34	20.57	0.27	-	
	C_4	18.36	0.31	18.37	0.64	-		-	
ω3	*n*-9, sat	31.95	11.90	31.95	9.04	31.95	8.66	31.94	6.62
	n-7	31.81	1.21	31.81	0.73	31.79	0.62	31.78	0.48
	n-6*c*	31.56	1.09	31.56	1.50	31.56	1.62	31.54	8.62
	n-6*t*	31.44	0.45	-		-		-	
	C_6	-		31.28	0.33	-		-	
allylic	*t*	32.64	4.52	32.64	2.05	32.64	2.96	32.64	2.18
	-	32.30	0.35	-		-		-	
	-	32.18	0.29	-		-		-	
	c	27.24	10.36	27.25	8.39	27.24	8.86	27.21	19.90
	c	27.20	9.88	27.21	7.73	-		-	
	cc	25.65	0.62	25.65	1.97	25.65	1.96	25.64	10.00
	cc	-		25.56	0.63	25.55	0.41	-	
olefinic (oleic, linoleic, α-linoleic)									
	Ln	-		131.92	0.36	131.94	0.34	-	
	L, Ln	130.17	1.31	130.19	1.87	130.20	2.24	130.21	8.67
	O, L	130.00	7.16	130.00	5.25	130.01	5.67	130.00	8.87
	O	129.69	6.38	129.70	4.49	129.70	4.24	129.70	2.86
	Ln	-		128.29	0.48	128.29	0.42	-	
	Ln	-		128.24	0.48	128.23	0.45	-	
	L	128.09	0.42	128.09	1.18	128.09	1.46	128.07	7.46
	L	127.91	0.45	127.91	1.14	127.90	1.44	127.89	7.81
	Ln	-		127.78	0.51	127.77	0.46	-	
	Ln	-		127.13	0.26	127.12	0.36	-	
olefinic (other)									
	8*t*	130.58	0.48	130.58	0.23	130.58	0.54	130.58	0.26
	9*t*	130.47	0.77	130.48	0.40	130.48	0.86	130.48	0.51
	10*t*	130.42	1.03	-		130.43	0.68	130.43	0.36
	11*t*	130.39	1.01	130.39	0.44	-		-	
	11*t*	130.29	1.11	130.29	0.46	-		-	
	10*t*	130.23	1.04	-		-		-	
	-	129.91	1.61	-		-		-	
	11*c*	129.82	1.40	129.82	0.67	129.82	0.72	129.83	0.68

Reference: [54]

Table 1.53

Composition (% mol) of a commercial emulsifier examined with three different solutions at different times.

	glycerol signals			C1 signals			C2 signals		
	(i)	(ii)	(iii)	(i)	(ii)	(iii)	(i)	(ii)	(iii)
monoacylglycerols	64	62	63	60	61	61	64	64	63
diacylglycerols	33	33	32	36	35	34	34	35	35
triacylglycerols	3	5	5	4	4	5	2	1	2

Reference: [50]

Table 1.54

Composition (% mol and % wt) of a commercial emulsifier compared with its specification.

	C1	glycerol		spec.
	% mol	% mol	% wt	% wt
1-monoacylglycerol	52.9	60.1	42.9	45.3
2-monoacylglycerol	-	4.6	3.3	
1,2-diacylglycerol	12.3	9.8	14.0	40.5
1,3-diacylglycerol	26.1	20.7	29.5	
triacylglycerol	8.7	4.8	10.3	13.2

Reference: [50]

Table 1.55

Composition (% mol) of a commercial sample of acetylated monoglyceride.

	monoglyceride	monoacetyl	diacetyl
C1	17	62	21
glycerol	18	58	24
$COCH_3$	(18)*	58	24
specification†	15	57	24

Reference: [55]
* average of two previous values
† also other components 4%

Table 1.56
Chemical shifts for propylene glycol mono- and diesters and composition (% mol) of two commercial samples.

	chemical shift			sample A		sample B		
	primary	sec.	diester	1-	2-	1-	2-	1,2-
C1	173.99	173.99	[173.51 173.24]	-	-	78		22
C2	34.23	34.58	[34.51 34.19]	69	31	57	22	21
C3	24.99	25.04	[25.03 24.97]	66	34	-	-	-
P1	69.46	65.92	65.42	[71	29	64	20	16
P2	66.13	71.77	67.98					
P3	19.2	16.25	16.5	72	28	61	23	16

Reference: [55]

Table 1.57
The proportion of 18:1 and 18:2 acids in individual sunflower seeds by ^{1}H and ^{13}C nmr spectroscopy and by gas chromatography.

	seed 1		seed 2		seed 3	
	18:1	18:2	18:1	18:2	18:1	18:2
^{1}H	96.0	4.0	80.6	19.4	21.5	78.5
^{13}C	100	-	80.0	20.0	17.1	82.9
GC	95.8	4.2	80.1	19.9	21.4	78.6

Reference: [132]

Table 1.58
The proportion of monoene, 18:2, and 18:3 acids in groups of 4 rape seeds by ^{13}C nmr spectroscopy.

	rape 2	rape 3	rape 4	rape 5	rape 6	rape 7
monoene*	66.1	61.8	77.4	75.3	69.6	75.4
18:2	21.7	25.1	13.2	13.5	18.5	13.2
18:3	12.1	13.1	9.4	11.2	11.9	11.4

Reference: [132]
* 18:1, 20:1, and 22:1

Table 1.59
Chemical shifts (ppm, peak centre) for glycerol tripalmitate in liquid and solid states.

		liquid	solid (polymorph)		
			α	β′	β
Glycerol	1′	61.9	61.9	64.9	59.6
	2′	68.8	67.4	68.8	67.8
	3′	61.9	61.9	64.9	61.2
C1	1′			173.3	173.4
	2′	172.7	171.9	172.5	172.9
	3′			173.3	174.6
C2		33.9	33.8	34.0	35.3
C3		25.0	26.0	26.1	27.0
C4-13		~29.5	32.7	32.8	34.0
C14 (ω3)		31.8	33.8	34.0	35.3
C15 (ω2)		22.5	24.1	23.9	25.5
C16 (ω1)		13.9	14.0	14.3	15.5

Reference: [21]

Chapter 2

PREPARATION OF ESTER DERIVATIVES OF FATTY ACIDS FOR CHROMATOGRAPHIC ANALYSIS

William W. Christie

The Scottish Crop Research Institute, Invergowrie, Dundee, Scotland DD2 5DA

A. INTRODUCTION

The technique of gas chromatography (GC) revolutionized the study of lipids by making it possible to determine the complete fatty acid composition of a lipid in a very short time [53]. For this purpose, the fatty acid components of lipids are converted to the simplest convenient volatile derivative, usually methyl esters, although other esters may be preferred for specific purposes. The preparation of such esters has therefore become by far the most common type of chemical reaction for lipid analysts. On the other hand, there are other chromatographic techniques, notably high-performance liquid chromatography (HPLC) [52], where alternative derivatives, such as those with UV chromophores, are better. Picolinyl esters or pyrrolidide derivatives have special properties for GC-mass spectrometry (MS) of fatty acids [128].

Although fatty acids can occur in nature in the free (unesterified) state, they are most often found as esters, linked to glycerol, cholesterol or long-chain aliphatic alcohols, and as amides in sphingolipids. The physical state of the lipids can vary. For example, they can be isolated as pure lipid classes or remain as a mixed lipid extract. It may be desirable to effect a reaction while lipids are still in the tissue matrix or on a chromatography adsorbent. There is no single esterification procedure which can be applied in all of these circumstances. The purpose of this review is to detail the principles behind the more important esterification and transesterification procedures, and to discuss their advantages and disadvantages and their applications to various classes of lipid. The various procedures are discussed with reference to the preparation of methyl esters from the more common C_{14} to C_{22} fatty acids mainly, in the free state or bound to lipids by ester or amide bonds. The preparation of esters from short-chain or unusual fatty acids is discussed in separate sections.

There has been a number of reviews of aspects of the topic [24,72,166,310], and the author has described esterification procedures for fatty acids on past occasions [49,52,53]; the latter reviews should be consulted for detailed recipes. The topic has also been reviewed from the standpoint of the synthetic organic chemist [129]. Many misconceptions continue to be published, and it is hoped that this new survey of the methodology and discussion of the reaction mechanisms will help to eliminate these in future. The situation has not been helped by the publication of recommended methods by various societies, that may have been tested rigorously in quality control laboratories with gram quantities of seed oils, but are inappropriate for medical laboratories say where microgram amounts only of samples, containing high proportions of polyunsaturated fatty acids, may be available. In the latter circumstance, traces of water adsorbed to glassware, in solvents or in the atmosphere, can be especially troublesome with many otherwise reliable procedures.

B. ACID-CATALYSED ESTERIFICATION AND TRANSESTERIFICATION

1. General mechanism

Carboxylic acids may be esterified by alcohols in the presence of a suitable acidic catalyst as illustrated in Scheme 2.1. The initial step is protonation

$$R{-}C({=}O){-}OH \underset{}{\overset{H^+}{\rightleftharpoons}} \underset{(1)}{R{-}C({=}O){-}\overset{+}{O}H_2} \overset{R'OH}{\rightleftharpoons} \underset{(2)}{R{-}C({=}O){-}\overset{+}{O}(H){-}R'} \overset{-H^+}{\rightleftharpoons} \underset{(3)}{R{-}C({=}O){-}OR'}$$

Scheme 2.1. Acid-catalysed esterification of fatty acids.

of the acid to give an oxonium ion (**1**) which can undergo an exchange reaction with an alcohol to give the intermediate (**2**), and this in turn can lose a proton to become an ester (**3**). Each step in the process is reversible but in the presence of a large excess of the alcohol, the equilibrium point of the reaction is displaced so that esterification proceeds virtually to completion. However, in the presence of water, which is a stronger electron donor than are aliphatic alcohols, formation of the intermediate (**2**) is not favoured and esterification will not proceed fully.

Ester exchange or transesterification occurs under similar conditions

(Scheme 2.2). In this instance, initial protonation of the ester is followed

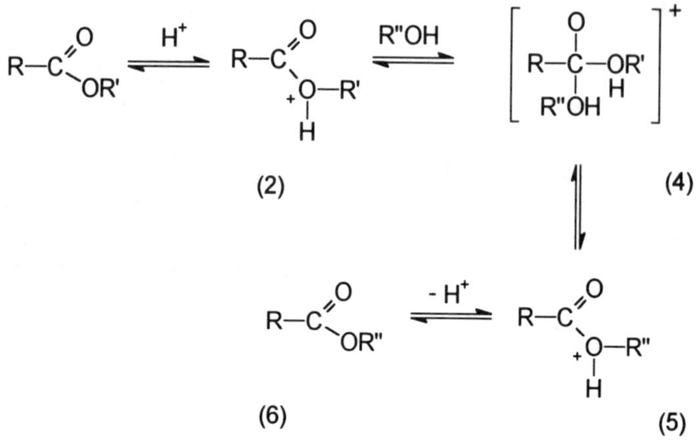

$$R{-}C(=O){-}OR' \overset{H^+}{\rightleftharpoons} R{-}C(=O){-}\overset{+}{O}(H){-}R' \;(2) \overset{R''OH}{\rightleftharpoons} \left[R{-}C(O)(R''\overset{H}{O}H){-}OR'\right]^+ \;(4)$$

$$\rightleftharpoons R{-}C(=O){-}\overset{+}{O}(H){-}R'' \;(5) \overset{-H^+}{\rightleftharpoons} R{-}C(=O){-}OR'' \;(6)$$

Scheme 2.2. Acid-catalysed transesterification of lipids.

by addition of the exchanging alcohol to give the intermediate (**4**), which can be dissociated *via* the transitions state (**5**) to give the ester (**6**). Again, each step is reversible and in the presence of a large excess of the alcohol, the equilibrium point of the reaction is displaced so that the product is almost entirely the required ester (**6**). Water must once more be excluded as it would produce some hydrolysis by dissociation of an intermediate analogous to (**4**) ($R'' = H$) to a free acid. The preferred conditions for acid-catalysed esterification of carboxylic acids or transesterification of existing esters are therefore a large excess of the appropriate alcohol and absence of water. While it may be possible to obtain water-free conditions simply by adding anhydrous sodium sulphate to the reaction medium [256], a better practice in general is to operate with dry reagents and glassware.

A critical practical point is the choice of acid as catalyst. This must facilitate the reaction but should not cause unwanted side-effects. In principle, the methodology can be used with any alcohol component, but in practice it is limited to those alcohols which can be eliminated from the reaction medium by selective evaporation, *i.e.* methanol to perhaps pentanol.

2. *Methanolic hydrogen chloride*

The most frequently cited reagent for the preparation of methyl esters is 5% anhydrous hydrogen chloride in methanol, prepared by bubbling dry gaseous hydrogen chloride into dry methanol. Gaseous hydrogen chloride is available commercially in cylinders or can be prepared when needed by dropping concentrated sulphuric acid onto fused ammonium chloride or into concentrated hydrochloric acid in a Kipp's apparatus [100]. The stability of the reagent was studied by Kishimoto and Radin [183], who found that half the titratable acid was lost at room temperature in six weeks, presumably by reaction between the acid and methanol to give methyl chloride and water. Similar findings were obtained with 1-butanol-hydrogen chloride [125]. In practice, the small amount of water formed does not affect the esterifying reaction significantly and the reagent has a useful shelf-life of about two weeks at room temperature or longer if refrigerated. An alternative method for rapid preparation of the reagent has been described [14] in which acetyl chloride is added to a large excess of dry methanol (Scheme 2.3). Methyl acetate is formed as a by-product

$$CH_3OH + CH_3COCl \longrightarrow CH_3COOCH_3 + HCl$$

Scheme 2.3. Preparation of methanolic hydrogen chloride *via* acetyl chloride.

but does not interfere with the reaction at the concentrations suggested; reagent prepared in this way is stable for about one week. It has been suggested that methanolic hydrogen chloride might be prepared by adding ammonium chloride and sulphuric acid to methanol [127].

In a typical esterification procedure using methanolic hydrogen chloride, the lipid sample is dissolved in at least a 100-fold excess of the reagent and the solution is refluxed for about two hours or is held at 50°C overnight (30 minutes at 50°C will suffice for free acids alone). At the end of this time, water is added and the required esters are extracted thoroughly into an appropriate solvent such as diethyl ether, hexane or light petroleum. The solvent layer is washed with dilute potassium bicarbonate solution to remove excess acid and dried over anhydrous sodium or magnesium sulphate (or anhydrous calcium chloride) and the esters are recovered after removal of the solvent by evaporation under reduced pressure on a rotary film evaporator or in a gentle stream of nitrogen. The reaction may also be performed in a sealed tube so that higher temperatures and shorter reaction times are possible. Longer reaction times are required as the molecular weight of the alcohol is increased.

All fatty acids are esterified at approximately the same rate by methanolic hydrogen chloride [126], so there are unlikely to be differential losses of specific fatty acids during the esterification step. On the other hand, special precautions are necessary to ensure quantitative recovery of short-chain esters and these are discussed in Section I.1. Certain classes of simple lipids, for example, cholesterol esters and triacylglycerols, are not soluble in methanolic hydrogen chloride alone and an inert solvent must be added to effect solution before the reaction will proceed. Benzene was once used frequently for the purpose, but the greater awareness of its toxicity now precludes its use. The author [49] has found that toluene, chloroform (ethanol-free), tetrahydrofuran and methyl-*tert*-butyl ether are all equally effective; the reaction is slower in methyl acetate, although this solvent has also been recommended [90].

While it has been claimed [163] that spurious components, which may interfere with GC analyses, are formed in hydrogen chloride-methanol solutions, these have not apparently been found by others. The author [49] has observed that such artefacts may be formed, apparently from the methanol, with a variety of acidic catalysts if superheating of the solution is allowed to occur in the presence of oxygen. With normal refluxing, under nitrogen especially, artefact formation is minimal.

Esterification will proceed with aqueous hydrochloric acid as catalyst if dimethoxypropane (7) is added to the reaction medium as a water scavenger (Scheme 2.4) [214]. Applications of this reagent to the

$$\underset{(7)}{(CH_3)_2C(OCH_3)_2} + H_2O \xrightarrow{H^+} (CH_3)_2CO + 2CH_3OH$$

Scheme 2.4. Dimethoxypropane as a water scavenger.

esterification of free fatty acids [287] and transesterification of triacylglycerols [232,327] have been described. In a variation of the procedure, glycerol was determined at the same time as the methyl esters [231]. The method has a major disadvantage, however, in that large amounts of coloured polymeric by-products are formed from the dimethoxypropane [47,232,311,329], and these often interfere with the subsequent GC analysis of the methyl esters.

Free fatty acids can be esterified with methanolic hydrogen chloride under mild conditions if they are first adsorbed onto a strongly-basic anion-exchange resin [139,343]. In one procedure [139], other lipids were eluted from the column first with hexane, leaving the free acids behind for selective methylation; in another, it was demonstrated that it was even possible to methylate free acids in the presence of other lipids by related methodology (see Section I.5) [62]. Unfortunately, such procedures are time-consuming and tedious, especially as it is necessary to wash the resin first with large volumes of solvents to remove contaminants prior to the reaction [194]. In addition, recoveries of individual fatty acids were found to be uneven [139]. An alternative method for selective methylation of free acids in the presence of other lipids using methanolic HCl [206] required very precise timing and manipulation of the reactants and appears to have limited applicability.

Some evidence [140] has been presented that addition of cupric acetate to methanolic hydrogen chloride facilitates the esterification reaction, but the improvement appears to be marginal.

In summary, hydrogen chloride in methanol (or another alcohol) can be used to esterify free fatty acids or to transesterify fatty acids linked by ester bonds to glycerol or cholesterol. While it has been employed to transesterify amide-linked fatty acids, it is not the best reagent for this purpose (see Section I.3). It could probably be claimed to be the best general-purpose esterifying agent available. The main disadvantage is the comparatively long reflux time needed for complete reaction to be achieved. As with other acid catalysts, it is not suited for certain fatty acids with sensitive functional groups, such as epoxyl, cyclopropane or cyclopropene rings (see Section I.2).

3. Methanolic sulphuric acid

Although much higher concentrations are sometimes used, a solution of 1 to 2% concentrated sulphuric acid in methanol has almost identical properties to 5% methanolic hydrogen chloride, and is very easy to prepare. Indeed in one method [280,359,360], extraction of the tissue and esterification were carried out in isopropanol to which neat sulphuric acid was added (following removal of water from the extract) to catalyse formation of isopropyl esters for GC analysis (see also Section K). Esters of other alcohols, such as butanol [101], have been prepared similarly. The same reaction times are usually recommended for H_2SO_4- and HCl-methanol. Lie Ken Jie and Yan-Kit [208] showed that free acids were esterified especially rapidly with the former and a microwave oven as an energy source. Inert solvents must again be added to effect solution of simple lipids.

Free fatty acids were esterified very rapidly by heating in 10% sulphuric acid in methanol until the reflux temperature was reached [290], but this procedure cannot be recommended for polyunsaturated fatty acids, as sulphuric acid is a strong oxidizing agent. There are reports that very long reflux times (up to six hours) [124,174], excessive sulphuric acid concentrations (20%) [16] or high temperatures (170°C) [122,277] will lead to the formation of coloured by-products and the destruction of polyenoic fatty acids. With the dilute reagent and moderate temperatures, however, there is no evidence for side-effects, and under such conditions the reagent was approved by the Instrumental Committee of the American Oil Chemists' Society [10]. Of course, it has the same drawbacks as other acidic catalysts with sensitive fatty acids (see Section I.2).

McGinnis and Dugan [222] described a modification of the reaction in which the sulphuric acid complex of the lipid was formed in diethyl ether solution at -60°C first, before it was decomposed with anhydrous methanol to give the required methyl esters. The procedure was shown to be applicable to the direct methylation of lipids in biological materials (see Section K) [86]. Although the reaction is rapid, the practical manipulations required are complex and the method has rarely been used. Others have described the use of a pre-column, impregnated with sulphuric acid, at the head of a GC column, where the sample in methanol solution is esterified directly [171]. Such a technique would appear to be rather hazardous for the column packing!

4. Boron trifluoride-methanol

The Lewis acid, boron trifluoride, in the form of its coordination complex with methanol is a powerful acidic catalyst for the esterification of fatty acids. For example, esterification of free fatty acids was completed in two minutes with 12 to 14% boron trifluoride in methanol under reflux [236]. Morrison and Smith [259] showed that the reagent could be used to transesterify most lipid classes (inert solvent must again be added to effect solution of simple lipids), although in general longer reaction times are necessary than with free fatty acids. For example, in this reagent cholesterol esters are transesterified in 45 minutes at 100°C in a sealed Teflon™-lined screw-top tube. Boron trifluoride can of course be used with other alcohols, and as examples ethyl [164], propyl [297] and butyl [154,155,164] esters have been prepared in this way. Methyl esters labelled with tritium in the methyl group have been prepared by esterification of fatty acids with boron trifluoride and ^{3}H-methanol [262]. The reaction has been accelerated by the use of microwave radiation (see also previous section) [23,73].

Unfortunately, boron trifluoride-methanol does have serious drawbacks. Lough [215] first reported that methoxy artefacts were produced from unsaturated fatty acids by addition of methanol across the double bond when very high concentrations of boron trifluoride in methanol (50%) were used. Although he later showed [216] that such artefacts were not necessarily formed with more normal concentrations of boron trifluoride in methanol and this was confirmed by Morrison and Smith [259], the warning was subsequently reiterated by others [65,97,185,190,235]. It is possible that the side-reactions are exacerbated by the presence of oxidised lipids [65]. In addition, it has been reported that sample size is critical with substantial losses sometimes occurring with samples of less than 200 mg [314]. There is some evidence that artefact formation is most likely with aged reagents [97]. The author has had similar experiences in his own laboratory and has heard much anecdotal evidence in support. Although Klopfenstein [187] could not confirm the suggested explanations for artefact formation, he confirmed the existence of the problem. Boron trifluoride-methanol suffers from the same disadvantages as other acidic reagents with fatty acids with labile functional groups (see Section I.2 below), although there is a suggestion that it produces by-products quantitatively in some instances and that this may be of analytical value [185]. Troublesome by-products are also known to be formed from some antioxidants commonly added to lipid extracts (see Section L below).

Solutions of boron trifluoride in methanol obtained commercially should therefore be checked carefully before use and periodically in use. The reagent has a limited shelf-life at room temperature and should be kept refrigerated. If such precautions are taken, the reagent may be a useful one in some circumstances. Problems are less likely to arise in quality control laboratories say, where there is a high through-put and large samples are the norm, than in medical laboratories where microgram amounts of lipids may be all that is available. The reagent has the blessing of the American Oil Chemists' Society [11] and of IUPAC [151] amongst others. It is certainly highly popular, but possibly because it is one of the few such reagents that can be purchased from commercial suppliers. In view of the many known side reactions and the high acid content in comparison to other analogous reagents, this author would not consider using it in his own laboratory.

The reagent is of value for the oxidative fission of ozonides with simultaneous methylation in a method for double bond location in fatty acids [2,5].

Boron trichloride in methanol can be used in a similar manner to prepare methyl esters [1,38,39], although the reaction is slower than when boron trifluoride is the catalyst. Klopfenstein [187] established that artefact formation was much less of a problem with boron trichloride in methanol, and others [40] have shown that it does not cause disruption of cyclopropane fatty acids (see Section I.2).

5. *Other acidic catalysts*

Aluminium trichloride appeared to be as effective as boron trifluoride as a catalyst for transesterification [307], but it has not been tested with a wide range of samples. Phenyl esters of fatty acids have been prepared by acid-catalysed esterification with *p*-toluenesulphonic acid as catalyst [152]. Similarly, a strong cation-exchange resin (presumably with chemically-bonded phenylsulphonic acid moieties) in a fixed bed has been used as part of an HPLC system for post-column transesterification of lipids [294].

C. BASE-CATALYSED TRANSESTERIFICATION

1. *General mechanism*

Esters, in the presence of base such as an alcoholate anion (**8**) (Scheme 2.5), form an anionic intermediate (**9**) which can dissociate back to the

$$R{-}C(=O){-}OR' + {}^{-}OR'' \rightleftharpoons \left[R{-}C(O^{-})(OR'){-}OR'' \right] \rightleftharpoons R{-}C(=O){-}OR'' + {}^{-}OR'$$

(8) (9) (10)

Scheme 2.5. Base-catalysed transesterification of lipids.

original ester or form the new ester (**10**). In the presence of a large excess of the alcohol from which the anion (**8**) was derived, the equilibrium point of the reaction will be displaced until virtually the sole product is the new ester (**10**). On the other hand, an unesterified fatty acid is converted to a carboxylate ion, $RCOO^-$, in a basic solution, and this is not subject to nucleophilic attack by alcohols or bases derived from them because of its negative charge. *Transesterification* can therefore occur by this mechanism with basic catalysis but *esterification* cannot.

In the presence of water, the intermediate (**9**, $R'' = H$) will dissociate irreversibly to the free acid. For base-catalysed transesterification of existing esters, it is necessary therefore to have a large excess of the new alcohol and an absence of water from the reaction medium.

2. *Sodium and potassium methoxide catalysts*

The most useful basic transesterifying agents are 0.5 to 2*M* sodium or potassium methoxide in anhydrous methanol, prepared by dissolving the clean metals in anhydrous methanol (the reaction is strongly exothermic). Potassium hydroxide at similar concentrations in methanol is occasionally used, but this is not recommended, since appreciable hydrolysis of lipids to free fatty acids can occur if the least trace of water is present, especially if the reaction is prolonged [107,142]. Methanolic sodium and potassium methoxide are stable for several months at refrigeration temperature but they eventually deteriorate, with precipitation of the bicarbonate salt by reaction with atmospheric carbon dioxide, and sometimes with the formation of other by-products which may interfere in GC analyses. Artefact formation is minimized and the shelf-life of the reagent improved if oxygen-free methanol is used in its preparation [49]. At equivalent molar concentrations with the same lipid samples, potassium methoxide effects complete esterification more quickly than does sodium methoxide, which is in turn more rapid than potassium hydroxide [49,218]. Because of the dangers inherent in handling metallic potassium, which has a very high heat of reaction with methanol, the author prefers to use sodium methoxide in methanol in his own laboratory. It should be noted that, in common with all strongly alkaline solutions, these reagents should be handled with caution.

Sodium methoxide in methanol effects transesterification of glycerolipids much more rapidly than is sometimes realised, and although reflux times of as long as 6 hours have on occasion been recommended in the literature, it has been shown that triacylglycerols can be completely transesterified in 2 to 5 minutes and phosphatidylcholine in only 1 minute at room temperature [226,227]. The reaction is slower with alcohols of higher molecular weight, *i.e.* taking up to 60 minutes with hexanol [102]. As with acidic catalysis, inert solvents must be added to dissolve simple lipids before methanolysis will proceed. Benzene has been used frequently for the purpose, but is no longer recommended for safety reasons, and the author [49] has found that the reaction is as fast or faster with dry toluene, dichloromethane (not with phospholipids), methyl-*tert*-butylether and tetrahydrofuran, and a little slower in diethyl ether, hexane or dimethyl carbonate. Early reports that this last solvent accelerates transesterification by participation in the reaction have now been refuted [59]. While chloroform is frequently recommended as a suitable solvent, the author [49] has noted that it reacts with sodium methoxide to give a precipitate of sodium chloride, presumably with generation of dichlorocarbene which has the potential to react with the double bonds of unsaturated fatty acids (although this does not appear to have been demonstrated in practice) [178]. Acetone is unsuitable also.

In a typical transesterification reaction, the lipid sample, dissolved if necessary in sufficient toluene or other solvent to ensure it remains in solution, is reacted with a 100-fold excess of 0.5 to 2*M* sodium methoxide at 50°C (refluxing will speed up the reaction but is not usually necessary). Triacylglycerols are completely transesterified in 10 minutes and phosphoglycerides in 5 minutes, although cholesterol esters require 60 minutes, under these conditions. At the end of the appropriate time, dilute acid is added to neutralise the sodium methoxide and so minimise the risk of hydrolysis occurring, and the required methyl esters are recovered by solvent extraction as with acid-catalysis. As an alternative for the micro-scale, a method can be recommended in which the reaction is carried out in an inert solvent with the minimum of methanolic sodium methoxide and with methyl acetate added to suppress the competing hydrolysis reaction [51].

A novel technique was described by Dutton and colleagues [34,74] in which methylation was carried out in a micro-reactor at the head of a GC column immediately prior to analysis. A heated sodium hydroxide-impregnated pre-column has also been used [172]. DePalma [76] adapted a laboratory robotic system for a similar purpose, while others have described the use of potassium methoxide/celite in small columns for methylation [228,304]. Recently, alumina treated with methanol has been incorporated into a supercritical fluid extraction-chromatography system to effect methylation on-line [181].

A reagent consisting of sodium borohydride with sodium hydroxide in methanol has been suggested as a means of reducing hydroperoxides to hydroxides and minimising over-oxidation during transesterification [283,284]. Hydroperoxides *per se* are not conveniently or safely methylated by any of the usual procedures (see Section I.2) [338].

Sodium methoxide in methanol is therefore a valuable reagent for rapid transesterification of fatty acids linked by ester bonds to alcohols (*e.g.* cholesterol, glycerol). It will not esterify free fatty acids or transesterify amide-bound fatty acids in sphingolipids. Also, unlike acidic catalysts, it will not liberate aldehydes from plasmalogens. Under normal conditions, no isomerization of double bonds in polyunsaturated fatty acids occurs [102,157], but prolonged or careless use of basic reagents can cause some alterations of fatty acids [17]. As examples, ethyl [102,108,164], propyl [102,109,122], butyl [102,109,298], isobutyl [328], isopentyl [328], hexyl [102] and phenylethyl [56] esters have been prepared by reaction with sodium in the appropriate alcohol.

3. *Organic base catalysis*

Pyrolysis of quaternary ammonium salts of free fatty acids was used as a method of methylation for some years (see Section E below), before it was realised that quaternary ammonium hydroxides could catalyse transesterification of lipids as part of relatively simple procedures for preparing methyl esters for GC analysis. The first use of such a procedure was in fact a quaternary ammonium ion exchange resin in the methoxide form [77], that was reported to be particularly suitable for transesterification of oils containing hydroxy fatty acids with conjugated double bond systems (dimorphecolic and lesquerolic acids) (see also Section I.2). However, the first reagent suited to use in homogeneous solution to be described was 0.2*M* methanolic (*m*-trifluoromethylphenyl)trimethylammonium hydroxide [219]. The glyceride sample in benzene was converted to the methyl esters in 30 minutes at room temperature, and it was possible to inject an aliquot of the reaction mixture directly onto the GC column without further work-up. If any free acids were present, they were converted to methyl esters in the heated injection port. A minor drawback is that *N*,*N*-dimethyl(*m*-trifluoromethyl)aniline, which is formed as a by-product, elutes as a distinct peak after the solvent front. The same type of reagent has been used by others to produce methyl [61,338], benzyl [337] and pentafluorobenzyl [291,339] esters from a variety of different lipid classes and fatty acids, including those derived from hydroperoxides. Tetramethylammonium hydroxide has been used as catalyst in a similar way [89,230,237,248] as has tetramethylguanidine hydroxide [302].

The most useful reagent of this type now appears to be trimethylsulphonium hydroxide ($(CH_3)_3SOH$), prepared ideally under conditions that guarantee the elimination of water [43,44,92], since it reacts very rapidly with lipids and the only by-product, dimethyl sulphide, is highly volatile [43,44,367] (see also Section E below). It was observed [92] that triacylglycerols and phospholipids were transesterified at a rate only a little slower than with sodium methoxide, but cholesterol esters were methylated very slowly. Polyunsaturated fatty acids, such as those of fish oils, were methylated safely. Although the reagent has seen limited use so far [233,264,265,303], it obviously has great potential.

D. DIAZOMETHANE AND RELATED REAGENTS

1. Diazomethane and methyl ester preparation

Diazomethane (**11**) reacts rapidly with unesterified fatty acids to give methyl esters (Scheme 2.6), but does not effect transesterification of other

$$R{-}C(=O){-}OH + CH_2N_2 \longrightarrow R{-}C(=O){-}OCH_3 + N_2$$

(11)

Scheme 2.6. Reaction of diazomethane with a fatty acid.

lipids (see also Section I.5 below). The reaction is not instantaneous, however, as has sometimes been assumed, unless a little methanol is present as a catalyst [70,110,301]. The reagent is generally prepared in ethereal solution by the action of alkali on a nitrosamide, *e.g.* *N*-methyl-*N*-nitroso-*p*-toluenesulphonamide (**12**) (Diazald™, Aldrich Chemical Co., Milwaukee, U.S.A.) in the presence of an alcohol (Scheme 2.7). A large-

$$CH_3C_6H_4SO_2N(NO)CH_3 + ROH \xrightarrow{KOH} CH_2N_2 + CH_3C_6H_4SO_3R + H_2O$$

(12) (11)

Scheme 2.7. Preparation of diazomethane from *N*-methyl-*N*-nitroso-*p*-toluenesulphonamide (**12**) (Diazald™).

scale practical procedure has been described [35], but simple small-scale preparations, in which special glassware is used and only enough diazomethane for immediate needs is prepared, are to be preferred [67,207,301,346]. An alternative procedure involving the reaction of hydrazine hydrate, chloroform and alkali has also been recommended [67]. Diazomethane has been used to prepare methyl esters labelled with ^{14}C [301,318] or tritium [189] in the methyl group.

Solutions of diazomethane in diethyl ether (with a little alcohol) are stable for short periods if stored in the dark over potassium hydroxide pellets at refrigeration temperatures. If they are kept too long, polymeric by-products are formed which interfere with GC analyses [122,258,301]. It has been claimed that loss of polyunsaturated fatty acids will occur by addition of carbene, formed by decomposition of diazomethane, to double bonds [122] or by related mechanisms [329]. Although others [301] were unable to detect this effect with a conventional range of fatty acids, it was found with α,β-unsaturated and α-keto acids [31]. There have been reports of etherification of hydroxy fatty acids with diazomethane, either with the absence of a small amount of methanol from the medium or in the presence of acidic or basic catalysts when reaction was prolonged [110,137], but this does not appear to be a problem otherwise. There is also evidence that diazomethane can cause some transesterification by prolonged irradiation under UV light in the presence of traces of acidic or basic catalysts [271].

Diazomethane is potentially explosive and great care must be exercised in its preparation; in particular, apparatus with ground-glass joints and strong light must be avoided. The reagent is toxic and is liable to cause development of specific sensitivity; nitrosamides used in the preparation of diazomethane must be handled with care as they are potential carcinogens. Small-scale procedures are recommended [67,207,301,346], since if sensible precautions are taken, the risks to health are slight, while methyl esters are prepared rapidly with virtually no artefact formation. Suitable micro-equipment is available from commercial sources.

2. Preparation of UV-absorbing and other derivatives

Pentafluorodiazoalkanes have been used to prepare derivatives from unesterified fatty acids that are suitable for GC with sensitive electron-capture detection [135], and benzyldiazomethane has been used to prepare benzyl esters [132,186].

Related methods have been used to prepare UV-absorbing derivatives for analysis by HPLC. For example, 9-anthryldiazomethane in diethyl ether, methanol, ethyl acetate or acetone solution gave good yields of anthrylmethyl esters in some hands [26,147,268,366], but others have preferred alternative methods (see Section G.2). 1-Pyrenyldiazomethane has been used also for derivative preparation [269].

E. PYROLYSIS OF TETRAMETHYLAMMONIUM SALTS OF FATTY ACIDS FOR THE PREPARATION OF METHYL ESTERS

Robb and Westbrook [289] first showed that tetramethylammonium salts of unesterified fatty acids in aqueous solution could be pyrolysed to form methyl esters in the heated (330 to 365°C) injection port of a gas chromatograph, but they were unable to obtain quantitative recovery of individual components of mixtures. However, by drying the samples carefully prior to analysis and using solid injection into the gas chromatograph [82,83] or by improvements in the design of the injector [21], others were able to obtain quantitative yields of methyl esters. The former adaptation was shown to be especially suited to the preparation of methyl esters from the mono- and dibasic acids obtained by oxidative cleavage of unsaturated fatty acids [84]. Although the method was used initially with saturated components, unsaturated fatty acids could also be esterified if the reaction conditions were controlled rigorously [83].

Subsequently greatly improved results were obtained by changing to trimethyl(*m*-trifluoromethylphenyl)ammonium hydroxide [105,184,220] or trimethylphenylammonium hydroxide [41,353] as the catalyst, when even polyunsaturated fatty acids in the free form could be analysed successfully.

Trimethylsulphonium and trialkylselenonium hydroxides could be even better, since they were found to decompose at lower temperatures (approximately 200°C) and the by-products produced (dimethyl sulphide from the former as illustrated in Scheme 2.8) are relatively volatile and do

$$RCOOH + (CH_3)_3S^+ OH^- \longrightarrow RCOO^- S^+(CH_3)_3 \xrightarrow{250^0 C} RCOOCH_3 + (CH_3)_2S$$

Scheme 2.8. Methylation of fatty acids by pyrolysis of the salt formed from trimethylsulphonium hydroxide (TMSH).

not interfere with the GC analysis [44,367]. However, some loss of polyunsaturated fatty acids was found with trimethylsulphonium hydroxide [92]. In some of these applications, samples were hydrolysed to the free acids prior to esterification, although it is now realized that the same catalysts can effect transesterification directly (see Section C.3 above).

The method may be of value for the analysis of unesterified fatty acids in aqueous solution as there is no extraction step during which selective losses of specific components may occur. It also appears to have value for selective esterification of free acids in the presence of other lipids (see Section I.5). On the other hand, this approach cannot be used with modern on-column injection techniques for gas chromatography.

F. PREPARATION OF ESTERS AND AMIDES *VIA* ACTIVATED FATTY ACIDS

1. Acid halides

Acid chlorides and anhydrides react with alcohols to give esters and with amines to give amides under appropriate conditions. The preparation and properties of acid chlorides and anhydrides have been reviewed [112]. Activation of fatty acids in this way is of special utility in the synthesis of esters of glycerol, for example for the synthesis of triacylglycerols or phosphoglycerides with specific fatty acids in the various positions [117]. More relevant to the topic of this review, such methodology can be of value for the synthesis of derivatives where the alcohol component has a relatively high molecular weight (six carbon atoms or more) for chromatographic analysis. Methods of this kind are preferred for the preparation of simple amine derivatives, but they do not have any advantages over other procedures described here for preparation of esters from short-chain aliphatic alcohols.

The use of acid chlorides for glycerolipid synthesis was reviewed by Mattson and Volpenhein [233a]. The mildest method of preparation is probably to react the acids with oxalyl chloride [233a]; the reaction is slow (up to 3 days at room temperature) but chlorides of polyunsaturated fatty acids can be prepared safely. Thionyl chloride will react much more quickly, and though it can cause isomerisation of double bonds if used carelessly, this is not a problem if the reaction time is confined to 1 minute [128]. In addition, fatty acid chlorides have been prepared under gentle conditions by reacting an acid with triphenylphosphine and carbon tetrachloride [134].

In practice to prepare an ester, the acid chloride is reacted with the required alcohol under strictly anhydrous conditions in the presence of a basic catalyst, traditionally pyridine or triethylamine but more often *N,N'*-dimethyl-4-aminopyridine or 4-pyrrolidinopyridine nowadays, to mop up the acid produced (Scheme 2.9). As an example, *L*-menthyl esters were

$$RCOCl + R'OH \xrightarrow{\text{pyridine}} RCOOR' + HCl$$

Scheme 2.9. Preparation of esters *via* acid chlorides.

prepared by reacting menthol with the appropriate acid chloride [3,4], in order to resolve optically active fatty acids by GC. Acid chlorides, prepared by a rapid reaction *in situ* with thionyl chloride, were favoured by Harvey [128] for the preparation of picolinyl (3-hydroxymethylpyridinyl) esters of fatty acids for analysis by GC/MS.

Similar methodology has been used for the synthesis of simple amide derivatives, such as those of *p*-methoxyaniline [134], naphthylamine [148], 9-aminophenanthrene [149] and aniline [197].

2. *Fatty acid anhydrides*

Fatty acid anhydrides, when used in the same way as acid chlorides, are reported to give fewer by-products. They have often been used with the tetraethylammonium salt of the same acid in the synthesis of lipids *per se* [68,200,201], but mixed acid anhydrides of fatty acids and trifluoroacetic acid are much more useful for the preparation of derivatives for chromatography [54,58,113,272]. With the latter, the mixed anhydride is prepared by reaction with trifluoroacetic anhydride immediately before use. Heptafluorobutyric anhydride has been utilised in a similar manner to prepare trichloroethyl esters [323]. While the presence of a base, such as *N,N*-dimethyl-4-aminopyridine, to neutralise the acid by-product is not essential, it appears to improve the yield when the reaction is carried out on the sub-milligram scale [54,58].

3. *Imidazolides*

Apart from the fact that preparation of esters *via* acid chlorides and anhydrides involves two steps, such procedures suffer from the disadvantages that they cannot be used for fatty acids with acid-labile functional moieties, *e.g.* epoxyl groups, and they tend to give poor yields with small samples (less than 1 milligram) when the slightest trace of atmospheric moisture is present, a situation encountered commonly in chromatographic analysis. A useful alternative, which gives high yields in terms of the fatty acid components [37,114], is then to activate the latter by reacting them with 1,1′-carbonyldiimidazole (**13**) to form an imidazolide (**14**), which is reacted immediately (without isolation) with the alcohol in the presence of a base to give the required ester (**15**) (Scheme

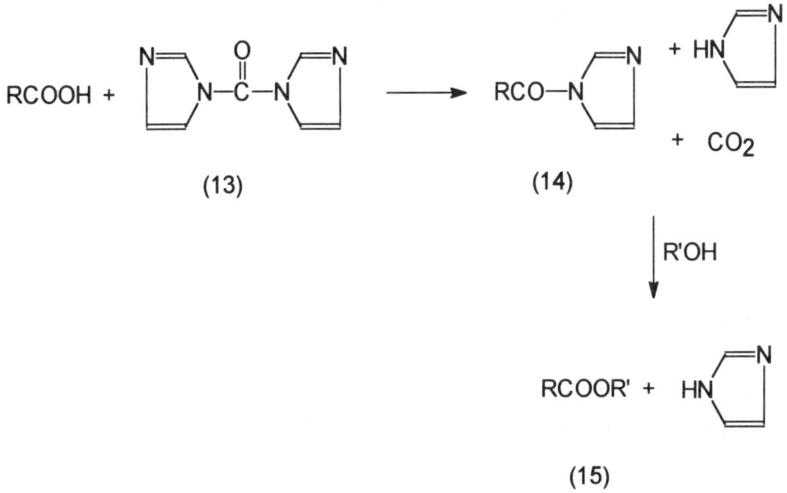

Scheme 2.10. Preparation of esters *via* imidazolide derivatives of fatty acids.

2.10). This procedure has been used in the preparation of picolinyl ester derivatives from epoxy [22] and thia [57] fatty acids, for example. It has also been employed for selective methylation of the free fatty acid fraction from plasma in the presence of other lipids (see also Section I.5) [188]. 1,1′-Carbonyldiimidazole is very sensitive to moisture and must be stored in a desiccator.

Dicyclohexylcarbodiimide (**16**) is a related reagent for effecting the coupling of a fatty acid and alcohol as shown in Scheme 2.11. It gives

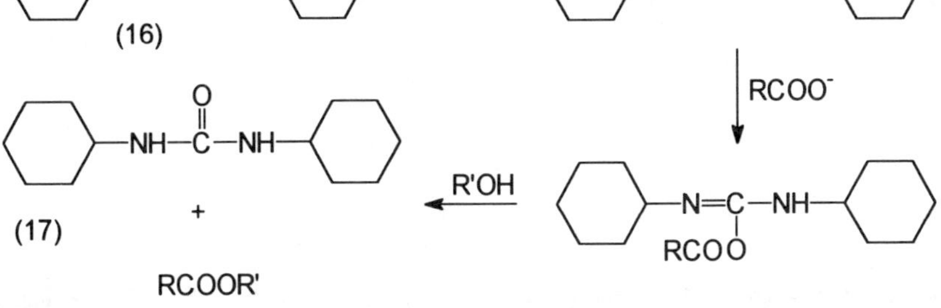

Scheme 2.11. Preparation of esters with dicyclohexylcarbodiimide (**16**) as the coupling agent.

quantitative esterification even in the presence of moisture, a considerable virtue when sub-milligram quantities of fatty acids must be derivatized. However, some drawbacks have been reported [129]. The *N,N*-dicyclohexylurea (**17**), produced as a by-product, is not easily removed from the reaction medium, and the reagent is potentially carcinogenic and must be handled with great care. The reaction has been used most often for the preparation of synthetic glycerolipids [117] rather than for simple derivatives, but it has been employed for the synthesis of dansyl-ethanolamine derivatives of fatty acids [293].

2-Nitrophenylhydrazide derivatives of long-chain fatty acids have been prepared by coupling the free acids with 2-nitrophenylhydrazine hydrochloride in the presence of dicyclohexylcarbodiimide or 1-ethyl-3-(3-dimethylaminopropyl)carbodiimide hydrochloride [249-254]. The same type of reaction was used as part of a post-column HPLC detection system to prepare fluorescent hydrazine derivatives for HPLC [156].

4. Other coupling reagents

Amide derivatives of monodansylcadaverine (**18**) have been prepared by coupling the amine and fatty acids in the presence of diethylphosphorocyanidate (**19**) (Scheme 2.12) [202,203]. The reaction

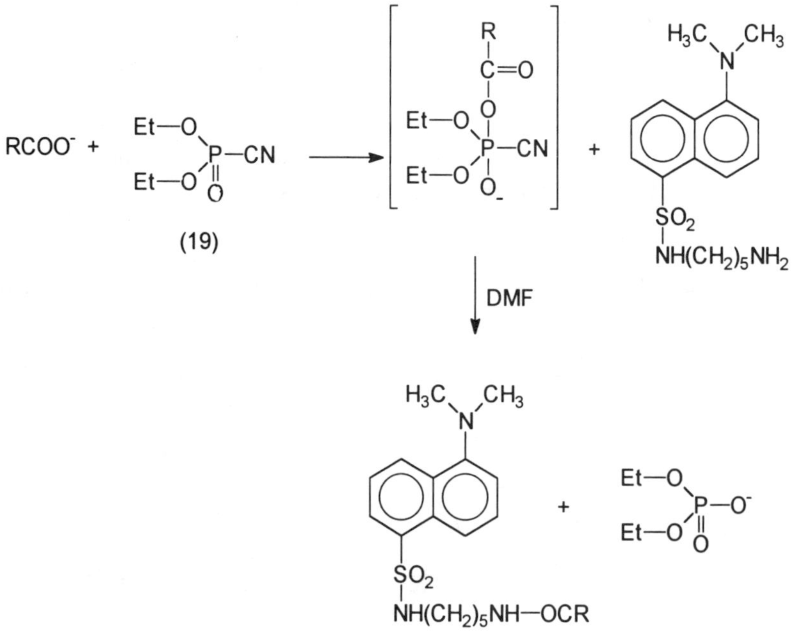

RCOO⁻ + (Et—O)₂P(=O)—CN (19) → [R—C(=O)—O—P(O⁻)(OEt)₂—CN] + dansylcadaverine (H_3C–N–CH_3 naphthalene, SO_2–$NH(CH_2)_5NH_2$)

↓ DMF

dansyl–SO_2–$NH(CH_2)_5NH$—OCR + (Et—O)₂P(=O)—O⁻

Scheme 2.12. Preparation of esters with diethylphosphorocyanidate (**19**) as the coupling agent.

occurred rapidly under simple practical conditions, but the coupling reagent is not yet readily available from commercial sources.

An alternative method consists in reacting the fatty acid with the appropriate alcohol in the presence of 2-bromo-1-methylpyridinium iodide (**20**) as catalyst (Scheme 2.13) [296]. The reaction proceeds *via* an

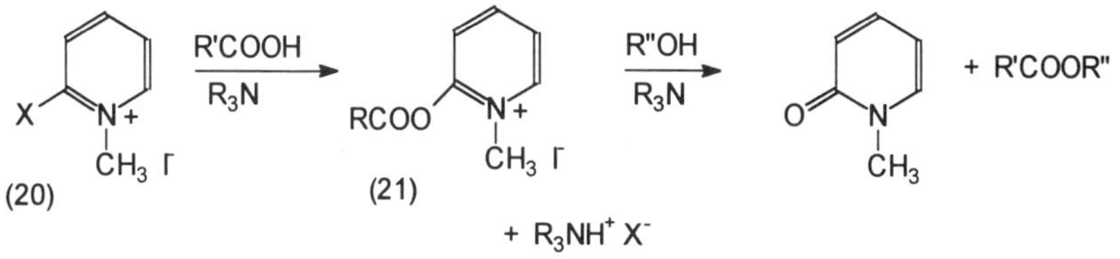

X N+ CH_3 I^- (20) —R'COOH / R_3N→ RCOO N+ CH_3 I^- (21) + $R_3NH^+X^-$ —R"OH / R_3N→ O N CH_3 + R'COOR"

Scheme 2.13. Preparation of esters with 2-bromo-1-methylpyridinium iodide (**20**) as the coupling agent.

intermediate 2-acyloxy-1-methylpyridinium iodide (**21**), which is subjected to nucleophilic attack by the alcohol in the presence of a tertiary amine to mop up the hydrogen iodide produced. As activation of the fatty acid and esterification takes place in one step, there are obvious advantages, which should be explored further. The procedure was used for the preparation of anthrylmethyl esters of fatty acids, where it was reportedly superior to the use of anthryldiazomethane (see Section D.2 above) for the purpose [28-30,211]. 2-Chloro-1-methylpyridinium iodide has been used similarly to prepare *N*-(4-aminobutyl)-*N*-ethyl-isoluminol esters for HPLC with fluorescence detection [368].

G. REACTION OF ALKYL OR ARYL HALIDES WITH A BASE

1. Derivatives for gas chromatography

Methyl esters can be prepared quantitatively by reaction of the silver salts of fatty acids with methyl iodide [104]. Although the reaction was said to be especially suitable for samples containing short-chain fatty acids [104,161], and an application to the analysis of milk fats was described, better methods are now available (see Section I.1). The method was subsequently applied to the preparation of benzyl esters, but a number of by-products were formed which interfered with the reaction [160].

In essence, the reaction involves S_N2 attack of a basic anion on an alkyl halide in a highly polar solvent system such as dimethylformamide (see Scheme 2.14 below). Other basic catalysts have been employed for the preparation of methyl esters including the organic catalysts phenyltrimethylammonium hydroxide [111] and tetramethylammonium hydroxide [121,351], sodium and potassium hydroxide [63,162], potassium carbonate [115], potassium hydroxide [79] and potassium ion-crown ether complexes (see next section) [133]. Similarly, benzyl esters have been prepared by reaction of potassium carboxylates with benzyl bromide and catalysed by a crown ether in acetonitrile solution [300]. Analogous methods have been used to prepare pentafluorobenzyl esters, *i.e.* by reacting the free acid with pentafluorobenzylbromide and triethylamine in acetonitrile solution, for analysis by electron capture GC [71,120,223,315,362]. The reaction has not been widely used but may be a useful alternative to diazomethane for the preparation of esters from free acids in the presence of other lipids.

A related method has been used to prepare methyl esters from the fatty acids of sphingolipids [162]. In this instance, the hydroxyl groups of the 2-hydroxy fatty acids present were methylated simultaneously (see also Section I.3 below) [63,162].

2. *Phenacyl esters and other derivatives for high-performance liquid chromatography*

Phenacyl esters and related derivatives, which absorb strongly in the UV region of the spectrum, are of particular value for analysis by means of HPLC. The method of preparation is in essence identical to that described above, but the reaction must be optimised for minimum use of the reagents, partly because of cost and partly because of the difficulty of removing the excess from the medium. Crown ethers, *i.e.* cyclic polyethers such as 1,2,7,10,13,16-hexaoxacyclooctadecane ("18-crown-6"), complex strongly with potassium ions and effectively solubilize them in aprotic solvents such as toluene, acetonitrile, methylene chloride and carbon tetrachloride, thus facilitating the reaction and simplifying the isolation of the esters produced. For example, phenacyl esters have been prepared by reaction of phenacyl bromide and a fatty acid in the presence of potassium carbonate and a crown ether in acetonitrile solution [88]. As an alternative, phenacyl esters (**23**) can be prepared under mild conditions by reaction of phenacyl bromide (**22**) and a fatty acid in the presence of triethylamine in acetone solution (Scheme 2.14) [36,361]. The procedure of

$$C_6H_5-C(=O)-CH_2Br + RCOOH + R'_3N \longrightarrow C_6H_5-C(=O)-CH_2OOCR + R'_3NH^+ Br^-$$

(22) (23)

Scheme 2.14. Preparation of phenacyl esters of fatty acids.

Wood and Lee [361] can be recommended. Some stereomutation of double bonds in phenacyl esters was observed when attempts were made to purify them by TLC [361].

Methoxyphenacyl [242], bromophenacyl [165], naphthacyl [64,79,165], methylanthryl [191], methylisatin [116], 4-bromomethyl-7-methoxycoumarin [87,198,357,358], 4-bromomethyl-7-acetoxycoumarin

[171,330-332] and other [267,364,365] esters have been prepared in the same way. The list is by no means complete. *p*-Nitrobenzyl and other UV-absorbing derivatives of fatty acids were prepared in an analogous manner by reaction with the *p*-toluenesulphonate derivatives of the appropriate alcohols in the presence of a basic catalyst [98], and bromophenacyl esters have been prepared *via* trifluoromethanesulphonates [150].

An alternative technique for solubilising an anion for reactions of this type is to use a phase-transfer-mediated reaction catalysed by tetrabutylammonium ions, which can at the same time extract the fatty acids from an aqueous medium. Thus, free fatty acids have been extracted and simultaneously converted to phenacyl esters by this means [99,369], and methyl, 4-nitrobenzyl [244,245], anthracyl, acetylfluorene [8], methylacridine [336] and *N*-(9-acridinyl)-acetamido [9] esters have been prepared similarly.

H. ALTERNATIVE METHODS FOR THE PREPARATION OF ESTERS, AMIDES AND OTHER FATTY ACID DERIVATIVES

1. Dimethylsulphate with dicyclohexylamine

Methyl esters can be prepared from unesterified fatty acids in a basic medium by heating them with dimethylsulphate in the presence of dicyclohexylamine for 15 to 60 minutes [317]. The method may be useful for esterifying fatty acids where diazomethane is not suitable or for fatty acids with functional groups that might be altered under acidic conditions. Unfortunately, the procedure does not appear to have been tested with a wide range of different fatty acids. Ethyl esters can be prepared in a similar way.

2. Preparation of esters with alkyl formamide derivatives

Carboxylic acids react with alkyl formates in dimethylformamide solution to give esters in good yield, and the method was used to prepare a variety of different esters [325]. In practice, alkyl chloroformates (**24**) are now the preferred reagents since they react very rapidly with acids under mild conditions in the presence of *N*,*N*-dimethyl-4-aminopyridine (DMAP) (Scheme 2.15) [180]. Others carried out the reaction in a solution of

$$\mathrm{RCOOH + ClCOOR' + Et_3N \xrightarrow{DMAP} RCOOR' + CO_2}$$

(24)

Scheme 2.15. Preparation of esters by reaction of acids with chloroformates.

acetonitrile, the alcohol and pyridine [144,146], though other basic catalysts, such as 3-picoline, *N*-methylpiperidine and dimethylaminopyridine appeared to be preferable [344]. When the method was applied to 2-hydroxy fatty acids, simultaneous alkylation of the hydroxyl group occurred [144]; this did not happen when the hydroxyl group was in other positions. The reaction was found to be less sensitive to traces of moisture than is the case in many other procedures [146,180,344]. As it is carried out in a mildly basic medium and is very rapid, the procedure may repay further study for esterification of fatty acids with labile functional groups. The use of chloroformates for derivatization has been reviewed by Husek [145].

3. *Trimethylsilyl esters*

Trimethylsilyl esters of fatty acids have been prepared for GC analysis, but they do not appear to be as useful as the conventional alkyl esters for the purpose, as they are sensitive to the presence of traces of water. In addition, non-polar silicone liquid phases must be used in the GC columns and these tend to have poorer resolving powers than the more usual polyesters. The common silylating agents, hexamethyldisilazane and trichloromethylsilane for example, can be used for the preparation of trimethylsilyl esters [60,81,94,173,196,229]. Woollard [362] has investigated procedures for preparing *tert*-butyldimethylsilyl esters of fatty acids.

4. *Preparation of pyrrolidides from esters*

Pyrrolidide derivatives of fatty acids are of value for structural identification of fatty acids by GC/MS [12,128]. Strongly nucleophilic bases such as pyrrolidine will react with esters (methyl, glyceryl, *etc.*) in the presence of acetic acid at 100°C (for 1 hour) to form pyrrolidides (**25**) (Scheme 2.16) [13]. However, the reaction has been accelerated

$$RCOOCH_3 + HN\langle\square\rangle \longrightarrow RCO{-}N\langle\square\rangle + CH_3OH$$

(25)

Scheme 2.16. Preparation of pyrrolidide derivatives from methyl esters.

appreciably by the use of microwave radiation [73]. In addition, pyrrolidides have been prepared from free fatty acids *via* trimethylsilyl esters in a one-pot reaction [340].

4,4-Dimethyloxazoline derivatives also have useful properties in locating functional groups by MS, but quantitative preparation requires heating the methyl esters with 2-amino-2-methylpropanol at 180°C overnight [95].

5. *Preparation of hydroxamic acid and related derivatives*

Hydroxamic acid derivatives of fatty acids have been prepared for HPLC analysis by the nickel-catalysed reaction of a fatty acid and hydroxylamine hydrochloride [123] and by reacting the fatty acid with hydroxylamine perchlorate [119]. Isopropylidene hydrazides have been prepared by hydrazinolysis followed by acetonation [6].

Imidazole derivatives of fatty acids for HPLC analysis were synthesised by reaction with 9,10-diaminophenanthrene, but strenuous reaction conditions were required [212].

6. Reaction with Grignard reagents

Fatty acid esters react with Grignard reagents, such as ethyl magnesium bromide (**26**), to form tertiary alcohols (**27**) (Scheme 2.17). The reaction

$$RCOOR' + 2CH_3CH_2MgBr \longrightarrow R(CH_3CH_2)_2COH$$

(26) (27)

Scheme 2.17. Reaction of fatty acids with a Grignard reagent, ethyl magnesium bromide (**26**).

was first used to convert the fatty acid components of the wax esters from jojoba oil to tertiary alcohols, the primary alcohols being released as such [282]. The two groups of compounds could be analysed simultaneously by GC in this form. More recently, it has been shown that the reaction can be used with a variety of fats and oils, especially those containing short-chain fatty acids as in milk fat [281]. Different alkyl moieties were employed to give theoretical response factors for flame-ionization detection that were close to unity, even with butyric acid. Although the reaction products are very different from the esters that have traditionally been favoured for GC analysis, the methodology is simple and may have wide applicability; it might repay further study.

I. SPECIAL CASES

1. Short-chain fatty acids

Much attention has been given to the preparation and analysis of esters of short-chain fatty acids by GC, largely because of their occurrence in milk fats. Reviews of the problems in such analyses have been published [53,158]. Short-chain fatty acids, in the free state or esterified to glycerol, can be converted completely to methyl esters by any of the reagents described above, but quantitative recovery from the reaction medium may not be achieved unless special precautions are taken. Losses can occur at several stages in any procedure. Short-chain fatty acid esters (methyl especially) are volatile and may be lost selectively on refluxing the esterification medium, they are more soluble in water than longer-chain esters and can be lost in an aqueous extraction step, or they may be distilled off when the extracting solvent is evaporated. Selective losses can also occur if non-saponifiable impurities have to be removed by sublimation or thin-layer chromatography (TLC) purification.

Losses occurring during refluxing of solutions can be avoided by carrying out the reaction in a sealed vessel or using procedures that work satisfactorily at room temperature. Losses of short-chain esters during aqueous extractions can never be eliminated entirely but they can be kept to a minimum with care. Some of the factors involved were studied by Dill [78], who found that recoveries could be improved greatly by salting out the esters. Hydrocarbon solvents, such as hexane, pentane or light petroleum gave better recoveries than did diethyl ether or benzene. Others [174] obtained excellent recoveries of short-chain esters by extracting with a solvent mixture of docosane and petroleum ether, but the docosane could interfere with the subsequent GC analysis of the esters. Careful removal of excess solvents at low temperatures on a rotary film evaporator is better than using a stream of nitrogen and will keep losses of short-chain esters down, but these cannot be eliminated completely.

The best esterification procedures for short-chain fatty acids are those in which heating of the reagents is avoided and in which stages involving aqueous extraction and solvent removal are absent. Many analysts, including the author, have found that the alkaline transesterification procedures of Christopherson and Glass [59] and of Luddy *et al.* [217] are the most convenient and quantitative for obtaining short-chain methyl esters from lipids, as no aqueous extraction or solvent removal steps are necessary. In the latter procedure, the lipid sample was transesterified at 65°C with a small amount of 0.4*M* potassium methoxide in a sealed tube. On cooling, carbon disulphide was added and excess methanol was removed by adding anhydrous calcium chloride. An aliquot of the supernatant solution was injected directly into the gas chromatograph. If the sample contained a high proportion of free fatty acids, these were esterified with boron trifluoride in methanol after the transesterification reaction had been completed. In the method of Christopherson and Glass [59], the fat sample was dissolved in 19 volumes of petroleum ether in a stoppered-tube and one volume of 2*M* sodium methoxide in methanol was added. Transesterification was complete in a few minutes at room temperature and an aliquot of the reaction mixture was again injected directly onto the GC column. If appreciable amounts of free fatty acids were present in the sample, they could be methylated when the transesterification reaction was finished by adding 5 volumes of 10% hydrogen chloride in methanol and leaving the reaction mixture for a further hour at room temperature before injection onto a packed GC column.

Injection of reaction media containing basic and acidic esterification catalysts directly onto GC columns shortens their working lives. The top few centimetres of packed columns can be replenished periodically, while lengths of deactivated tubing or "retention gaps" ahead of capillary columns will protect the latter. This can be a small price to pay for the speed, simplicity and accuracy of these procedures.

The method of Christopherson and Glass has been evaluated and recommended by Badings and De Jong [20]. Others have suggested minor modifications in which acid was added to neutralise the sample prior to GC analysis [25,53] or in which methyl acetate was added to suppress the competing hydrolysis reaction [51,53]. Basic transesterification-esterification with a quaternary ammonium salt has also been considered [230]. Of course, it is necessary that the GC conditions be optimised for the analysis otherwise the value of these transesterification procedures will be negated [20,25,53].

Quantitative recovery of esters of short-chain acids is less of a problem if esters of higher molecular weight alcohols than methanol are prepared. For example, propanol, butanol or 2-chloroethanol may be used, but great care is still necessary. For example, a procedure utilising boron trifluoride-butanol has been recommended [154,155] and the Grignard reaction described above (Section H.6) appears to have some promise [281]. In addition to the combined methods just described, free short-chain fatty acids can be esterified with diazomethane in diethyl ether and the reaction medium injected directly onto the GC column. Alternatively, the use of alkyl chloroformates could be considered (Section H.2). Methods requiring hydrolysis prior to esterification are not recommended.

HPLC methods have also been developed for the analysis of milk fatty acids. Benzyl esters of milk fat were prepared by transesterification with potassium benzylate [55], and butyrate in milk fat was determined in the form of the phenethyl ester [56]. As an alternative, milk fat was saponified with potassium hydroxide, solvent was removed from the alkaline solution and the potassium salts were solubilized with a crown ether for conversion to phenacyl or related esters [179]. A similar method was described for the analysis of the free fatty acid fraction in butter fat [288].

2. *Fatty acids with unusual structures*

If the appropriate precautions are taken, most of the methods described above can be used to esterify unsaturated fatty acids with two to six methylene-interrupted double bonds without causing stereomutation or double bond migration, *i.e.* they can be safely applied to virtually all fatty acids of animal origin with the possible exception of certain of the eicosenoids. Plant lipids, on the other hand, may contain fatty acids with a variety of different functional moieties, such as cyclopropene rings, conjugated unsaturation (especially of concern when adjacent to other functional groups), or epoxyl groups, that can be altered chemically by certain esterification catalysts [19,312]. Some knowledge of the chemistry and potential occurrence of these fatty acids is necessary, therefore, before a decision can be taken as to which of the available methods of esterification is likely to be most appropriate in each instance.

The occurrence and chemistry of cyclopropene and cyclopropane fatty acids have been reviewed [48,306]. Both types of functional group may be destroyed by acidic conditions, but triacylglycerols and other lipids containing such acids can be transesterified safely with basic reagents. Unesterified fatty acids can be methylated with diazomethane and presumably by alkyl chloroformates (Section H.2) or *via* the imidazolides (Section F.3). Boron trifluoride-methanol complex, although it is acidic, was reportedly suitable for esterifying cyclopropene fatty acids without causing any alteration to the functional group [185]. However, this reagent reacts with cyclopropane fatty acids, which are often found with cyclopropene fatty acids in seed oils, with the addition of methanol across the ring carbons [75,243,247], a property which can be of diagnostic value [243,247]. Methanolic-hydrogen chloride has similar drawbacks [199,345], but boron trichloride-methanol has been used safely [40].

Epoxy fatty acids are widely distributed in seed oils, and their occurrence and chemical reactivity have been reviewed [193]. Fatty acids of this type are very sensitive to acidic conditions and they react with opening of the oxirane ring. For example, hydrogen chloride adds across the ring to form halogen hydrins, and boron trifluoride-methanol adds methanol across the ring to give a methoxy-hydroxy product which is potentially useful for quantitative analysis of epoxy fatty acids in natural oils [185]. Epoxy acids are not harmed by basic conditions under normal circumstances, however, and seed oils containing these acids can be transesterified safely with alkaline reagents. Unesterified epoxy acids can be methylated with diazomethane and they have been converted to picolinyl esters *via* the imidazolides (Section F.3) [22].

The occurrence and chemistry of fatty acids with conjugated unsaturation have been reviewed [138]. Such conjugated polyenoic acids as α-eleostearic (9-*cis*,11-*trans*,13-*trans*-octadecatrienoic) acid underwent stereomutation and double bond migration, when esterified with methanolic hydrogen chloride but not when reacted with boron trifluoride-methanol [185], possibly because of the shorter reaction time necessary, or when transesterified with basic reagents. On the other hand, methanol containing boron trifluoride or other strong acids was reported to cause addition of methanol to conjugated diene systems [190]. Conjugated fatty acids with hydroxyl groups adjacent to the double bond system are especially liable to rearrangement, and double bond migration, dehydration and ether formation may all occur. For example, dimorphecolic (9-hydroxyoctadeca-10-*trans*,12-*trans*-dienoic) and related acids were dehydrated to conjugated trienoic acids by strongly acidic conditions [313]. With methanol and hydrogen chloride or boron trifluoride, methoxy dienes were formed [185,286], and they were also produced on reaction with diazomethane [77]. Similarly, hydroperoxides formed during autoxidation and containing conjugated double bond systems were destroyed completely under conditions of acidic methylation and a substantial portion can be lost with base-catalysed esterification

[335]. Simultaneous reduction and transesterification with sodium borohydride in methanol may help in analysis of these compounds, although the hydroperoxide group *per se* is lost [283,284]. α-Hydroxy acetylenic fatty acids, in contrast to their ethylenic analogues, were resistant to acid-catalysed dehydration [118].

Fatty acids with double bonds or keto groups in position 2 were reported to react with diazomethane with addition of a methylene group [31].

3. *Sphingolipids and other N-acyl lipids*

In sphingolipids, fatty acids are joined by an amide rather than an ester linkage to a long-chain base. The structures, occurrence and biochemistry of these compounds have been reviewed [167,352,356]. In a comprehensive analysis, it may be necessary to prepare not only the methyl esters of the component fatty acids but also the free sphingoid bases in a pure state. A method that is suitable for the first purpose may not always be suited to the second, and more than one procedure may have to be used for complete analysis of the constituents. In addition, the *N*-acylated lipid, *N*-acyl-phosphatidylethanolamine, has been found in brain and in a variety of plant tissues, and *N*-acyl-phosphatidylserine has been described; *N*-acyl amino acid derivatives have been found in several species of bacteria.

Sphingolipids and other *N*-acyl lipids are transesterified by vigorous acid-catalysed methanolysis, but *O*-methyl ethers of the long-chain bases may be formed as artefacts [46,240,347] or inversion of configuration of the functional groups at C(2) or C(3) of the bases may occur [238]. Anhydrous methanolic hydrogen chloride gave approximately 75% recovery of methyl esters from sphingomyelin after 5 hours under reflux [170], but large amounts of by-products are known to be formed from the sphingoid bases with this reagent [46,182]. In contrast, methanol containing concentrated hydrochloric acid gave much better recoveries of methyl esters and smaller amounts of artefacts from the long-chain bases [103,106,170], especially when the reagents were heated in sealed tubes in the absence of air [106,261]. All acid-catalysed transesterification procedures produce some artefacts from sphingoid bases [240], but these need not interfere with the GC analysis of methyl esters, which are easily cleaned up by TLC or column procedures in any case (see Section L) [49]. If the organic bases are not required for analysis, quantitative recovery of methyl esters can be obtained from sphingolipids by refluxing in methanol-concentrated hydrochloric acid (5:1, v/v) for five hours [170] or by heating with the reagents in a stoppered tube at 60°C for 24 hours [49]. Others have suggested the use of hydrochloric acid in aqueous butanol at 85°C for 80 minutes, followed by hydrolysis of the resultant butyl esters under mild conditions and esterification *via* pyrolysis of the quaternary ammonium salt [221]. Similarly, hydrolysis with 0.5*M* hydrochloric acid in acetonitrile-water followed by methylation has been reported to give excellent results [18]. It has been claimed that boron trifluoride-methanol reagent can transesterify amide-bound fatty acids in sphingolipids comparatively rapidly [259,260], but others [91] found that a reaction time of 15 hours at 90°C was required for complete transesterification of sphingomyelin with boron trifluoride (15%) or sulphuric acid (1 or 2.5*M*) in methanol. Such prolonged heating of unsaturated fatty acids in these reagents may not be advisable (see Sections B.4 and L).

Sphingolipids are not transesterified in the presence of mild basic catalysts and this property has been utilized in the bulk preparation of sphingolipids to remove *O*-acyl impurities [322,349]. Basic transesterification is also of value for the quantitative release of *O*-acyl-bound fatty acids alone from sphingolipids and other *N*-acyl lipids containing both *O*-acyl and *N*-acyl fatty acids [270]. Prolonged treatment with aqueous alkali (up to 8 hours at reflux) hydrolysed sphingolipids completely (sphingomyelin must first be enzymatically dephosphorylated with phospholipase C before chemical hydrolysis), and quantitative yields of artefact-free long-chain bases were obtained [168,239,257,305]. Polyunsaturated fatty acids, which might possibly be harmed by such vigorous conditions of hydrolysis, are not normally found in significant amounts in sphingolipids so the free fatty acids released can be safely recovered and esterified by an appropriate method. Alkaline hydrolysis methods are therefore to be preferred when both the fatty acid and long-chain base components are required for analysis. Fatty acids obtained in this way can then be methylated by any of the methods described above for free acids. The hydroxyl group of 2-hydroxy acids can be methylated at the same time if reaction with methyl iodide and a base is used (see Section G.1) [63,162].

4. Sterol esters

It is well documented (but often rediscovered) that sterol esters are transesterified much more slowly than are glycerolipids. The same reagents can be employed but much longer reaction times or higher temperatures are necessary. Reaction times of an hour or more with sodium methoxide in methanol and an inert solvent at 25 to 50°C have been quoted typically [51,333,370]. While irreversible hydrolysis was found to occur under certain conditions [333], leading to a suggestion that acid-catalysed methylation should follow basic catalysis, the unwanted side effects could be avoided by the simple expedient of adding a little methyl acetate to suppress hydrolysis [51].

Acid-catalysed methylation procedures must be avoided with sterol esters (or lipid extracts containing sterols) as they react with the other constituent of the lipid, *i.e.* cholesterol [130,176,192,234,259,309] or phytosterols [320]. Dehydration, methoxylated and other products are formed that interfere with the analysis of polyunsaturated fatty acids by means of GC. In practice, broad peaks tended to emerge after the fatty acids of interest, often during subsequent analyses. Cholesterol itself can interfere in this way [130] and is best removed from methyl ester preparations by mini-column or TLC procedures prior to GC analysis (see Section L).

It is less well known that wax esters also are transesterified rather slowly by most reagents.

5. *Selective esterification of free fatty acids in the presence of other lipids*

The free or unesterified fatty acid fraction in plasma and other tissues is of special metabolic importance, so some effort has been expended to devise methods for selective esterification of this lipid class without effecting transesterification of other lipids that might be present. In general, the topic can be considered in terms of the preparation of derivatives for either GC (usually methyl esters) or HPLC (UV-absorbing or fluorescent derivatives). Methods involving selective extraction of free fatty acids or isolation by chromatography prior to esterification are not considered here.

Diazomethane [301] (see Section D.1) was one of the first methods to be used for selective methylation of free fatty acids, and it has been employed in analyses of this lipid fraction from plasma, when it has been reported to give reliable results [195,209,263,274]. It has also been utilised in a method for simultaneous extraction and methylation of free acids in plasma [274]. However, routine use has probably been discouraged in clinical laboratories because of the toxicity problems or occasional reports of side-reactions. Preparation of methyl esters by reaction of free acids with *N*,*N'*-carbonyldiimidazole and methanol also gave good results [188], but the method does not appear to have been followed by others for no clear reason. Reaction with free fatty acids from plasma and dimethylacetal-dimethylformamide (Section H.2) was judged to be unsatisfactory, because of deleterious effects on the GC column [263], and methyl iodide-potassium carbonate "proved to be less than completely reliable". On the other hand, the latter reaction did appear to give acceptable results when a crown ether was added as catalyst [133] or with aqueous potassium hydroxide as the base [7]. Pyrolysis of the fatty acid salt of trimethyl(trifluoro-*m*-tolyl)ammonium hydroxide was also reported to work satisfactorily when other lipids were present [7,169]. A variation on the last procedure was used to extract and derivatize the free fatty acid fraction of vegetable oils [354].

Selective methylation of free fatty acids, after adsorption onto a strong anion exchange resin, has been effected with methanolic hydrogen chloride

[62], and this method has been adapted to the analysis of free acids in milk fat [266,316]. Others have attempted to utilise the fact that free acids are esterified much more rapidly than other lipids are transesterified as a means of selective methylation of the former [206,329]. While such methods may work satisfactorily under ideal circumstances, there is no margin for error.

A quite different and novel approach involved transesterification of glycerolipids to benzyl esters with benzoyl alcohol and (*m*-trifluoromethylphenyl)trimethyl ammonium hydroxide as catalyst, with pyrolysis of the salt of the acids to the methyl esters in the injection port of the gas chromatograph [337]. Others used sodium methoxide in methanol to transesterify the ester-linked lipids, then subjected the methyl esters and the unchanged free acids to GC analysis [278].

J. PREPARATION OF ESTERS IN THE PRESENCE OF ADSORBENTS FOR THIN-LAYER CHROMATOGRAPHY

It is a common practice in lipid analysis to fractionate samples by TLC so that the fatty acid composition of each lipid class can be determined. In conventional procedures, lipid classes are separately eluted from the TLC adsorbent before methylation, but it has been shown that methyl esters can be prepared *in situ* on the TLC adsorbent without such preliminary extraction. The elimination of this step reduces the risk of loss of sample, particularly of phospholipids which are strongly adsorbed, and of contamination of the sample by traces of impurity in the large volumes of solvent that otherwise might be necessary for elution.

Most of the common acid- or base-catalysed procedures have been adapted for this purpose. After the components have been located by spraying the developed plate with a solution of a suitable dye, such as 2′,7′-dichlorofluorescein, the methylating agent can be sprayed onto the whole plate or pipetted gently onto the individual spots. The reaction can be facilitated by warming the plate, and the esters are eventually obtained for analysis by elution from the adsorbent by a less polar solvent than might otherwise have been required to extract the original lipid component. Alternatively, the silica gel on which the lipid is adsorbed can be scraped from the TLC plate into a suitable vessel and the esterification reaction carried out in a similar manner to that when no adsorbent is present.

For example, potassium hydroxide in methanol (12%) was pipetted onto TLC plates to transesterify phospholipids [175], and this procedure was used successfully in other laboratories [295,341]. Neutral lipids have been transesterified similarly by spraying the TLC plates with 2*M* sodium methoxide in methanol [270,342]. Others [50,143,285,324] found that procedures involving spraying TLC plates with reagents were messy and wasteful of reagents, and that esterification could be carried out more conveniently by scraping the bands into a test-tube or flask to which the esterifying reagent was added. The author once made some use of a procedure of this type but eventually found that variable amounts of water bound to the silica gel caused some hydrolysis so that the results were unreliable. This is in accord with a report that the ratio of the weight of lipid to that of silica gel must be higher than 4:1000 for consistent results [45]. In addition, there is ample evidence that cholesterol esters are esterified only slowly and unevenly in such systems [50,93]. Satisfactory results were apparently obtained when the silica gel bands were dried *in vacuo* over phosphorus pentoxide prior to adding sodium methoxide solution [321], but this additional step negates some of the potential advantages of this approach to methylation.

Boron trifluoride-methanol has been used to methylate unesterified fatty acids in contact with TLC adsorbents [50,136]. (Although diazomethane in diethyl ether has been sprayed onto TLC plates for the same purpose, this is much too hazardous a procedure to be recommended). Methanolic sulphuric acid [96,225], boron trifluoride-methanol [143,307], methanolic hydrogen chloride [45] and aluminium chloride-methanol [307] have been used for acid-catalysed transesterification of most lipid classes in the presence of silica gel from TLC plates. Methanol containing hydrochloric acid was found to be necessary for transesterification of sphingomyelin (see Section I.3) [50]. Again, problems were encountered mainly with cholesterol esters [50,93].

K. SIMULTANEOUS EXTRACTION FROM TISSUES AND TRANSESTERIFICATION

A common requirement of lipid analysts is to determine the total content of fatty acids and the overall fatty acid composition of a tissue of animal, plant or microbial origin. To minimize the labour involved and some potential sources of error, there have been many attempts to combine extraction of the lipids from the tissue with methylation of the fatty acid components for GC analysis. An internal standard, usually an odd-chain fatty acid, can be added to enable the total lipid content as well as the fatty acid composition to be determined; this can also compensate for any partial hydrolysis that may occur. The results have been somewhat variable, depending on the nature of the tissue, especially that of the matrix and its water content, since this has an important bearing on the yield of methyl esters. Lyophilization of the tissue has sometimes been undertaken to minimize this problem. As with other transesterification procedures, it is usually advisable to add an inert solvent such as toluene to solubilize triacylglycerols.

Most analysts have used acidic media for simultaneous extraction and transesterification, since these are affected less by small amounts of water. For example, as long ago as 1963, it was shown that the fatty acids of bacteria could be extracted efficiently and methylated with boron trichloride in methanol [1], a finding confirmed later [85]. A distinctive low temperature method, utilising methanol-sulphuric acid in diethyl ether, was used to methylate lipids in animal tissues [86]. Cereal grains of various kinds were ground to a powder and the lipids transesterified directly with 10% boron trichloride [27] or 1 to 2% sulphuric acid in methanol [69,348]. 5% Methanolic hydrogen chloride was used similarly in experiments with ruminant feeds, digesta and faeces [273], and with added toluene for various food materials [334]. Rat serum and brain tissues were first dehydrated by reaction with dimethoxypropane; excess solvents were removed by evaporation (so artefact formation (see Section B2) was not troublesome), before the tissue was extracted and methylated with methanolic hydrogen chloride [192]. Alternatively, lipoprotein fractions from plasma were lyophilized prior to methylation with boron trifluoride-methanol [299]. On the other hand, Lepage and Roy [204,205] have suggested that such a step is not necessary. Provided that the water content of the medium was not allowed to rise above 5%, much better recoveries of fatty acids were obtained by extraction/transesterification with methanol-hydrogen chloride-benzene than with standard extraction procedures. Plasma, milk, adipose tissue, bile, liver and other tissues were analysed, and the method has been adapted for the specific methylation of the free fatty acid fraction of plasma [206]. The validity of this procedure was confirmed by others and it was applied to further types of sample [319,350]. A related procedure has been used with green leaf tissue [42].

In an interesting alternative approach, isopropanol or hexane-isopropanol mixtures were used to extract the lipids from such materials as cheese, other dairy products and erythrocytes [280,359,360], anhydrous sodium sulphate was used to remove the water and then concentrated sulphuric acid was added directly to the solution as a catalyst to effect transesterification with formation of isopropyl esters for GC analysis.

Base-catalysed methods have been little used for this specific purpose, but satisfactory results were obtained with small samples of materials with a low water content, such as crushed rapeseed [141], soy meal, cheese and buttermilk blends by means of direct methylation with sodium methoxide [213]. Similarly, trimethylphenylammonium hydroxide has been utilised to transesterify plasma extracts for the determination of phytanic acid [61]. Whole bacterial cells have been subjected to methylation or pyrolytic methylation with some success, *i.e.* by reaction with tetramethylammonium hydroxide [89,246] or trimethylsulphonium hydroxide [264,265].

Free fatty acids in plasma have been transesterified directly and selectively by methylation with diazomethane in methanol [274].

As an alternative, many analysts have used either acidic or basic treatments to improve the extractabilty of fatty acids from tissue matrices before proceeding to methylation, but this is out with the scope of the present review.

L. ARTEFACTS OF ESTERIFICATION PROCEDURES

Vigilance must be exercised continuously to detect contamination of samples by impurities in the reagents or from any other extraneous source. Middleditch [241] has compiled an exhaustive treatise on compounds that can be troublesome in chromatographic analyses (including mass spectral information). In addition, the samples themselves may be a source of contaminants or artefacts, for example from the fatty acids *per se* (see Section I.2), endogenous cholesterol, other sterols or their esters (see Section I.4), and other lipids. Thus, dimethyl acetals are formed from plasmalogens by the reaction of acidic transesterifying reagents, and these tend to elute just ahead of the corresponding esters (generally C_{16} and C_{18}) on GC. All solvents (including water) and reagents should be of the highest grade and may have to be distilled before use to remove non-volatile impurities, especially when preparing very small quantities of esters. Extraneous substances can be introduced into samples from a variety of sources, for example filter papers, soaps, hair preparations, tobacco smoke and laboratory grease, and care must be taken to recognise and avoid such contaminants [15,210,292].

Phthalate esters, used as plasticisers, are probably the most common contaminants and are encountered whenever plastic ware of any kind is in contact with solvents, lipid samples, reagents and even distilled water [32,66,153,210,275,276,279]. They can enter preparations *via* the brief contact between disposable pipette tips and reagents. In addition, phthalate esters can interact with transesterification reagents to give mono- and sometimes dimethyl esters, basic catalysts reacting more rapidly than acids [308]. In GC analysis, the precise point of elution relative to fatty acid derivatives is dependent on the nature of the phthalate ester and the stationary phase, but typically is in the same range as the C_{18} to C_{22} fatty acids. In HPLC, the precise elution point is dependent on the mode of chromatography, and the problem is especially troublesome with UV detection systems.

Samples containing polyunsaturated fatty acids should be handled under nitrogen whenever possible, and antioxidants such as 2,6-di-*tert*-butyl-4-methylphenol (BHT) may be added to solvents and reagents to minimize autoxidation [363]. In GC analyses, BHT emerges as a sharp peak which can interfere with the analysis of methyl myristate with packed columns but not usually when fused silica capillaries are used; in HPLC, it often emerges close to the solvent front where it can be a nuisance with UV detection. Boron trifluoride is known to interact with BHT to produce methoxy derivatives as artefacts that co-elute with methyl pentadecanoate or hexadecenoate on GC analysis [33,80,131,255].

This does not appear to be a problem with other acidic reagents.

Ethyl esters were found in methyl ester preparations, when chloroform containing ethanol as a stabilizer was employed as a solvent to facilitate transesterification [159]. During GC analysis *per se*, artefact peaks have been produced when traces of basic transesterification reagents, remaining in samples, were introduced into GC columns [326].

Non-lipid contaminants or non-ester by-products of esterification reactions, for example cholesterol, can be removed by preparative TLC on plates coated with silica gel G and developed in a solvent system of hexane-diethyl ether (9:1, v/v), though there may be some selective loss of short-chain esters if these are present. BHT migrates with the solvent front and is closely followed by the required esters. Alternatively, a short column of silicic acid or Florisil™, in a disposable Pasteur pipette say, can be used and the esters recovered by elution with hexane-diethyl ether (95:5, v/v). Alumina column chromatography has been recommended for the same purpose [234]. Dimethyl acetals formed from plasmalogens have been separated from methyl esters by preparative TLC with toluene [224] or dichloroethane [355] as mobile phase, when esters migrate more rapidly.

M. THE CHOICE OF REAGENTS – A SUMMARY

Methyl esters are by far the favourite derivatives for GC analysis of fatty acids, although a case can be made for isopropyl or butyl esters in some circumstances. One of the first considerations when deciding which procedure to adopt for preparing ester derivatives for GC is the lipid composition of the samples to be analysed. If these are free fatty acids alone, mixtures containing significant amounts of free acids or mixtures of unknown composition which might contain free acids, then acid-catalysed procedures are generally preferred. The newer quaternary ammonium reagents, which can be used both for esterification and transesterification appear to be promising alternatives. On the other hand, alkaline transesterification procedures are so rapid and convenient that they must be considered for mixed lipid samples which contain no unesterified fatty acids or for single lipid classes containing ester-bound fatty acids. If a single method is required for use with all routine lipid samples, hydrogen chloride (5%) or sulphuric acid (2%) in methanol, despite the comparatively long reaction times needed, is probably the best general purpose reagent available. The author, does not find it inconvenient to have more than one method in routine use in his own laboratory, *i.e.* sulphuric acid in methanol for those samples containing free fatty acids, and sodium methoxide in methanol for all other lipid classes with the exception of sphingolipids (for which special procedures are in any case necessary (see Section I.3)).

A further consideration when deciding on a reagent is the fatty acid composition of the samples to be esterified. If short-chain fatty acids (Section I.1) or others of unusual structure (Section I.2) are present, appropriate methods must be adopted. Polyunsaturated fatty acids must always be handled with care and should not be subjected to more vigorous conditions than are necessary. It should be recognised that reagents which are perfectly satisfactory when used under optimised conditions can be destructive to fatty acids if used carelessly.

Similar restrictions apply in the preparation of derivatives for HPLC. Here there is no consensus as to which is the best derivative from a chromatographic standpoint, and the choice may be determined to some extent by the nature of the detector available to the analyst.

ACKNOWLEDGEMENT

This review is published as part of a programme funded by the Scottish Office Agriculture and Fisheries Dept.

ABBREVIATIONS

BHT, 2,6-di-*tert*-butyl-4-methylphenol; DMAP, *N,N*-dimethyl-4-aminopyridine; GC, gas chromatography; HPLC, high-performance liquid chromatography; MS, mass spectrometry; TLC, thin-layer chromatography.

REFERENCES

1. Abel,K., de Schmertzing,H. and Peterson,J.I., *J. Bact.*, **85**, 1039-1044 ((1963).
2. Ackman,R.G., *Lipids*, **12**, 293-296 (1977).
3. Ackman,R.G., Hooper,S.N., Kates,M., Sen Gupta,A.K., Eglinton,G. and MacLean,I., *J. Chromatogr.*, **44**, 256-261 (1969).
4. Ackman,R.G., Kates,M. and Hansen,R.P., *Biochim. Biophys. Acta*, **176**, 673-674 (1969).
5. Ackman,R.G., Sebedio,J.-L. and Ratnayake,W.N., *Methods Enzymol.*, **72**, 253-276 (1981).
6. Agrawal,V.P. and Schulte,E., *Anal. Biochem.*, **131**, 356-359 (1983).
7. Allen,K.G., MacGee,J., Fellows,M.E., Tornheim,P.A. and Wagner,K.R., *J. Chromatogr.*, **309**, 33-42 (1984).
8. Allenmark,S. and Chelminska-Bertilsson,M., *J. Chromatogr.*, **456**, 410-416 (1988).
9. Allenmark,S., Chelminska-Bertilsson,M. and Thompson,R.A., *Anal. Biochem.*, **185**, 279-285 (1990).
10. American Oil Chemists' Society, Instrumental Committee, *J. Am. Oil Chem. Soc.*, **43**, 10A (1966).
11. American Oil Chemists' Society, *Official and Tentative Methods of the American Oil Chemists' Society, 3rd Edition*, Champaign, IL., Method Ce 2-66 (1973).
12. Andersson,B.A., *Prog. Chem. Fats Other Lipids*, **16**, 279-308 (1978).
13. Andersson,B.A. and Holman,R.T., *Lipids*, **9**, 185-190 (1974).
14. Anon., *Gas-Chrom. Newsletter* (Applied Science Laboratories Inc.), Vol. 11, No. 4., p.2. (1970).
15. Appelqvist,L.-A., *Arkiv. Kemi*, **28**, 551-570 (1968).
16. Archibald,F.M. and Skipski,V.P., *J. Lipid Res.*, **7**, 442-445 (1966).
17. Ast,H.J., *Anal. Chem.*, **35**, 1539-1540 (1963).
18. Aveldano,M.I. and Horrocks,L.A., *J. Lipid Res.*, **24**, 1101-1105 (1983).

19. Badami,R.C. and Patil,K.B., *Prog. Lipid Res.*, **19**, 119-153 (1982).
20. Badings,H.T. and De Jong,C., *J. Chromatogr.*, **279**, 493-506 (1983).
21. Bailey,J.J., *Anal. Chem.*, **39**, 1485-1489 (1967).
22. Balazy,M. and Nies,A.S., *Biomed. Environm. Mass Spectrom.*, **18**, 328-336 (1989).
23. Banerjee,P., Dawson,G. and Dasgupta,A., *Biochim. Biophys. Acta*, **1110**, 65-74 (1992).
24. Bannon,C.D., Breen,G.J., Craske,J.D., Hai,N.T., Harper,N.L. and O'Rourke,K.L., *J. Chromatogr.*, **247**, 71-89 (1982).
25. Bannon,C.D., Craske,J.D. and Hilliker,A.E., *J. Am. Oil Chem. Soc.*, **62**, 1501-1507 (1985).
26. Barker,S.A., Monti,J.A., Christian,S.T., Benington,F. and Morin,R.D., *Anal. Biochem.*, **107**, 116-123 (1980).
27. Barnes,P.C. and Holaday,C.E., *J. Chromatogr. Sci.*, **10**, 181-183 (1972).
28. Baty,J.D., Pazouki,S. and Dolphin,J., *J. Chromatogr.*, **395**, 403-411 (1987).
29. Baty,J.D., Willis,R.G. and Tavendale,R., *Biomed. Mass Spectrom.*, **12**, 565-569 (1985).
30. Baty,J.D., Willis,R.G. and Tavendale,R., *J. Chromatogr.*, **353**, 319-328 (1986).
31. Bauer,S., Neupert,M. and Spiteller,G., *J. Chromatogr.*, **309**, 243-259 (1984).
32. Baumann,A.J., Cameron,R.E., Kritchevsky,E. and Rouser,G., *Lipids*, **2**, 85-86 (1967).
33. Beaumelle,B.D. and Vial,H.J., *J. Chromatogr.*, **356**, 187-194 (1986).
34. Bitner,E.D., Davison,V.L. and Dutton,H.J., *J. Am. Oil Chem. Soc.*, **46**, 113-117 (1969).
35. Boer,T.J. de and Backer,H.J., in *Organic Synthesis, Coll. Vol.* 4, pp. 250-253 (1963) (edited by N. Rabjohn, John Wiley & Sons, London).
36. Borch,R.F., *Anal. Chem.*, **47**, 2438-2439 (1975).
37. Boss,W.F., Kelley,C.J. and Landsberger,F.R., *Anal. Biochem.*, **64**, 289-292 (1975).
38. Brian,B.L. and Gardner,E.W., *Appl. Microbiol.*, **15**, 1499-1500 (1967).
39. Brian,B.L. and Gardner,E.W., *J. Bact.*, **96**, 2181-2182 (1968).
40. Brian,B.L., Gracy,R.W. and Scholes,V.E., *J. Chromatogr.*, **66**, 138-140 (1972).
41. Brondz,I. and Olsen,I., *J. Chromatogr.*, **576**, 328-333 (1992).
42. Browse,J., McCourt,P.J. and Somerville,C.R., *Anal. Biochem.*, **152**, 141-145 (1986).
43. Butte,W., *J. Chromatogr.*, **261**, 142-145 (1983).
44. Butte,W., Eilers,J. and Kirsch,M., *Anal. Letts.*, **15A**, 841-850 (1982).
45. Carreau,J.P. and Dubacq,J.P., *J. Chromatogr.*, **151**, 384-390 (1978).
46. Carter,H.E., Nalbandov,O. and Tavormina,P.A., *J. Biol. Chem.*, **192**, 197-207 (1951).
47. Castell,J.D. and Ackman,R.G., *Canad. J. Chem.*, **45**, 1405-1410 (1967).
48. Christie,W.W., in *Topics in Lipid Chemistry*. Vol. 1, pp. 1-49 (1970) (edited by F.D. Gunstone, Logos Press, London).
49. Christie,W.W., in *Topics in Lipid Chemistry*. Vol. 3, pp. 171-197 (1972) (edited by F.D. Gunstone, Paul Elek, London).
50. Christie,W.W., *Analyst (London)*, **97**, 221-223 (1972).
51. Christie,W.W., *J. Lipid Res.*, **23**, 1072-1075 (1982).
52. Christie,W.W., *High-Performance Liquid Chromatography and Lipids*, Pergamon Press, Oxford, 1987.
53. Christie,W.W., *Gas Chromatography and Lipids*, Oily Press, Ayr, 1989.
54. Christie,W.W., Brechany,E.Y. and Stefanov,K., *Chem. Phys. Lipids*, **46**, 127-135 (1988).
55. Christie,W.W., Connor,K. and Noble,R.C., *J. Chromatogr.*, **298**, 513-515 (1984).
56. Christie,W.W., Connor,K., Noble,R.C., Shand,J.H. and Wagstaffe,P.J., *J. Chromatogr.*, **390**, 444-447 (1987).
57. Christie,W.W., Lie Ken Jie,M.S.F., Brechany,E.Y. and Bakare,O., *Biol. Mass Spectrom.*, **20**, 629-635 (1991).
58. Christie,W.W. and Stefanov,K., *J. Chromatogr.*, **392**, 259-265 (1987).
59. Christopherson,S.W. and Glass,R.L., *J. Dairy Sci.*, **52**, 1289-1290 (1969).
60. Churacek,J., Drahokoupilova,M., Matousek,P. and Komarek,K., *Chromatographia*, **2**, 493-499 (1969).
61. Cingolani,L., *J. Chromatogr.*, **419**, 475-478 (1987).
62. Ciucanu,I. and Kerek,F., *J. Chromatogr.*, **257**, 101-106 (1983).
63. Ciucanu,I. and Kerek,F., *J. Chromatogr.*, **284**, 179-185 (1984).
64. Cooper,W.J. and Anders,M.W., *Anal. Chem.*, **46**, 1849-1852 (1974).
65. Coppock,J.B.M., Daniels,N.W.R. and Eggitt,P.W.R., *J. Am. Oil Chem. Soc.*, **42**, 652-656 (1965).
66. Crosby,D.G. and Aharonson,N., *J. Chromatogr.*, **25**, 330-335 (1966).
67. Crotte,C., Mule,A. and Planche,N.E., *Bull. Soc. Chim. Biol.*, **52**, 108-109 (1970).

68. Cubero Robles,E. and Van den Berg,D., *Biochim. Biophys. Acta*, **187**, 520-526 (1969).
69. Dahmer,M.L., Fleming,P.D., Collins,G.B and Hildebrand,D.F., *J. Am. Oil Chem. Soc.*, **66**, 543-548 (1989).
70. Dalgliesh,C.E., Horning,E.C., Horning,M.G., Knox,K.L. and Yarger,K., *Biochem. J.*, **101**, 792-810 (1966).
71. Daneshvar,M.I. and Brooks,J.B., *J. Chromatogr.*, **433**, 248-256 (1988).
72. Darbre,A., in *Handbook of Derivatives for Chromatography*, pp. 36-103 (1978) (edited by K. Blau & G.S. King, Heyden & Son, London).
73. Dasgupta,A., Banerjee,P. and Malik,S., *Chem. Phys. Lipids*, **62**, 281-291 (1992).
74. Davison,V.L. and Dutton,H.J., *J. Lipid Res.*, **8**, 147-149 (1967).
75. Dawidowowicz,E.A. and Thompson,T.E., *J. Lipid Res.*, **12**, 636-637 (1971).
76. DePalma,R.A., *J. Chromatogr. Sci.*, **25**, 219-222 (1987).
77. Diamond,M.J., Knowles,R.E., Binder,R.G. and Goldblatt,L.A., *J. Am. Oil Chem. Soc.*, **41**, 430-433 (1964).
78. Dill,C.W., *J. Dairy Sci.*, **49**, 1276-1279 (1966).
79. Distler,W., *J. Chromatogr.*, **192**, 240-246 (1980).
80. Dodge,J.T. and Phillips,G.B., *J. Lipid Res.*, **8**, 667-675 (1967).
81. Donike,M., Hollmann,W. and Stratmann,D., *J. Chromatogr.*, **43**, 490-492 (1969).
82. Downing,D.T., *Anal. Chem.*, **39**, 218-220 (1967).
83. Downing,D.T. and Greene,R.S., *Anal. Chem.*, **40**, 827-828 (1968).
84. Downing,D.T. and Greene,R.S., *Lipids*, **3**, 96-100 (1968).
85. Drucker,D.B., *Microbios*, **5**, 109-112 (1972).
86. Dugan,L.R., McGinnis,G.W. and Vahedra,D.V., *Lipids*, **1**, 305-315 (1966).
87. Dunges,W., *Anal. Chem.*, **49**, 442-445 (1977).
88. Durst,H.D., Milano,M., Kikta,E.J., Connelly,S.A. and Grushka,E., *Anal. Chem.*, **47**, 1797-1801 (1975).
89. Dworzanski,J.P., Berwald,P.and Meuzelaar,H.L.C., *Appl. Environm. Microbiol.*, **56**, 1717-1724 (1990).
90. Eberhagen,D., *Z. Anal. Chem.*, **212**, 230-238 (1965).
91. Eder,K., Reichlmayr-Lais,A.M. and Kirchgessner,M., *J. Chromatogr.*, **607**, 55-67 (1992).
92. El-Hamdy,A.H. and Christie,W.W., *J. Chromatogr.*, **630**, 438-441 (1993).
93. Epps,D.E. and Kaluzny,M.A., *J. Chromatogr.*, **343**, 143-148 (1985).
94. Esposito,G.G., *Anal. Chem.*, **40**, 1902-1902 (1968).
95. Fay,L. and Richli,U., *J. Chromatogr.*, **541**, 89-98 (1991).
96. Feldman,G.L. and Rouser,G., *J. Am. Oil Chem. Soc.*, **42**, 290-293 (1965).
97. Fulk,W.K. and Shorb,M.S., *J. Lipid Res.*, **11**, 276-277 (1970).
98. Funazo,K., Tanaka,M., Yasaka,Y., Takigawa,H. and Shono,T., *J. Chromatogr.*, **481**, 211-219 (1989).
99. Furangen,A., *J. Chromatogr.*, **353**, 259-271 (1986).
100. Furniss,B.S., Hannaford,A.J., Smith,P.W.G. and Tatchell,A.R. (Editors), *Vogel's Textbook of Practical Organic Chemistry* (*5th Edition*), Longman Scientific & Technical, Harlow, 1989.
101. Gander,G.W., Jensen,R.G. and Sampugna,J., *J. Dairy Sci.*, **45**, 323-328 (1962).
102. Gauglitz,E.J. and Lehman,L.W., *J. Am. Oil Chem. Soc.*, **40**, 197-198 (1963).
103. Gaver,R.C. and Sweeley,C.C., *J. Am. Oil Chem. Soc.*, **42**, 294-298 (1965).
104. Gehrke,C.W. and Goerlitz,D.F., *Anal. Chem.*, **35**, 76-80 (1963).
105. Gerhardt,K.G. and Gehrke,C.W., *J. Chromatogr.*, **143**, 335-344 (1977).
106. Gilliland,K.M. and Moscatelli,E.A., *Biochim. Biophys. Acta*, **187**, 221-229 (1969).
107. Glass,R.L., *Lipids*, **6**, 919-925 (1971).
108. Glass,R.L., Jenness,R. and Troolin,H.A., *J. Dairy Sci.*, **48**, 1106-1109 (1965).
109. Glass,R.L. and Troolin,H.A., *J. Dairy Sci.*, **49**, 1469-1472 (1966).
110. Graff,G., Anderson,L.A., Jaques,L.W. and Scannell,R.T., *Chem. Phys. Lipids*, **53**, 27-36 (1990).
111. Greeley,R.H., *J. Chromatogr.*, **88**, 229-233 (1974).
112. Grimm,R.A., in *Fatty Acids*, pp. 218-235 (1979) (edited by E.H. Pryde, American Oil Chemists' Society, Champaign, IL).
113. Grossert,J.J., Ratnayake,W.M.N. and Swee,T., *Canad. J. Chem.*, **59**, 2617-2620 (1981).
114. Grover,A.K. and Cushley,R.J., *J. Labelled Compounds Radiopharm.*, **16**, 307-313 (1978).
115. Grunert,A. and Bassler,K.H., *Z. Anal. Chem.*, **267**, 342-346 (1973).
116. Gubitz,G., *J. Chromatogr.*, **187**, 208-211 (1980).

117. Gunstone,F.D., Harwood,J.L. and Krog,N., in *The Lipid Handbook*, pp. 287-320 (1986) (edited by F.D. Gunstone, J.L. Harwood and F.B. Padley, Chapman & Hall, London).
118. Gunstone,F.D. and Sealy,A.J., *J. Chem. Soc.*, 5772-5778 (1963).
119. Gutnikov,G. and Streng,J.R., *J. Chromatogr.*, **587**, 292-296 (1991).
120. Gyllenhaal,O., Brotell,H. and Hartvig,P., *J. Chromatogr.*, **129**, 295-302 (1976).
121. Haan,G.J., van der Heide,S. and Wolthers,B.G., *J. Chromatogr.*, **162**, 261-271 (1979).
122. Hadorn,H. and Zuercher,K., *Mitt. Lebensmittelunters. Hyg.*, **58**, 236-258 (1967).
123. Hamilton,R.J., Mitchell,S.F. and Sewell,P.A., *J. Chromatogr.*, **395**, 33-46 (1987).
124. Hansen,R.P. and Smith,J.F., *Lipids*, **1**, 316-321 (1966).
125. Hardy,J.P., Kerrin,S.L. and Manatt,S.L., *J. Org. Chem.*, **38**, 4196-4200 (1973).
126. Hartman,L., *J. Am. Oil Chem. Soc.*, **42**, 664 (1965)
127. Hartman,L. and Lago,R.C.A., *Lab. Practice*, **22**, 475-476 (1973).
128. Harvey,D.J., in *Advances in Lipid Methodology - One*, pp. 19-80 (1992) (edited by W.W. Christie, Oily Press, Ayr).
129. Haslam,E., *Tetrahedron*, **36**, 2409-2433 (1980).
130. Hayes,L., Lowry,R.R. and Tinsley,I.J., *Lipids*, **6**, 65-66 (1971).
131. Heckers,H., Melcher,F.W. and Schloeder,U., *J. Chromatogr.*, **136**, 311-317 (1977).
132. Hintze,U., Roper,H. and Gercken,G., *J. Chromatogr.*, **87**, 481-489 (1973).
133. Hockel,M., Dunges,W., Holzer,A., Brockerhoff,P. and Rathgen,G.H., *J. Chromatogr.*, **221**, 205-214 (1980).
134. Hoffman,N.E. and Liao,J.C., *Anal. Chem.*, **48**, 1104-1106 (1976).
135. Hofmann,U., Holzer,S. and Meese,C.O., *J. Chromatogr.*, **508**, 349-356 (1990).
136. Holloway,P.J. and Challen,S.B., *J. Chromatogr.*, **25**, 336-346 (1966).
137. Holloway,P.J. and Deas,A.H.B., *Chem. Ind.* (*London*), 1140 (1971).
138. Hopkins,C.Y., in *Topics in Lipid Chemistry*. Vol. 3, pp. 37-87 (1972) (edited by F.D. Gunstone, Paul Elek Ltd, London).
139. Hornstein,I., Alford,J.A., Elliot,L.E. and Crowe,P.F., *Anal. Chem.*, **32**, 540-542 (1960).
140. Hoshi,M., Williams,M. and Kishimoto,Y., *J. Lipid Res.*, **14**, 599-601 (1973).
141. Hougen,F.W. and Bodo,V., *J. Am. Oil Chem. Soc.*, **50**, 230-234 (1973).
142. Hubscher,G., Hawthorne,J.N. and Kemp,P., *J. Lipid Res.*, **1**, 433-438 (1960).
143. Husek,P., *Z. Klin. Chem. Klin. Biochem.*, **6**, 627-630 (1969).
144. Husek,P., *J. Chromatogr.*, **547**, 307-314 (1991).
145. Husek,P., *LC-GC International*, **5(9)**, 43-49 (1992).
146. Husek,P., Rijks,J.A., Leclercq,P.A. and Cramers,C.A., *J. High Resolut. Chromatogr.*, **13**, 633-638 (1990).
147. Ichinose,N., Nakamura,K., Shimizu,C., Kurokura,H. and Okamoto,K., *J. Chromatogr.*, **295**, 463-469 (1984).
148. Ikeda,M., Shimada,K. and Sakaguchi,T., *J. Chromatogr.*, **272**, 251-259 (1983).
149. Ikeda,M., Shimada,K., Sakaguchi,T. and Matsumoto,U., *J. Chromatogr.*, **305**, 261-270 (1984).
150. Ingalls,S.T., Minkler,P.E., Hoppel,C.L. and Nordlander,J.E., *J. Chromatogr.*, **299**, 365-376 (1984).
151. International Union of Pure and Applied Chemistry (IUPAC), *Standard Methods for the Analysis of Oils, Fats and Derivatives, 6th Edition, Applied Chemistry Division, Commission on Oils, Fats and Derivatives, Part* 1 (*Sections* 1 *and* 2), Pergamon Press, Oxford, Method 2.301, Section 3 (1979).
152. Isaiah,N.H., Subbarao,R. and Aggarwal,J.S., *J. Chromatogr.*, **43**, 519-522 (1969).
153. Ishida,M., Suyama,K. and Adachi,S., *J. Chromatogr.*, **189**, 421-424 (1980).
154. Iverson,J.L. and Sheppard,A.J., *J. Assoc. Off. Anal. Chem.*, **60**, 284-288 (1977).
155. Iverson,J.L. and Sheppard,A.J., *Food Chem.*, **21**, 223-234 (1986).
156. Iwata,T., Inoue,K., Nakamura,M. and Masatoshi,Y., *Biomed. Chromatogr.*, **6**, 120-123 (1992).
157. Jamieson,G.R. and Reid,E.H., *J. Chromatogr.*, **20**, 232-239 (1969).
158. Jensen,R.G., Quinn,J.G., Carpenter,D.L. and Sampugna,J., *J. Dairy Sci.*, **50**, 119-126 (1967).
159. Johnson,A.R., Fogerty,A.C., Hood,R.L., Kozuharov,S. and Ford,G.L., *J. Lipid Res.*, **17**, 431-432 (1976).
160. Johnson,C.B., *Anal. Biochem.*, **71**, 594-596 (1976).
161. Johnson,C.B. and Wong,E., *J. Chromatogr.*, **109**, 404-408 (1975).
162. Johnson,S.B. and Brown,R.E., *J. Chromatogr.*, **605**, 281-286 (1992).

163. Johnston,P.V. and Roots,B.I., *J. Lipid Res.*, **5**, 477-478 (1964).
164. Jones,E.P. and Davison,V.L., *J. Am. Oil Chem. Soc.*, **42**, 121-126 (1965).
165. Jordi,H.C., *J. Liqu. Chromatogr.*, **1**, 215-230 (1978).
166. Jupille,T., *J. Chromatogr. Sci.*, **17**, 161-167 (1979).
167. Kanfer,J.N. and Hakomori,S.-I. (editors), *Sphingolipid Biochemistry* (*Handbook of Lipid Research Vol.* 3) (1983) (Plenum Press, New York).
168. Karlsson,K.A., *Acta Chem. Scand.*, **19**, 2425-2427 (1965).
169. Kashyap,M.L., Mellies,M.J., Brady,D., Hynd,B.A. and Robinson,K., *Anal. Biochem.*, **107**, 432-435 (1980).
170. Kates,M., *J. Lipid Res.*, **5**, 132-135 (1964).
171. Kato,A., Tomita,H. and Yamaura,Y., *Chem. Ind.* (*London*), 302-303 (1971).
172. Kato,A. and Yamaura,Y., *Chem. Ind.* (*London*), 1260 (1970).
173. Kaufman,M.L., Friedman,S. and Wender,I., *Anal. Chem.*, **39**, 1011-1014 (1967).
174. Kaufmann,H.P. and Mankel,G., *Fette Seifen Anstrichm.*, **65**, 179-184 (1963).
175. Kaufmann,H.P., Radwan,S.S. and Ahmad,A.K.S., *Fette Seifen Anstrichm.*, **68**, 262-268 (1966).
176. Kawamura,N. and Taketomi,T., *Japan J. Exp. Med.*, **43**, 157-162 (1973).
177. Kelly,R.A., O'Hara,D.S. and Kelley,V., *J. Chromatogr.*, **416**, 247-254 (1987).
178. Kenney,H.E., Komanowsky,D., Cook,L.L. and Wigley,A.N., *J. Am. Oil Chem. Soc.*, **41**, 82-85 (1964).
179. Kihara,K., Rokushika,S. and Hatano,H., *Bunseki Kagaku*, **33**, 647-652 (1984).
180. Kim,S., Kim,Y.C. and Lee,J.I., *Tetrahedron Letts.*, **24**, 3365-3368 (1983).
181. King,J.W., France,J.E. and Snyder,J.M., *Fres. J. Anal. Chem.*, **344**, 474-478 (1992).
182. Kishimoto,Y. and Radin,N.S., *J. Lipid Res.*, **1**, 72-78 (1965).
183. Kishimoto,Y. and Radin,N.S., *J. Lipid Res.*, **6**, 435-436 (1965).
184. Kishiro,K. and Yasuda,H., *Anal. Biochem.*, **175**, 516-520 (1988).
185. Kleiman,R., Spencer,G.F. and Earle,F.R., *Lipids*, **4**, 118-122 (1969).
186. Klem,H.-P., Hintze,U. and Gercken,G., *J. Chromatogr.*, **75**, 19-27 (1973).
187. Klopfenstein,W.E., *J. Lipid Res.*, **12**, 773-776 (1971).
188. Ko,H. and Royer,M.E., *J. Chromatogr.*, **88**, 253-263 (1974).
189. Koch,G.K. and Jurriens,G., *Nature*, **208**, 1312-1313 (1965).
190. Koritala,S. and Rohwedder,W.K., *Lipids*, **7**, 274 (1972).
191. Korte,W.D., *J. Chromatogr.*, **243**, 153-157 (1982).
192. Kramer,J.K.G. and Hulan,H.W., *J. Lipid Res.*, **17**, 674-676 (1976).
193. Krewson,C.F., *J. Am. Oil Chem. Soc.*, **45**, 250-256 (1968).
194. Kuchmak,K., *Lipids*, **2**, 192 (1967).
195. Kuksis,A., Kovacevic,N., Lau,D. and Vranic,M., *Fed. Proc.*, **34**, 2238-2241 (1975).
196. Kuksis,A., Myher,J.J., Marai,L. and Geher,K., *Anal. Biochem.*, **70**, 302-312 (1979).
197. Kusaka,T., Ikeda,M., Nakano,H. and Numajiri,Y., *J. Biochem.* (*Tokyo*), **104**, 495-497 (1988).
198. Lam,S. and Grushka,E., *J. Chromatogr.*, **158**, 207-214 (1978).
199. Lambert,M.A. and Moss,C.W., *J. Clin. Microbiol.*, **18**, 1370-1377 (1983).
200. Lapidot,Y., Barzilay,I. and Hajdu,J., *Chem. Phys. Lipids*, **3**, 125-134 (1969).
201. Lapidot,Y. and Selinger,Z., *J. Am. Chem. Soc.*, **87**, 5522-5523 (1965).
202. Lee,Y.-M., Nakamura,H. and Nakajima,T., *Anal. Sci.*, **5**, 681-685 (1989).
203. Lee,Y.-M., Nakamura,H. and Nakajima,T., *J. Chromatogr.*, **515**, 467-473 (1990).
204. Lepage,G. and Roy,C.C., *J. Lipid Res.*, **25**, 1391-1396 (1984).
205. Lepage,G. and Roy,C.C., *J. Lipid Res.*, **27**, 114-120 (1986).
206. Lepage,G. and Roy,C.C., *J. Lipid Res.*, **29**, 227-235 (1988).
207. Levitt,M.J., *Anal. Chem.*, **45**, 618-620 (1973).
208. Lie Ken Jie,M.S.F. and Yan-Yit,C., *Lipids*, **23**, 367-369 (1988).
209. Lin,S.-N. and Horning,E.C., *J. Chromatogr.*, **112**, 465-482 (1975).
210. Lindgren,F.T., Nichols,A.V., Freeman,N.K. and Wills,R.D., *J. Lipid Res.*, **3**, 390-391 (1962).
211. Lingeman,H., Hulshoff,A., Underberg,W.J.M. and Offermann,F.B.J.M., *J. Chromatogr.*, **290**, 215-222 (1984).
212. Lloyd,J.B.F., *J. Chromatogr.*, **189**, 359-373 (1980).
213. Long,A.R., Massie,S.J. and Tyznik,W.J., *J. Food Sci.*, **53**, 940-942 (1988).
214. Lorette,N.B. and Brown,J.H., *J. Org. Chem.*, **24**, 261-262 (1959).
215. Lough,A.K., *Biochem. J.*, **90**, 4C-5C (1964).

216. Lough,A.K., *Nature*, **202**, 795 (1964).
217. Luddy,F.E., Barford,R.A., Herb,S.F. and Magidman,P., *J. Am. Oil Chem. Soc.*, **45**, 549-552 (1968).
218. Luddy,F.E., Barford,R.A. and Riemenschneider,R.W., *J. Am. Oil Chem. Soc.*, **37**, 447-451 (1960).
219. McCreary,D.K., Kossa,W.C., Ramachandran,S. and Kurtz,R.R., *J. Chromatogr. Sci.*, **16**, 329-331 (1978).
220. MacGee,J. and Allen,K.G., *J. Chromatogr.*, **100**, 35-42 (1974).
221. MacGee,J. and Williams,M.G., *J. Chromatogr.*, **205**, 281-288 (1981).
222. McGinnis,G.W. and Dugan,L.R., *J. Am. Oil Chem. Soc.*, **42**, 305-307 (1965).
223. McGrath,L.T. and Elliot,R.J., *Anal. Biochem.*, **187**, 273-276 (1990).
224. Mahadevan,V., Viswanathan,C.V. and Lundberg,W.O., *J. Chromatogr.*, **24**, 357-363 (1966).
225. Mancha Perello,M., *Grasas Aceites*, **18**, 231 (1967).
226. Marinetti,G.V., *Biochemistry*, **1**, 350-353 (1962).
227. Marinetti,G.V., *J. Lipid Res.*, **7**, 786-788 (1966).
228. Marmer,W.N., *Lipids*, **13**, 835-839 (1978).
229. Martin,G.E. and Swinehart,J.S., *J. Gas Chromatogr.*, **6**, 533-539 (1968).
230. Martinez-Castro,I., Alonso,L. and Juarez,M., *Chromatographia*, **21**, 37-40 (1986).
231. Mason,M.E., Eager,M.E. and Waller,G.R., *Anal. Chem.*, **36**, 587-590 (1964).
232. Mason,M.E. and Waller,G.R., *Anal. Chem.*, **36**, 583-586 (1964).
233. Matter,L., Schenker,D., Husmann,H. and Schomburg,G., *Chromatographia*, **27**, 31-36 (1989).
233a. Mattson,F.H. and Volpenhein, R.A., *J. Lipid Res.*, **3**, 281-296 (1962).
234. Matusik,E.J., Reeves,V.B. and Flanagan,V.P., *Anal. Chim. Acta*, **166**, 179-188 (1984).
235. Medina,I., Aubourg,S., Gallardo,J.M. and Perez-Martin,R., *Int. J. Food Sci. Technol.*, **27**, 597-601 (1992).
236. Metcalfe,L.D. and Schmitz,A.A., *Anal. Chem.*, **33**, 363-364 (1961).
237. Metcalfe,L.D. and Wang,C.N., *J. Chromatogr. Sci.*, **19**, 530-535 (1981).
238. Michalec,C., *J. Chromatogr.*, **31**, 643-645 (1967).
239. Michalec,C. and Kolman,Z., *J. Chromatogr.*, **31**, 632-635 (1967).
240. Michalec,C. and Kolman,Z., *J. Chromatogr.*, **34**, 375-381 (1968).
241. Middleditch,B.S., *Analytical Artifacts* (*Journal of Chromatography Library - Vol.* 44) (1989) (Elsevier, Amsterdam).
242. Miller,R.A., Bussell,N.E. and Ricketts,C., *J. Liqu. Chromatogr.*, **1**, 291-304 (1978).
243. Minnikin,D.E., *Lipids*, **7**, 398-403 (1972).
244. Minnikin,D.E., Dobson,G., Goodfellow,M., Draper,P. and Magnusson,M., *J. Gen. Microbiol.*, **131**, 2013-2021 (1985).
245. Minnikin,D.E., Dobson,G., Parlett,J.H., Datta,A.K., Minnikin,S.M. and Goodfellow,M., in *Topics in Lipid Research*, pp. 139-143 (1986) (edited by R.A. Klein and B. Schmitz, Royal Society of Chemistry, London).
246. Minnikin,D.E., Minnikin,S.M., O'Donnell,A.G. and Goodfellow,M., *J. Microbiol. Meth.*, **2**, 243-249 (1984).
247. Minnikin,D.E. and Polgar,N., *Chem. Commun.*, 312-314 (1967).
248. Misir,R., Laarveld,B. and Blair,R., *J. Chromatogr.*, **331**, 141-148 (1985).
249. Miwa,H., Hiyama,C. and Yamamoto,M., *J. Chromatogr.*, **321**, 165-174 (1985).
250. Miwa,H. and Yamamoto,M., *J. Chromatogr.*, **421**, 33-41 (1987).
251. Miwa,H. and Yamamoto,M., *J. Chromatogr.*, **523**, 235-246 (1990).
252. Miwa,H., Yamamoto,M. and Momose,T., *Chem. Pharm. Bull.*, **28**, 599-605 (1980).
253. Miwa,H., Yamamoto,M. and Nishida,T., *Clin. Chim. Acta*, **155**, 95-102 (1986).
254. Miwa,H., Yamamoto,M., Nishida,T., Nunoi,K. and Kikuchi,M., *J. Chromatogr.*, **416**, 237-245 (1987).
255. Moffat,C.F., McGill,A.S. and Anderson,R.S., *J. High Resolut. Chromatogr.*, **14**, 322-326 (1991).
256. Molnar-Perl,I. and Pinter-Szakacs,M., *J. Chromatogr.*, **365**, 171-182 (1986).
257. Morrison,W.R. and Hay,J.D., *J. Chromatogr.*, **202**, 460-467 (1970).
258. Morrison,W.R., Lawrie,T.D.V. and Blades,J., *Chem. Ind.* (*London*), 1534-1535 (1961).
259. Morrison,W.R. and Smith,L.M., *J. Lipid Res.*, **5**, 600-608 (1964).
260. Moscatelli,E.A., *Lipids*, **7**, 268-271 (1973).
261. Moscatelli,E.A. and Isaacson,E., *Lipids*, 4, 550-555 (1969).

262. Mounts,T.L. and Dutton,H.J., *J. Labelled Compounds*, **3**, 343-345 (1967).
263. Mueller,H.W. and Binz,K., *J. Chromatogr.*, **228**, 75-93 (1982).
264. Muller,K.-D., Husmann,H. and Nalik,H.P., *Zbl. Bakt.*, **274**, 174-182 (1990).
265. Muller,K.-D., Husmann,H., Nalik,H.P. and Schomburg,G., *Chromatographia*, **30**, 245-248 (1990).
266. Needs,E.C., Ford,G.D., Owen,A.J., Tuckley,B. and Anderson,B., *J. Dairy Res.*, **50**, 321-339 (1983).
267. Netting,A.G. and Duffield,A.M., *J. Chromatogr.*, **336**, 115-123 (1984).
268. Nimura,N. and Kinoshita,T., *Anal. Letts*, **13**, 191-202 (1980).
269. Nimura,N., Kinoshita,T., Yoshida,T., Uetake,A. and Nakai,C., *Anal. Chem.*, **60**, 2067-2070 (1988).
270. Oette,K. and Doss,M., *J. Chromatogr.*, **32**, 439-450 (1968).
271. Ord,W.O. and Bamford,P.C., *Chem. Ind. (London)*, 2115-2116 (1967).
272. Oswald,E.O., Piantadosi,C., Anderson,C.E. and Snyder,F., *Lipids*, **1**, 241-246 (1966).
273. Outen,G.E., Beever,D.E. and Fenlon,J.S., *J. Sci. Food Agric.*, **27**, 419-425 (1976).
274. Pace-Asciak,C.R., *J. Lipid Res.*, **30**, 451-454 (1989).
275. Pascal,J.C. and Ackman,R.G., *Comp. Biochem. Physiol.*, **51B**, 111-113 (1976).
276. Pascaud,M., *Anal. Biochem.*, **18**, 570-572 (1967).
277. Peisker,K.V., *J. Am. Oil Chem. Soc.*, **41**, 87-89 (1964).
278. Penttila,I., Huhtikangas,A., Herranen,J., Eskelinen,S. and Moilanen,O., *Annals Clin. Res.*, **16**, 13-17 (1984).
279. Perkins,E.G., *J. Am. Oil Chem. Soc.*, **44**, 197-199 (1967).
280. Peuchant,E., Wolff,R., Salles,C. and Jensen,R., *Anal. Biochem.*, **181**, 341-344 (1989).
281. Pina,M., Montet,D., Graille,J., Ozenne,C. and Lamberet,G., *Rev. Franc. Corps Gras*, **38**, 213-218 (1991).
282. Pina,M., Pioch,D. and Graille,J., *Lipids*, **22**, 358-361 (1987).
283. Piretti,M.V., Pagliuca,G. and Vasina,M., *Anal. Biochem.*, **167**, 358-361 (1987).
284. Piretti,M.V., Pagliuca,G. and Vasina,M., *Chem. Phys. Lipids*, **47**, 149-153 (1987).
285. Pohl,P., Glasl,H. and Wagner,H., *J. Chromatogr.*, **49**, 488-492 (1970).
286. Powell,R.G., Smith,C.R. and Wolff,I.A., *J. Org. Chem.*, **32**, 1442-1446 (1967).
287. Radin,N.S., Hajra,A.K. and Akahori,Y., *J. Lipid Res.*, **1**, 250-251 (1960).
288. Reed,A.W., Deeth,H.C. and Clegg,D.E., *J. Assoc. Off. Anal. Chem.*, **67**, 718-721 (1984).
289. Robb,E.W. and Westbrook,J.J., *Anal. Chem.*, **35**, 1644-1647 (1963).
290. Rogozinski,M., *J. Gas Chromatogr.*, **2**, 136-137 (1964).
291. Rosenfeld,J.M., Hammerberg,O. and Orvidas,M.C., *J. Chromatogr.*, **378**, 9-16 (1986).
292. Rouser,G., Kritchevsky,G., Whatley,M. and Baxter,C.F., *Lipids*, **1**, 107-112 (1966).
293. Ryan,P.J. and Honeyman,T.W., *J. Chromatogr.*, **312**, 461-466 (1984).
294. Saglione,T.V. and Hartwick,R.A., *J. Chromatogr.*, **454**, 157-167 (1988).
295. Saha,S. and Dutta,J., *Lipids*, **8**, 653-655 (1973).
296. Saigo,K., Usui,M., Kikuchi,K., Shimada,E. and Mukaiyama,T., *Bull. Chem. Soc. Japan.*, **50**, 1863-1866 (1977).
297. Salwin,H. and Bond,J.F., *J. Assoc. Off. Anal. Chem.*, **52**, 41-47 (1969).
298. Sampugna,J., Pitas,R.E. and Jensen,R.G., *J. Dairy Sci.*, **49**, 1462-1463 (1966).
299. Sattler,W., Puhl,H., Hayn,M., Kostner,G.M. and Esterbauer,H., *Anal. Biochem.*, **198**, 184-190 (1991).
300. Schatowitz,B. and Gercken,G., *J. Chromatogr.*, **409**, 43-54 (1987).
301. Schlenk,H. and Gellerman,J.L., *Anal. Chem.*, **32**, 1412-1414 (1960).
302. Schuchardt,U. and Lopes,O.C., *J. Am. Oil Chem. Soc.*, **65**, 1940-1941 (1988).
303. Schulte,E. and Weber,K., *Fat Sci. Technol.*, **91**, 181-183 (1989).
304. Schwartz,D.P., *Microchem. J.*, **22**, 457-462 (1977).
305. Scribney, *Biochim. Biophys. Acta*, **125**, 542-547 (1966).
306. Sebedio,J.L. and Grandgirard,A., *Prog. Lipid Res.*, **28**, 303-336 (1989).
307. Segura,R., *J. Chromatogr.*, **441**, 99-113 (1988).
308. Shantha,N.C. and Ackman,R.G., *J. Chromatogr.*, **587**, 263-267 (1991).
309. Shapiro,I.L. and Kritchevsky,D., *J. Chromatogr.*, **18**, 599-601 (1965).
310. Sheppard,A.J. and Iverson,J.L., *J. Chromatogr. Sci.*, **13**, 448-452 (1975).
311. Shimasaki,H., Phillips,F.C. and Privett,O.S., *J. Lipid Res.*, **18**, 540-543 (1977).
312. Smith,C.R., *Prog. Chem. Fats other Lipids*, **11**, 137-177 (1970).
313. Smith,C.R., Wilson,T.L., Melvin,E.H. and Wolff,I.A., *J. Am. Chem. Soc.*, **82**, 1417-1421 (1960).

314. Solomon,H.L., Hubbard,W.D., Prosser,A.R. and Sheppard,A.J., *J. Am. Oil Chem. Soc.*, **51**, 424-425 (1974).
315. Sonesson,A., Larsson,,L. and Jimenez,J., *J. Chromatogr.*, **417**, 366-370 (1987).
316. Spangelo,A., Karijord,O., Svensen,A. and Abrahamsen,R.K., *J. Dairy Sci.*, **69**, 1787-1792 (1986).
317. Stodola,F.H., *J. Org. Chem.*, **29**, 2490-2491 (1964).
318. Stoll,A., Rutschmann,J., von Wartburg,A. and Renz,J., *Helv. Chim. Acta*, **39**, 993-999 (1956).
319. Sukhija,P.S. and Palmquist,D.L., *J. Agric. Food Chem.*, **36**, 1202-1206 (1988).
320. Suyama,K., Hori,K. and Adachi,S., *J. Chromatogr.*, **174**, 234-238 (1979).
321. Svennerholm,L., Bostrom,K., Fredman,P., Jungbjer,B., Mansson,J.-E. and Rynmark,B.-M., *Biochim. Biophys. Acta*, **1128**, 1-7 (1992).
322. Sweeley,C.C., *J. Lipid Res.*, **4**, 402-406 (1963).
323. Sweeney,R.W., Beech,J., Whitlock,R.H. and Castelli,P.L., *J. Chromatogr.*, **494**, 278-282 (1989).
324. Szoke,K., Kramer,M. and Lindner,K., *Fette Seifen Anstrichm.*, **67**, 257-259 (1965).
325. Thenot,J.-P., Horning,E.C., Stafford,M. and Horning,M.G., *Anal. Letts.*, **5**, 217-223 (1972).
326. Timms,R.E., *Austral. J. Dairy Technol.*, **33**, 4-6 (1978).
327. Tove,S.B., *J. Nutr.*, **75**, 361-365 (1961).
328. Traitler,H., Richli,U., Winter,H., Kappeler,A.M. and Monnard,C., *J. High Resolut. Chromatogr., Chromatogr. Commun.*, **8**, 440-443 (1985).
329. Tserng,K.-Y., Kliegman,R.M., Miettinen,E.-L. and Kalhan,S.C., *J. Lipid Res.*, **22**, 852-858 (1981).
330. Tsuchiya,H., Hayashi,T., Naruse,H. and Takagi,N., *J. Chromatogr.*, **231**, 247-254 (1982).
331. Tsuchiya,H., Hayashi,T., Naruse,H. and Takagi,N., *J. Chromatogr.*, **234**, 121-130 (1982).
332. Tsuchiya,H., Hayashi,T., Sato,M., Tatsumi,M. and Takagi,N., *J. Chromatogr.*, **309**, 43-52 (1984).
333. Tuckey,R.C. and Stevenson,P.M., *Anal. Biochem.*, **94**, 402-408 (1979).
334. Ulberth,F. and Henninger,M., *J. Am. Oil Chem. Soc.*, **69**, 174-177 (1992).
335. Ulberth,F. and Kamptner,W., *Anal. Biochem.*, **203**, 35-38 (1992).
336. Van der Horst,F.A.L., Post,M.H., Holthuis,J.J.M. and Brinkman,U.A.Th., *J. Chromatogr.*, **500**, 443-452 (1990).
337. van Kuijk,F.J.G.M., Thomas,D.W., Konopelski,J.P. and Dratz,E.A., *J. Lipid Res.*, **27**, 452-456 (1986).
338. van Kuijk,F.G.J.M., Thomas,D.W., Stephens,R.J. and Dratz,E.A., *J. Free Radical Biol. Med.*, **1**, 215-225 (1985).
339. van Kuijk,F.G.J.M., Thomas,D.W., Stephens,R.J. and Dratz,E.A., *J. Free Radical Biol. Med.*, **1**, 387-393 (1985).
340. Vetter,W. and Walther,W., *J. Chromatogr.*, **513**, 405-407 (1990).
341. Viswanathan,C.V., Basilio,M., Hoevet,S.P. and Lundberg,W.O., *J. Chromatogr.*, **34**, 241-245 (1968).
342. Viswanathan,C.V., Phillips,F. and Lundberg,W.O., *J. Chromatogr.*, **38**, 267-273 (1968).
343. Vorbeck,M.L., Mattick,L.R., Lee,F.A. and Pederson,C.S., *Anal. Chem.*, **33**, 1512-1514 (1961).
344. Vreeken,R.J., Jager,M.E., Ghijsen,R.T. and Brinkman, U.A.T., *J. High Resolut. Chromatogr.*, **15**, 785-790 (1992).
345. Vulliet,P., Markey,P.S. and Tornabene,T.G., *Biochim. Biophys. Acta*, **348**, 299-301 (1974).
346. Walker,M.A., Roberts,D.R. and Dumbroff,E.B., *J. Chromatogr.*, **241**, 390-391 (1982).
347. Weiss,B., *Biochemistry*, **3**, 1288-1293 (1964).
348. Welch,R.W., *J. Sci. Food Agric.*, **26**, 429-435 (1975).
349. Wells,M.A. and Dittmer,J.C., *J. Chromatogr.*, **18**, 503-511 (1965).
350. Welz,W., Sattler,W., Leis,H.-J. and Malle,E., *J. Chromatogr.*, **526**, 319-329 (1990).
351. West,J.C., *Anal. Chem.*, **47**, 1708-1709 (1975).
352. Wiegandt,H. (editor), *Glycolipids* (*New Comprehensive Biochemistry, Vol.* 10) (1985) (Elsevier, Amsterdam).
353. Williams,M.G. and MacGee,J., *J. Chromatogr.*, **234**, 468-471 (1982).
354. Williams,M.G. and MacGee,J., *J. Am. Oil Chem. Soc.*, **60**, 1507-1509 (1983).
355. Winterfeld,M. and Debuch,H., *Hoppe Seyler's Z. Physiol. Chem.*, **345**, 11-21 (1966).

356. Witting,L.A. (editor), *Glycolipid Methodology* (1976) (American Oil Chemists' Society, Champaign).
357. Wolf,J.H. and Korf,J., *J. Chromatogr.*, **436**, 437-445 (1988).
358. Wolf,J.H. and Korf,J., *J. Chromatogr.*, **502**, 423-430 (1990).
359. Wolff,R.L. and Castera-Rossignol,F.M., *Rev. Franc. Corps Gras*, **34**, 123-132 (1987).
360. Wolff,R.L. and Fabien,R.J., *Lait*, **69**, 33-46 (1989).
361. Wood,R. and Lee,T., *J. Chromatogr.*, **254**, 237-246 (1983).
362. Woollard,P.M., *Biomed. Mass Spectrom.*, **10**, 143-154 (1983).
363. Wren,J.J. and Szczepanowska,A.D., *J. Chromatogr.*, **15**, 404-410 (1964).
364. Yamaguchi,M., Hara,S., Matsunaga,R., Nakamura,M. and Ohkura,Y., *J. Chromatogr.*, **346**, 227-236 (1985).
365. Yamaguchi,M., Matsunaga,R., Fukuda,K., Nakamura,M. and Ohkura,Y., *Anal. Biochem.*, **155**, 256-261 (1986).
366. Yamaki,K. and Oh-ishi,S., *Chem. Pharm. Bull.*, **34**, 3526-3529 (1986).
367. Yamauchi,K., Tanabe,T. and Kinoshita,M., *J. Org. Chem.*, **44**, 638-639 (1979).
368. Yuki,H., Azuma,Y., Maeda,N. and Kawasaki,H., *Chem. Pharm. Bull.*, **36**, 1905-1908 (1988).
369. Zamir,I., *J. Chromatogr.*, **586**, 347-350 (1991).
370. Zubillaga,M.P. and Maerker,G., *J. Am. Oil Chem. Soc.*, **65**, 780-782 (1988).

Chapter 3

SIZE EXCLUSION CHROMATOGRAPHY IN THE ANALYSIS OF LIPIDS

M. Carmen Dobarganes and Gloria Márquez-Ruiz
Instituto de la Grasa y sus Derivados (C.S.I.C.), 41012-Seville, Spain

A. INTRODUCTION

Size exclusion chromatography (SEC) is a technique to separate compounds according to their molecular size and can be applied to almost any molecule, from those of around 100 molecular weight (MW) to those of several millions MW. The technique has been developed by two independent groups of scientists using different terminology and separate types of stationary phase. On one hand, biochemists applied the technique to analysis of biomolecules and defined it as *gel filtration*, normally utilizing macroporous cross-linked dextrans and acrylamides compatible with aqueous buffers. On the other hand, the term adopted by industrial and organic chemists for synthetic polymer characterization was *gel permeation* and referred to the use of rigid, polymer-based resins, such as polystyrene, which are stable in organic solvents.

Advances in this technique have benefited both scientific communities, first by the introduction of high speed columns of greater efficiency, and second, through the development of new stationary phases, which are more stable, chemically inert and manufactured with special care to decrease parallel effects other than molecular size separation. Hence, using a distinct terminology for each type of application no longer seems

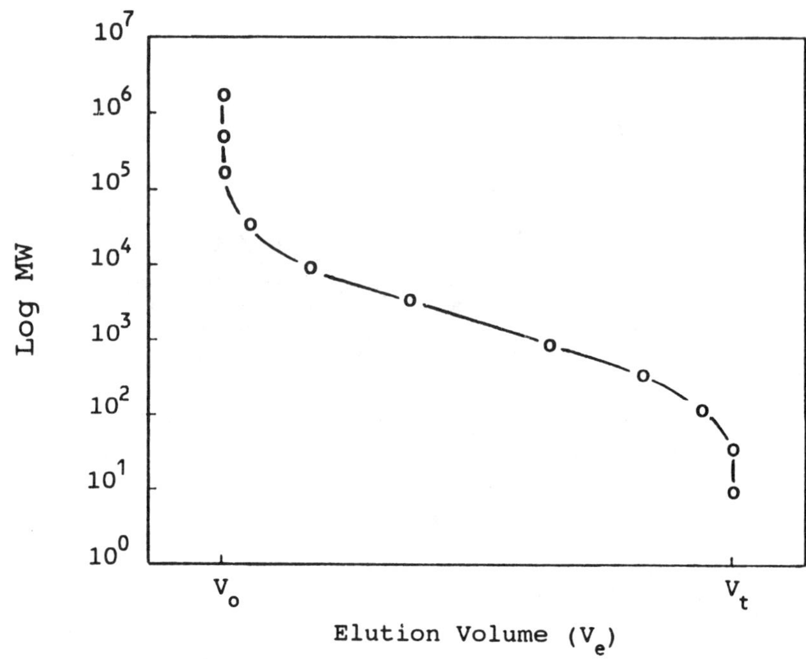

Fig. 3.1. Explanatory calibration curve for SEC.

justified and the technique will be referred to only as *size exclusion chromatography* (SEC) throughout this chapter.

SEC is carried out on columns packed with a porous material of a specific pore size and completely filled with the same solvent used to dissolve the sample. Depending on their respective sizes, smaller sample molecules can diffuse into the pores of the gel, partially or completely, while the larger ones are excluded and emerge first. Thus the elution order is inversely related to their molecular size or weight. Figure 3.1 shows a typical calibration graph where the empirical elution volumes are plotted versus the logarithms of molecular weight. Column calibration is normally carried out using standard molecules of known MW, such as polystyrene or globular proteins. These plots can provide accurate determinations of MW for molecules which adopt a similar conformation in solution to that of the standard. A species is eluted at a volume (V_e) exactly equal to the volume available to it in the column, following the general equation:

$$V_e = V_o + K_d V_i$$

where V_o is the volume external to the beads or void volume, V_i is the internal (pore) volume and K_d is the distribution or partition coefficient.

V_o represents the elution volume for completely excluded molecules and thereby $K_d = 0$, whereas V_T is the elution volume of a material experiencing total penetration of the gel pores, and it is equal to the sum of V_o and V_i (K_d = 1). When K_d is greater than 1, the separation by size is modified by some interaction such as adsorption by the stationary phase.

The most important dimensions are the size of the pores that access the internal volume and the size of the molecules to be separated. In practice, it is well-known that flow-rate and particle size in addition to packing methodology are essential factors for column efficiency [34]. A large variety of stationary phases is currently available commercially from different suppliers, who also provide valuable technical information and detailed instructions regarding the use of their products. There is now a choice of many different gels since, for a given chemical constitution, various fractionation ranges can be selected and within them, different particle sizes can be found. For a specific column, the primary rule is that smaller particles give higher efficiency, but then the resistance to flow tends to increase, resulting in a greater pressure drop. Suitable compromises between these factors must be found to achieve maximum resolution under the flow conditions required for each particular application. Mono-sized particles, spherical in shape, have been developed in order to ensure identical bead properties and uniform packing [123]. Such a strict control of both the particle size and pore-size distribution contributes considerably to obtaining columns with high efficiency and separation capacity [70].

Stationary phases for aqueous SEC are conveniently divided into two classes: derivatized glass and silica, and cross-linked gels. Among the latter, dextran, acrylamide and agarose polymers stand out. It has been noted that the main advantages of the phases based on porous glass or silica are their mechanical strength and homogeneous particle size and porosity. Nevertheless it is difficult to obtain a silica packing completely free of adsorptive properties. In contrast, adsorption effects are less pronounced in cross-linked gels and recovery of the sample is usually satisfactory [33].

For SEC in organic solvents, silica-based phases are also used although copolymers of styrene and divinyl benzene have become the most popular stationary phases. They are now available over a wide range of pore sizes, and these chromatographic columns are often used in series.

Nowadays SEC is a very well-accepted technique for characterization of either synthetic polymers or biopolymers. In fact, its applications continue to expand and are largely supported by the evident technical advances.

For those readers interested in different aspects of this topic, we strongly recommend *Size Exclusion Chromatography Fundamental Reviews*, published biannually in *Analytical Chemistry*, which cover fundamental developments and new applications in the SEC area during the review period and give short descriptions of those books, symposia and reviews dedicated to SEC [6,7].

SEC has found only limited use in lipid analysis being reduced to special applications. Hence it is not surprising that the previous reviews of SEC applications do not deal specifically with lipid separations, which are instead included in some general surveys of high-performance liquid chromatography (HPLC) in oils [2,3,22].

The most important applications, *i.e.* fat and oil polymerization, and lipoprotein analyses, have been developed by two distinct scientific communities, as mentioned earlier in this chapter, and these use different solvents, stationary phases, detection systems and even terminology. Accordingly, the rest of this chapter has been divided into two general sections, namely SEC in organic media and SEC in aqueous media.

B. SIZE EXCLUSION CHROMATOGRAPHY IN ORGANIC MEDIA

To better understand the present situation and the evolution of SEC development and utilization on lipid analysis, it would be helpful to keep in mind several aspects of this technique from a historical perspective. In this sense, some general points of greater interest could be summarized as follows.

The development of lipophilic gels applicable to the separation of organic compounds of low MW (fractionation range from 100-1500 MW) encouraged researchers to apply the SEC technique to the analysis of lipids for the first time [10,12,79]. This was especially observed upon the development of a hydroxypropyl derivative of cross-linked dextran by Pharmacia, and sold commercially under the trade designation of Sephadex™ LH-20. Then, the literature contained many comparisons between the new lipophilic phases and gels based on polystyrene, particularly Bio-Beads™ SX from Bio-Rad Laboratories and Styragel™ from Waters Associates. In most cases, effects others than exclusion by size were present to some extent, depending on the solvent selected, and these were frequently used deliberately to achieve better separations.

On the other hand, advances in those general chromatographic techniques which could be more suitable for lipid analysis, namely HPLC on normal and reversed phases, limited the potential applications based on size exclusion effects initially.

Yet development of high performance supports for SEC greatly benefited those phases based on polystyrene, especially in terms of good control of particle size and pore size distribution, leading to a wide choice of gels with different fractionation ranges, chemical stability, and mechanical stability, which enable high operating pressures and result in a shorter time for a given separation. Height equivalent to a theoretical plate (HETP) values as low as 0.006 mm have been reported [70] versus 0.4 to 0.2 mm for conventional columns [79] (using *o*-dichlorobenzene as test substance in both cases).

Accordingly, the stationary phases based on copolymers of styrene divinyl benzene have been to date the most promising and widely used.

They have been sold commercially under different trade designations, Bio-Beads™ SX from Bio-Rad Laboratories, Styragel™ from Waters Associates, PLGel™ from Polymer Laboratories, and Spherogel™ from Beckman, among others.

At present, SEC is ultimately of major interest for separating groups of lipid compounds varying in MW, and it is not applied to those separations of lipids based on differences in fatty acid chain length or degree of unsaturation. The methodologies proposed for lipid analysis have in common the stationary phase used, with different possibilities of pore size (50, 100, 500 and 1000 Å), number of columns used, selection of solvents (tetrahydrofuran (THF) being outstanding) and choice of detection system.

1. Polymers

(i) Polymerized triacylglycerols. SEC appears to be the most important technique for the analysis of polymerized triacylglycerols (TAG). This type of evaluation has been particularly useful in the area of heated or frying fats and oils, as polymerized TAG are quantitatively the most relevant group of compounds among those formed due to the combined action of oxygen and temperature. Polymerized TAG form a complex mixture of hundreds of individual components which vary in MW as the polymerization level increases from the original monomers to dimers, trimers, *etc.* Moreover, compounds of a wide polarity range may result from a process such as the possible oxidation of the fatty acyl groups, and joined by different types of linkages (C-C, C-O-C, C-O-O-C) to form dimers and so on. However, such variations in polarity are not reflected in significant MW changes, and SEC can then be considered as a good alternative to the previous methodologies based on urea non-adduct formation, solvent partition or distillation.

Preliminary results emphasised that a good separation of polymeric TAG was possible either in Sephadex™ LH-20 swelled in chloroform [95] or ethanol [1], BioBeads™ in benzene [1], chloroform [95], or Styragel™ in THF [124]. Later [113], it was demonstrated that in used frying fats a good correlation existed between the amount of polymerized TAG determined by HPSEC using BioBeads™ in dichloromethane and a refractive index detector and non-polar components separated by column chromatography. This indicated that the concentration of polymerized TAG was a reasonably good measurement of alteration in frying fats. Still, the major drawbacks of these methods were that long columns, low pressures and long analysis time were required for an acceptable resolution, because of the particular characteristics of the stationary phases available in terms of pore size distribution, and particle size and shape. Further developments to optimize chromatographic matrices have led to the column now normally used, which is approximately 30 cm x 0.8 cm i.d., with spherically shaped particles (5 or 10 μm) and a controlled pore size distribution.

Perrin *et al.* [96] were the first to propose HP columns of polystyrene divinyl benzene for the analysis of polymers. Three 50, 100 and 50-10^6Å mixed columns were used separately and in series. THF was selected as solvent and a refractometer as detector. Although efforts to obtain quantitative data were not made, some chromatograms shown in this short laboratory note illustrated the potential possibilities of the technique. This was later confirmed by Kupranycz *et al.* [71], who utilized the method to examine increasing alteration levels in oils. A differential refractometer and a series of three columns (1000, 500 and 100 Å Ultrastyragel) were operated at room temperature, and THF was again used as solvent. At the same time, White and Wang [129] evaluated different heated soybean oils using a variable wavelength UV/VIS detector. Two columns of 500 and 1000 Å were employed and methylene chloride served as the mobile phase. Three UV wavelengths (234, 254 and 270 nm) were tested initially, and because no advantage of one wavelength over another due to functional group selectivity was evident, 234 nm was selected on the basis of sensitivity. Chromatograms were thus simpler than when using a refractometer, as unchanged TAG, diacylglycerols (DAG) and fatty acids (FA) were not detected. Regardless of the lack of exact quantitative data, the authors concluded that some estimate of the relative quantities of high MW compounds formed could be given and suggested this method as a rapid test for use in the oil industry to compare stabilities of various frying oils.

Overall, good resolution has been achieved in polymer analysis in general in all laboratories, by using sample concentrations between 30-50 mg/mL and loops of 10-20 μL. The method stands out for its simplicity as it is only necessary to dilute the fat in the appropriate solvent before the chromatographic determination, which is performed with a single solvent in 10 to 30 min depending on the flow rate and number of columns used. Because of the complexity of the polymeric fractions, quantitative results are usually obtained from the calculation of Peak Area % divided by Total Area.

Because of the promising results obtained in the analysis of polymerized TAG, the IUPAC Commission on Oils, Fats and Derivatives has recently published the results of two inter-laboratory tests, carried out at an international level in 1986-87, in which ten and seventeen laboratories participated, each obtaining two test results for each sample. The main conclusion after evaluation of 8 different samples with varying polymer contents was that repeatability was in general high but reproducibility was rather poor for samples with low contents of polymers. Hence the Commission finally decided to adopt a general methodology with a limitation of application to samples containing 3% or more of polymerized

TAG [133]. The method proposed use of a single column of 30 cm x 0.77 cm i.d. packed with a high-performance spherical gel made of copolystyrene divinyl benzene of 5 μm, THF as the mobile phase and a refractive index detector with a sensitivity at full scale at least 1 x 10^{-4} of refractive index. The sample concentration suggested was 50 mg/mL for an injection valve with a 10 μl loop. The analysis time was about 10 min with a flow rate of 1 mL/min.

Quite a different approach has consisted in increasing sensitivity and widening the applications of polymer determination to any kind of fat or oil, heated or not [29]. The basis of the methodology is present in the complex evaluation of heated fats perfected by Gere [38]. The originality of the scheme relied on conveniently combining two methods proposed by IUPAC, namely the determination of polar compounds by adsorption chromatography [127] and the determination of polymerized triglycerides by HPSEC [133], which was applied to the polar fraction obtained by the former method. Undoubted advantages resulted from the scheme adopted in comparison with the analysis of the total fat sample; first, a substantial increase in possibilities for quantification of all the groups of altered compounds because of the effect of concentration, second, an independent determination of oxidized TAG monomers as a measurement of oxidative alteration (since non-altered TAG would elute in the first separation stage), and third, simultaneous evaluation of DAG as an indication of hydrolytic alteration, avoiding any overlap with TAG in the whole sample. Of course, this evaluation is time-consuming, as determination of the polar compounds takes approximately 2 hours. However, it is not proposed to analyse polymer content exclusively, but rather with the broader aim of obtaining fairly complete information on the type of degradation involved. It is thus suggested as a rapid complementary method for those laboratories where polar compound determination is already undertaken following the present regulation. Silica column chromatography used for the initial separation by polarity may be substituted for methods which require lower amounts of solvent and less time, such as solid phase extraction using NH_2 columns [57] or silica gel cartridges [115].

(ii) Polymerized fatty acids. Identical methodology to that applied to polymerized TAG has been used for the evaluation of polymerized FA or fatty acid methyl esters (FAME). Strictly speaking, this application was studied earlier. In fact, the first studies on the application of SEC to polymer analysis were dedicated to polymerized FA in the area of oleochemicals. This analysis was especially valuable for technical mixtures of monomeric, dimeric and trimeric FA derivatives, used as raw materials for the manufacture of polyamides, polyesters, *etc.* [8,20,61]. In this context, the study of Harris *et al.* [51] deserves particular mention as an excellent example of rigorous quantification. Detector linearity was

demonstrated over the concentration range used, response factors were determined by mixing pure dimers and trimers with known amounts of two internal standards (heptanoic acid and polystyrene) and the analytical method thus tested was applied to the analysis of typical commercial dimer acid products.

SEC was soon revealed as a promising technique for this type of application and the first studies were followed by a number of others devoted to analysis of TAG polymers, which also included evaluation of polymerized FA after transesterification or hydrolysis, using the same columns, solvents and even chromatographic conditions [1,95]. Another study worthy of note was by Jensen and Moller [62], who characterized polymerized FA in different raw materials used as commercial feed fats for poultry, pigs and ruminants, providing important data to better understand the actual concentrations and feeding consequences of polymerized FA present in feeds. They accomplished the analysis of 11 commercial feed fats and one sample of animal fat distillation residue, bleach earth oil and abused frying fat, using a Ultrastyragel™ 500 Å column and THF as eluant at a flow rate of 0.8 mL/min. Although trimeric acids were only detected in three samples, the dimeric fraction ranged from 3.7 to 13.7% so it seemed obvious that this latter fraction may have a marked effect on the feeding value.

A systematic study of the methodology has recently been presented by Christopoulou and Perkins [24]. The system used was composed of two styrene/divinylbenzene copolymer columns with toluene as the mobile phase and refractometry as the mode of detection. Quantification of monomer, dimer and trimer content in various samples correlated well with results obtained by gas liquid chromatography (GLC) analysis and gravimetically by SEC. Additionally, comparisons of SEC with other chromatographic techniques available for analysis of monomer, dimer and trimer FA, such as GLC and thin-layer chromatography with flame ionization detection (TLC-FID), have been carried out recently [19], and no significant differences were found between the quantitative results given by any of the methods tested.

Similarly to the comments on TAG evaluation above, analysis of FA monomers, dimers and polymers by SEC exclusively, and therefore based on molecular size only, may be rather poor when the major fraction of the sample is composed of nonpolar FA. Again it is possible to enhance the possibilities for the determination by separating two fractions by polarity prior to the SEC analysis [74]. Figure 3.2 shows a simple scheme that combines adsorption and size exclusion chromatographies for the determination of FAME derivatives in heated fats. The method proposed allowed quantification of unaltered FA in addition to four groups of degradation compounds, *i.e.* non-polar FA dimers, oxidized FA monomers, polar FA dimers and FA polymers, and was further applied to evaluate their respective digestibilities in nutritional studies [76]. Among

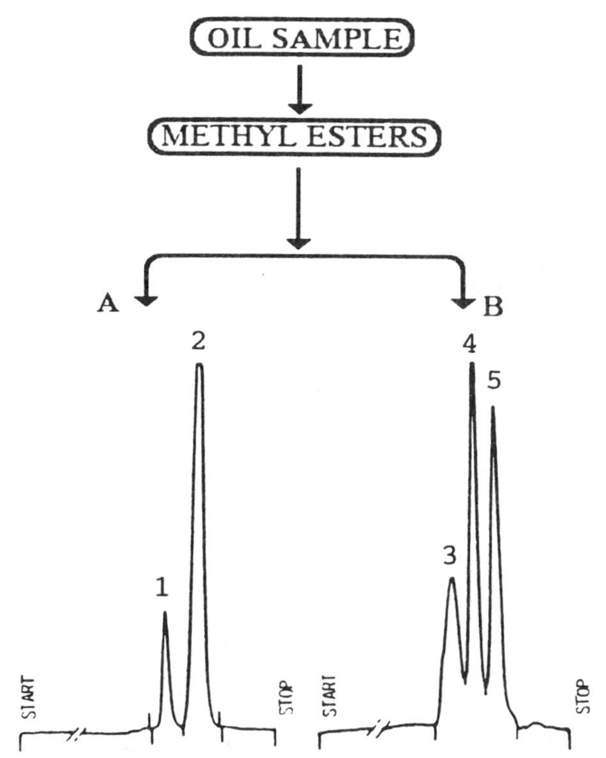

Fig. 3.2. Combination of adsorption and size exclusion chromatographies for the determination of fatty acid methyl ester derivatives in heated fats. A, non-polar fraction eluted with hexane-diethyl ether (88:12, v/v); B, polar fraction eluted with diethyl ether and methanol. Peaks: 1, non-polar dimers; 2, non-polar monomers; 3, polymers; 4, oxidized dimers; 5, oxidized monomers. (Reproduced by kind permission of the *Journal of Chromatography*, and redrawn from the original paper).

polymeric FA, the lowest digestibilities were found for nonpolar dimers (10.9% on average), whereas oxidized dimers and polymers showed higher apparent absorbability than expected, ranging from 22.7% to 49.6%.

2. *Partial glycerides*

Partial glyceride evaluation is of proven interest in many areas of chemistry and biochemistry. Control of food emulsifiers, technical oleochemicals, chemical and enzymatic reactions of hydrolysis or esterification and analysis of special seed oils are typical examples. Adsorption chromatography on TLC and HPLC, as well as GLC, may be used alternatively for this purpose although quantification is often tedious and inaccurate. Prior to the development of high-performance chromatographic supports, acceptable results were obtained in the early 1970s by Umbehend *et al.* [124] for the analysis of technical dipalmitin, using a BioBeads™ SX column, THF as eluant and a refractive index

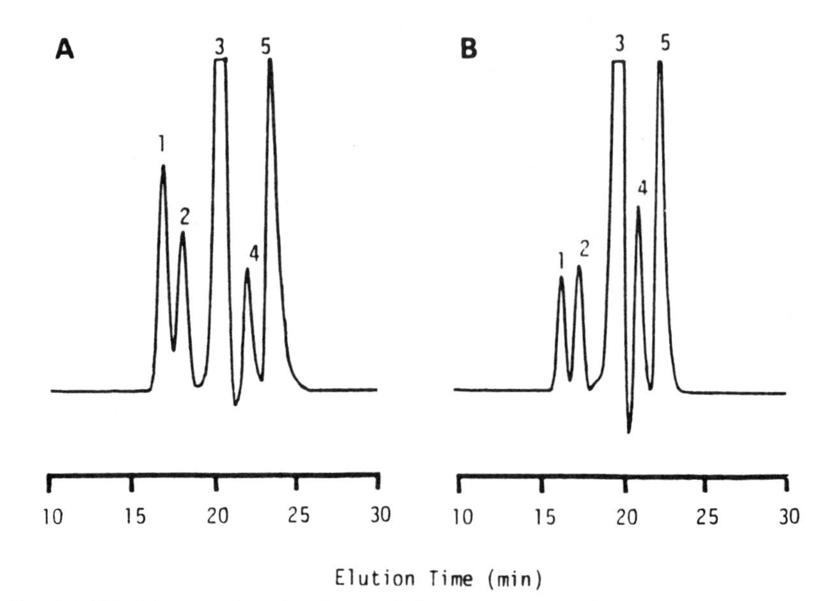

Fig. 3.3. HPSEC separation of safflower oil lipolysis mixtures after methylation of the free fatty acids with diazomethane . A, 15 min lipolysis, B, 30 min lipolysis. Peaks: 1, triacylglycerols; 2, diacylglycerols; 3, methyl esters; 4, monoacylglycerols; 5, monolaurin (internal standard). (Reproduced by kind permission of the authors and of *The Journal of American Oil Chemists' Society*).

detector. As was pointed out later [2,3], by selecting THF it is not only possible to employ a refractive index detector but also an infrared detector, if the windows in the spectrum of THF that permit the registration of the IR absorbance of the ester carbonyl and of the free alcohol group are used. In this latter case, small amounts of glycerol and water might be also detected.

The most complete study among HPSEC applications to the separation and quantification of FA, and mono-, di-, and TAG mixtures was accomplished by Christopoulou and Perkins [23]. Two columns packed with 5 μm styrene-divinylbenzene copolymer were connected in series, toluene was employed as eluant and lipid components were monitored by refractometry. A formula based on known values of correction factors for simple lipid components and the fatty acid composition of the sample allowed quantification of lipid mixtures. The relative standard deviation for each lipid component was below 5%, and they could be detected at the 0.05% level. Figure 3.3 displays a representative example of the efficacy obtained in the separation of safflower oil lipolysis mixtures after 15 and 30 min reaction. Interestingly, monoacylglycerols (MAG) and FA were not resolved unless FA were first methylated with diazomethane to improve the separation. An inversion in the elution volumes of MAG and FAME was observed, that was apparently associated with adsorption effects between the stationary phase and MAG. However, this was not found in other applications [29,57,96] using the same stationary phase, where MAG eluted between DAG and FA.

An analogous methodology was developed by Mittelbach and Trathnigg [77] for the analysis of FAME. They stressed the importance of controlling the content of partial glycerides in FAME used as diesel fuel for their influence on combustion characteristics. One Bondapak™ CN and two Ultrastyragel™ columns (100 Å + 500 Å) were connected, chloroform was selected as the mobile phase and a density detection system was used. Although good resolution was obtained, quantification was not easy.

The use of an evaporative light-scattering (mass) detector has recently been described by Hopia *et al.* [57] for the analysis of lipid classes including TAG, DAG, MAG and FA. These four groups of compounds were successfully separated but, unfortunately, optimization of the detector was difficult and the response was not linear over the weight range covered. This detector has a slightly superior sensitivity and is a good solution in the search for a lipid sensitive detector under gradient elution conditions. However, for quantitative purposes in HPSEC applications, a refractometer is simpler in that a single solvent is used and linear responses are normally obtained in the ranges of interest.

Lastly, HPSEC has been used for the evaluation of enzymatic hydrolysis *in vitro* of thermoxidized oils [75]. Parallel quantification of the increase of partial glycerides and fatty acids and the decrease of high-molecular weight compounds with hydrolysis time clearly showed the problems involved in the action of pancreatic lipase on complex glycerides.

3. Oxidized compounds

Lipid oxidation has been long recognized as a major problem in the storage of fatty foods [63]. The complexity of the chemistry involved is still obscure and so far no universal analytical test has emerged for the measurement of the oxidation level. The present methods give information about particular stages of the oxidation process and some are only applicable to certain lipid systems.

In recent years, SEC has also been applied to the evaluation of oxidative deterioration in foods, especially in the area of fish oils. Polyunsaturated fatty acids such as eicosapentaenoic (EPA) and docosahexaenoic acids (DHA) are particularly abundant in fish oils. On one hand, they may account for a lower incidence of coronary heart disease but, on the other, are readily oxidized and their ingestion may entail a risk of exposure to potentially harmful products. As a consequence, an increasing number of general studies on autoxidation of fish oils at low temperatures have been carried out lately, with special emphasis on determining those compounds which are possibly deleterious.

As to SEC applications to this area, a few papers are of particular interest. Firstly, Cho *et al.* [21] evaluated the extent of oxidation of ethyl esters of EPA and DHA in comparison to ethyl linoleate and linolenate, at 5C in the dark. Polar materials were fractionated by SEC on a BioBeads™ S-X3 column. Dimers and polymers thus isolated were further analysed by HPSEC on a column of Ultrastyragel™ 500 Å using dichloromethane as eluant with a refractive index detector. Results indicated that oxidation of ethyl EPA and DHA was very rapid after an induction period of 3-4 days while linoleate was oxidized after three weeks. Over 70% of polar material from oxidized EPA and DHA were dimers plus polymers. Whereas the peroxide value was not found to be a good indicator of oxidation, the determination of secondary products such as polymers was suggested by this study to be important in evaluating oxidative deterioration because of the instability of hydroperoxides.

It is worthy to note the study undertaken by Shukla and Perkins [117] on the presence of oxidation materials in encapsulated fish oils, which have become very popular in recent years as a health supplement to reduce the risk of cardiovascular disease. A series of polystyrene-divinyl benzene columns connected in the order 500 Å, 100 Å, and 100 Å and THF as mobile phase were used. The results supported the need for a close control of commercial fish oil capsules and the utility of HPSEC in this particular regard.

In the same direction, Burkow and Henderson [13] proposed SEC for the routine assessment of fish oil quality. They used a Ultrastyragel™ 500 Å column with dichloromethane (0.8 mL/min) as mobile phase. In experiments carried out at 35C in artificial daylight, the oils were rapidly oxidized, with values of polymer content reaching as high as 35 to 45% after 4 days. Later, these authors improved the separation by coupling three columns in series with effective fractionation ranges from 100-30,000 daltons. The new combination was applied to both TAG and hydrolysed samples. The detection system chosen was an evaporative light-scattering detector and glycerol was used as internal standard. The difficulties found in quantification led to the use of curvilinear calibration graphs to give the response-weight relationship between the polymers and the internal standard [14].

Oxidized TAG have also been proposed as a measurement of the level of oxidation in fats and fatty foods other than fish oils [93,94]. Following examination of HPSEC profiles and data of the polar compounds formed after storage of various fatty foods at different temperatures, it was evident that oxidized TAG were the group of compounds which experienced the most remarkable increase. The results obtained in oils extracted from roasted sunflower seeds after storage for 30, 60 and 90 days are presented in Table 3.1. A notable increase of oxidized TAG (polymers plus monomers) was observed while, in contrast, DAG and FA coming from hydrolysis remained at the initial levels. The analytical method was thus

Table 3.188
Polar compound distribution in oils from roasted sunflower seeds after storage at room temperature for 30, 60 and 90 days.

Polar compounds (mg/g oil)	Time (days)	Total TAG P	oxTAG M	DAG	FA
0	53	0.8	33.9	9.6	8.7
30	86	9.2	60.8	8.9	7.1
60	139	25.2	93.5	9.9	10.4
90	184	63.0	101.7	9.8	9.5

Abbreviations: TAG P, triacylglycerol polymers; oxTAG M, oxidized triacylglycerol monomers; DAG, diacylglycerols; FA, fatty acids.

suggested as an objective measurement of the total oxidation level in fats, as quantification included both the primary and secondary compounds formed.

Similar results were obtained by Hopia *et al.* [56] in a study in which TAG mixtures were autoxidized and the polar products analysed through HPSEC with light scattering detection, following preliminary elimination of nonpolar TAG by solid phase extraction.

4. Miscellaneous

Numerous applications of SEC to various groups of lipids and to minor lipid compounds were described in the 1970s. Neutral hydroxy lipids [16], phospholipids [4,66], oxidation products [91,98], unsaponifiable fractions [99], synthetic antioxidants [32], waxes [53], hydrocarbons [27], and fat-soluble vitamins [54] are only some examples of applications of different SEC methodologies that did not achieve general acceptability, in part due to the simultaneous development of other chromatographic techniques based on much more differentiating principles for lipid separation than that of molecular size exclusion.

Nonetheless, one of these former applications has endured to date and is attracting increasing attention, namely, the use of SEC as a lipid clean-up method prior to pesticide analysis by GLC. The technique was primarily found to be efficient and reproducible for lipid removal from fish extracts, using an automatic system, with a Bio-Beads™ S-X2 column and cyclohexane as mobile phase [58,120]. Since then, efforts have been directed to improve the technique as a competitive alternative to separations based on polarity. Attempts to eliminate lipids thoroughly for the clean-up of such samples [5,18,69,107] and technical developments to simplify manual operations associated with the loading systems [28] have been reported. Recently, a new automated system of on-line SEC-GLC is receiving considerable attention [43,44]. Advantages and limitations of the latter were discussed in detail by Grob and Klin [44], who applied the technique to analyses of olive oil, fat extracts of chicken and fish, and lettuce, using a 25 cm x 3 mm i.d. column packed with cross-linked

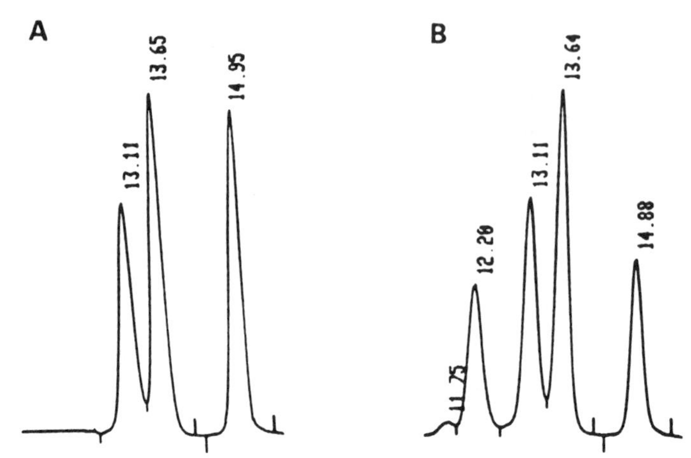

Fig. 3.4. Significant part of chromatograms of minor glyceridic compounds in an olive oil (A) before and (B) after refining. Retention times (min): 12.2, triacylglycerol dimers; 13.1, oxidized triacylglycerols; 13.6, diacylglycerols; 14.9, fatty acids and polar unsaponifiable fraction.

polystyrene. They found the technique saved considerable time in preparative work and was of improved reproducibility, while the main drawbacks were related to the necessity for smaller SEC columns for on-line coupling with GLC that limited sensitivity. Finally, the authors anticipated that on-line SEC-LC-GLC system would probably become the most successful concept in the future.

Another set of novel applications refers to different aspects of fat and oil characterization and quality evaluation. Recently, Husain *et al.* [60] reported the use of HPSEC for accurate determination of molecular weight averages in a number of oils and fats, and their binary mixtures. The authors discussed the conditions under which the method might be suitable for the determination of adulteration in certain oils and fats. SEC has been also proposed for characterization of virgin olive oils [39,40,41,42]. Refined olive oils showed dimer contents higher than 0.5% while, in contrast, no dimers were detected in virgin oil samples. These results were confirmed in our laboratory, in which a significant change of the ratio DAG/FA from virgin to refined oils was also reported [30]. Furthermore, we evaluated the possibilities of HPSEC for quality assessment of refined oils [31]. Quantification of polar compounds in crude oils and in samples taken at different stages of the refining process indicated that the quality of an initial crude oil could be deduced from the resulting refined oil by virtue of certain markers of the oxidative and hydrolytic alterations. An illustrative example of the above is presented in Figure 3.4, which shows polar fractions from an olive oil before and after refining. On one hand, compounds of higher MW than that of oxidized TAG monomers are formed as a consequence of the high temperature in the deodorization step. On the other hand, oxidized TAG and DAG remain after refining and so their level in the different samples is a measure of oxidation and hydrolysis, respectively, regardless of whether the oil is crude or refined. Finally, the peak including FA and the polar unsaponifiable fraction decreases due to the neutralization step.

SEC evaluation has been extended to lipid samples other than classical fats and oils. For example, a great deal of work is currently done on those low digestibility fats that can be heated at high temperature and hence used in baked and fried low caloric fatty foods. Among the fat substitutes proposed, sucrose polyesters (SPE), *i.e.* sucrose molecules esterified by 5 to 8 fatty acids, probably stands out. Birch and Crow [9] relied on SEC to determine SPE in faeces and diets, and they proposed the method for measurements of absorption of such materials, but based initially on radioactivity markers and hence not suitable for human subjects. The authors established the conditions best suited to quantify SPE as a single peak eluting before other components of the extracts, by using two μStyragel™ columns (500 and 1000 Å) in series, THF as eluant at a flow-rate of 1.5 mL/min and a refractive index detector. More recently, studies on the polymers formed at high temperature from SPE were carried out with a single 500 Å PLGel™ column (60 cm), and indicated not only that dimers and trimers were present but also mixed dimers when SPE were subjected to this thermal treatment in mixtures with TAG [37]. Susceptibility of SPE to thermal, oxidative and hydrolytic modifications is a current subject of study in our laboratory. Starting from purified samples derived from olive oil and following an analytical procedure based on adsorption chromatography and HPSEC, we have reported that oxidation of sucrose octaesters takes place more slowly than TAG with a similar FA composition. Given the different molecular structures of sucrose octaesters and TAG, which have eight and three acyl groups respectively, it was essential to apply the methodology to the methyl ester derivatives. In essence, monomeric and, in lower amounts, dimeric compounds were obtained at 100C after 12 hours [105]. In other experiments, the behaviours of sucrose octaoleate and linoleate were compared upon oxidation at 60C and 100C, and showed that differences in the degree of unsaturation of FA esterifying sucrose molecules had a greater influence on polymerization than did the temperature. Thus, sucrose octalinoleate oxidized at 60C during 12 h showed a higher level of dimers plus polymers (6.0%) than that found for sucrose octaoleate after 20 h at 100C [106].

A special application of promising utility in lipid analysis is micelle separation. Certain lipids may be formed into reversed micelles in nonpolar solvents. Clearly, micelles can be separated from typical lipids by SEC, being thus excluded because of their size. In a novel study, Shansky and Kaneda [116] applied SEC to soy lecithin, which had showed an anomalous chromatographic behaviour in THF, suggesting that formation of reverse micelles or aggregates of phospholipid components might have occurred. Calibration of the column was performed using polystyrene standards. A broad peak was observed at a relative molecular mass of about 8000 in two samples of soy lecithin. Phospholipid aggregates were not completely broken up by dilution through the chromatographic system until a sufficiently low concentration (≤ mg/mL) of soy lecithin was used.

C. SIZE EXCLUSION CHROMATOGRAPHY IN AQUEOUS MEDIA

In contrast to SEC applications in organic media, at present characterized by the use of a unique stationary phase intended to separate low MW groups of compounds, SEC in aqueous media is normally applied to complex molecules, molecular aggregates or particles of low stability and high MW. Conventional and high performance stationary phases based on silica or cross-linked polymer-based phases compete to give solutions to diverse problems [121]. Different aqueous buffers are used to preserve the structural integrity of the solute or to mimic physiological conditions. Another distinctive feature is that previous equilibration of the column with salt solutions or lipids may also be necessary in order to reduce interactions between the stationary phase and the solute. In this context, two major sets of studies are presently of growing importance, namely lipoprotein analysis and size estimation of micelles and vesicles.

1. Lipoproteins

Plasma lipoproteins are water-soluble macromolecular complexes of lipids (triglycerides, cholesterol and phospholipids) and specific proteins, referred to as apolipoproteins. They are classified according to their hydrated densities into five major classes: chylomicrons, very low density lipoproteins (VLDL), low density lipoproteins (LDL), high density lipoproteins (HDL) and very high density lipoproteins (VHDL). Most lipoproteins range from approximately 200,000 to 10,000,000 MW and can be fractionated into groups or classes by a variety of techniques, ultracentrifugal flotation or electrophoresis standing out [36,78]. Many efforts have been devoted to determining the factors that control the concentration and composition of lipoproteins, both in humans and in animal models, not least for their important role in the pathogenesis of atherosclerosis. Separation by density after ultracentrifugation is widely used, although density ranges may vary among animal species and contamination by other plasma constituents may occur [36,45]. SEC represents an alternative technique for lipoprotein separation and isolation in that size differs among classes. In general, it offers several advantages, *i.e.* the procedure is gentle and non-destructive, the size distribution of particles can be easily visualized during separation and the recovery is high.

Fractionation of lipoproteins by conventional agarose gels, available in bulk and distinguished by their relative economy, has been practised successfully by a number of investigators. Most notably, Rudel and coworkers [108,109,110] using agarose gel beads, BioGel™ A (BioRad, USA) consisting of 4 and 6% agarose, were able to separate human lipoproteins into three discrete elution peaks (VLDL plus chylomicrons, LDL and HDL). In addition, 4% agarose gels were used to sub-fractionate chylomicrons and VLDL [35,100,111,112]. However, limitations of this procedure are the long elution time (16-30 hours), large sample injection volume, poor reproducibility and the great dilution of the fractions obtained.

Hara *et al.* developed a simple and rapid method for lipoprotein analysis using HPLC with gel permeation columns from Toyo Soda Mfg. Co. (Tokyo). Reviews of their work are available [49,50,89]. Chemically modified silica gel (TSK™ SW type) or hydrophilic organic copolymer-based (TSK™ PW type) columns of 60 cm x 0.75 cm i.d. were used in combinations varying in pore size. Originally, TSK™ columns were developed for high-speed size exclusion of proteins; they can resist high pressure in aqueous systems and possess a large number of theoretical plates. TSK™ SW support, available in three grades based on pore size, allows separation between 5000 and 7,000,000 MW range whereas TSK™ PW support, available in six grades, cover a wider MW range (1000-8,000,000) and is ideal for separation of low MW compounds. However, within its limited separation range, resolution is better with the SW type column [64].

The analytical procedure was briefly as follows: serum was usually obtained and individual lipoprotein fractions or total lipoproteins isolated by ultracentrifugation and containing 20-100 μg of protein were applied onto the column in an HPLC apparatus, and the separation was monitored by absorbance at 280 nm. 0.15 M NaCl solution (pH 7.0) was used as the preferred eluent for the separation of serum lipoproteins. The flow rate of the mobile phase was 1.0 mL/min for three connected columns, 0.60 mL/min for two connected columns, and 0.50 mL/min for a single column. The best column system for the separation of all lipoprotein classes may be the combined system of TSK™ 4000 SW + TSK™ 3000 SW [86,87], although Carrol and Rudel [17] reported the best resolution of individual lipoprotein fractions to be possible with the 5000 PW + 4000 SW + 3000 SW system. However, the latter admitted that this was an expensive combination and chose a single 5000 PW column for most of their work, that offered acceptable resolution.

Additionally, Hara *et al.* developed direct quantification methods for cholesterol [47,83,84], triglycerides [48] and phospholipids [86,87] through selective detection in the post-column effluent by using enzymatic reaction reagents where the required loaded volume of normolipidemic human serum was 10, 20 and 50 μL, respectively. Recovery of the total lipoprotein fraction and individual lipoproteins from the TSK™ gel columns was examined on the basis of total cholesterol determination and was found to be over 90% for the total lipoprotein fraction, LDL, HDL_2 and HDL_3, and only under this value for chylomicron + VLDL (85.3%). Correlation between HPSEC and the sequential ultracentrifugal flotation method was very good except for the VHDL [88]. Overall, the main advantages of this procedure are the short analysis time (less than one hour), small sample amount required (less than 20 μL of serum), high resolution and good reproducibility.

TSK columns were found reliable also by other investigators, especially for being easily applied to the follow-up of individuals at risk for cardiovascular disease, either in adult or infant serum [15,68,126]. Similarly, they were shown to be effective for screening a large number of animals [17,130,131]. As examples of analysis of lipoprotein subclasses, Okazaki *et al.* determined cholesterol in HDL_2 and HDL_3 using two TSK™ 3000 SW columns [85], while Holmquist and Carlson [55] studied the HDL particle size heterogeneity and interconversion of small to large particles using three coupled TSK™ 3000 SW columns. However, in the range of higher particle size, after preliminary assays with various columns, Williams and Kushwaha [132] concluded that HPLC separation of plasma triglyceride-rich lipoproteins could be easier using a GF-450 column (25 cm x 0.94 cm i.d., DuPont, USA).

Apart from separation of lipoprotein classes, the availability of new high performance columns also invited applications to improving separation of the apolipoprotein constituents of lipoproteins, utilizing denaturants such as urea or guanidinium chloride. To quote some examples, Wehr *et al.* [128] resolved human serum apoproteins from a delipidated total HDL fraction on a MicroPak™ TSK 3000SW column at a low flow rate and sample load. To cover the wide molecular range (10,000 to 250,000) of the protein components of apoVLDL, Pfaffinger *et al.* [97] connected five TSK columns in series and isolated apoE in good yield in less than one hour. A further improvement was provided by Young *et al.* [134] who resolved delipidated human HDL (apoHDL) without the use of chemical denaturants or detergents in the eluent buffer.

In recent years, separation of plasma lipoproteins has gained in both simplicity and economy, as it is usually carried out on a single column of an extensively cross-linked agarose gel matrix, Superose™ 6 (Pharmacia Fine Chemicals, Sweden). Superose™ support was especially developed to meet the demands for faster separation and improved resolution using agarose-based media. The main advantages in comparison to conventional agarose gels are the great reduction in the separation time and a considerably higher rigidity. The analytical Superose™ HR 10/30 (10 mm i.d. x 30 cm) has a bead size of approximately 10 μm, whereas the Superose™ 6 prep grade, formerly Superose™ 6B, or Superose™ 6 HR 16/50 (16 mm i.d. x 56 cm) if supplied prepacked, is intended for preparative work and has a bead size of about 30 μm. The useful fractionation range is 5000-5,000,000 MW for globular proteins. Separation of plasma lipoproteins into three major classes was acceptable and achieved within two hours when using Superose™ 6B [25,46], although resolution did not match that obtained with the Toyo Soda columns. In this work, the mobile phase was 0.15 M sodium chloride containing EDTA and sodium azide, pH 7.2, the flow rate was 0.5 or 0.75 mL/min and the separation was monitored at 280 nm. Under such conditions, the HDL fraction eluted as a single peak, yet Clifton *et al.* [25] collected multiple fractions of HDL that coincided with the sub-populations identified by gradient gel electrophoresis. The method described is essentially a preparative one and enables a 20-fold increase in the amount of HDL separated on the column, thereby providing workable quantities of small HDL particles. Shorter elution times (around one hour) were achieved by using Superose™ 6 HR 10/30 [67]. These authors also reported the use of Nile red for single-stage lipoprotein evaluation, using a fluorescence detector and thus avoiding the lack of specificity associated with monitoring the absorbance.

From the literature of recent years, it seems that there has been less emphasis on the use of new analytical SEC packings and more attention placed on applicability and optimization of the existing procedures. As an alternative to the latest general scheme described, which consists of one ultracentrifugation step with the plasma prior to SEC on Superose™ columns, Van Gent *et al.* [125] have suggested an automated lipoprotein separation without that preliminary ultracentrifugal step. 2 mL of plasma were required and the column chosen in this case was Superose™ 6 HR 16/50. The elution scheme consisted of a low-flow separation of the lipoprotein complexes followed by a high-flow washout of the remaining plasma proteins. Another novel study of interest has been carried out by Nyyssönen and Salonen [82], who compared the use of two Superose™ 6 columns in series (total run time of 105 min) with density gradient ultracentrifugation and precipitation methods. Quite a good example of the broad utility of the system has been provided by Kieft *et al.* [65] who analysed lipoprotein profiles in humans and several species fed normal or cholesterol-containing diets. Figure 3.5 illustrates the influence of the diets tested on cholesterol distribution among lipoprotein classes for seven species. As with conventional agarose columns, these data can be used to estimate and compare the relative sizes of lipoprotein particles, with the advantages derived from the use of a single HPSEC column, only microlitre quantities of whole plasma and a short elution time (within 40 min).

It is clear that SEC is a useful and rapid quantitative method in the

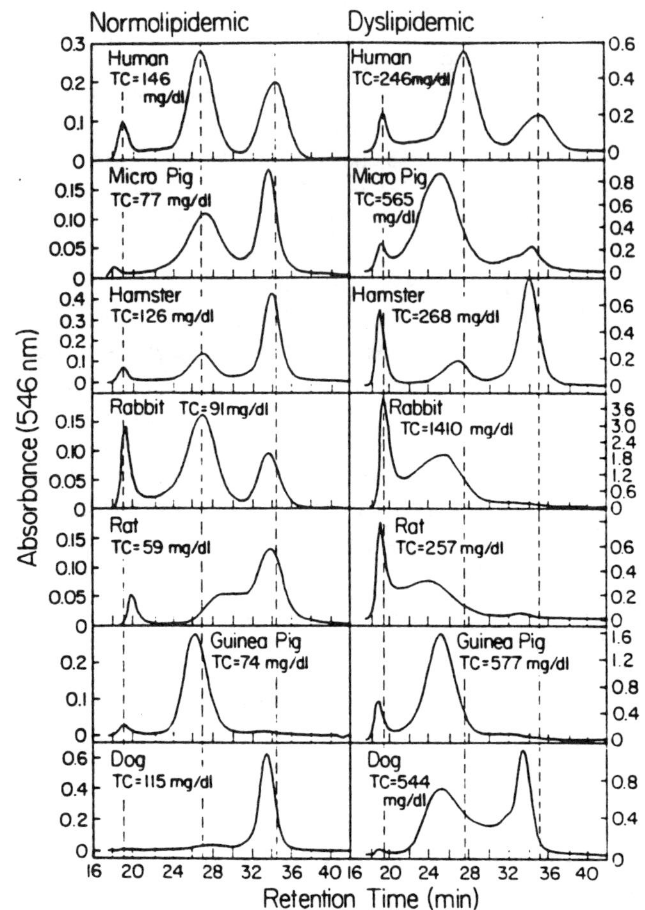

Fig. 3.5. Cholesterol distribution profiles for seven species with normal (normolipidemic) or abnormal (dyslipidemic) plasma total cholesterol concentrations. The vertical, dashed lines mark the elution peaks for human VLDL, LDL and HDL. (Reproduced by kind permission of the authors and of the *Journal of Lipid Research*).

lipoprotein field, and it continues to be applied to systematic studies of both genetic and dietary effects on lipoprotein phenotypes.

2. *Vesicles and micelles*

The average size and size-distribution appear to be among the major concerns in lipid vesicle characterization. Knowledge of vesicle size parameters is essential in many physical and biological investigations such as quantitative studies of membrane transport, ligand binding to receptors or pharmacological use of liposomes. Electron microscopy, dynamic light-scattering, analytical ultracentrifugation, nuclear magnetic resonance, sedimentation field flow fractionation and SEC are used for estimating vesicle diameter. Many of these approaches require large amounts of material, depend on expensive equipment or do not allow separations of mixed vesicle populations. However, SEC is simple, relatively inexpensive and has been shown to provide a suitable method for both fractionation and average size determination of biological vesicles [59,104]. In recent years, one of the most widely used exclusion gels for sizing and separation of phospholipid vesicles has been the acrylamide-based polymer Sephacryl™ S1000 from Pharmacia [81,101,102,114]. It possesses an exclusion diameter of about 3000 Å. Recovery of phospholipid vesicles of different size has been reported to be greater than 95% provided that the column has been pre-saturated with lipids, and the diameters obtained are in good agreement with those obtained by electron microscopy [102]. Satisfactory results were also provided by Rees *et al.* [101] for the characterization of unilamellar phospholipid vesicles, applicable to studies on fatty acid donor and acceptor properties of membranes and fatty acid binding proteins. Gel columns were first saturated with phospholipids and then fractionation of vesicles was performed with the same buffer used for vesicle preparation. Eluting particles were detected by monitoring turbidities at 350 nm.

Development of high performance supports has overcome the main limitations of SEC for routine analysis. Ollivon *et al.* [90] reported that the TSK™ G6000PW column was able to separate larger particles ($>$5000 Å) than the Sephacryl™ S1000, and, when used in combination with the TSK™ G5000PW column, gave more discrete separation of smaller particles (100 to 300 Å diameter). Moreover, the HPLC columns can be run significantly faster (10-20 min as opposed to several hours) and give more precise results than Sephacryl™ S1000. As is usual, columns were saturated with lipids, and the eluant was that used for vesicle preparation. Samples were detected by absorbance measurement at 254 nm. A TSK™ G 6000PW column has been applied recently to the study of stability of vesicles of egg phosphatidylcholine and phosphatidic acid ranging from 1300 to 6400Å in diameter. The results indicated that the apparent size variation of egg phosphatidylcholine vesicles observed over a week, was due mainly to their aggregation, which was reduced significantly by the introduction of a small amount of egg phosphatidic acid in the vesicle membrane [73].

A survey of the recent literature shows an increasing number of reports connected to the biological processes where not only vesicles but also micelles play a key role. Particularly relevant has been the use of SEC to characterize both size and composition of biliary micelles and vesicles. In this context, classical studies date from the 1960s [11,80,122], but most work has been done in the last decade. SEC has been used as the primary technique either to ascertain the effect of bile salt concentration on the size of mixed micelles [103] or, more often, to study cholesterol-transporting particles inasmuch as a non-micellar, bile salt-independent mode of cholesterol transport in human bile involving phospholipid vesicles has been described [118]. Studies have in common, first, the use of low performance stationary phases such as Sephadex™ G-100 and G-200 [72,103], Sephacryl™ S-300 [119], Sepharose™ CL-4B-200 [92], from Pharmacia, and Biogel™ P-200 stacked on top of Bio-Gel™ A-0.5m from Bio-Rad [52], secondly, column pre-equilibration with bile salts to preserve micellar integrity and, thirdly, further application of conventional techniques for lipid determination in the bile fractions isolated by SEC.

The introduction of high performance columns has considerably improved the technique, as recently proposed by Cohen *et al.* [26]. They described the use and validation of Superose™ 6 HR 10/30 for rapid, high resolution separation and sizing of co-existing simple micelles, mixed micelles and vesicles in bile. Fractionation of model biles composed of labelled lecithin, cholesterol and the sodium salt of taurocholate in varying concentrations and resuspended in aqueous buffer (0.15 M NaCl, pH 6-7) was carried out either on Superose™ 6 or Sephacryl™ S-300 pre-equilibrated with Na-taurocholate solution. Superose™ 6 gel fractionation of model biles supersaturated with cholesterol yielded high resolution separation of vesicles (640Å apparent diameter) from mixed micelles of bile salt-lecithin-cholesterol (80-100Å diameter) and simple Na-taurocholate micelles (22-30Å diameter). Compared with Superose™, Sephacryl™ yielded particle separations with lower resolution and speed (30 hours against one hour).

ABBREVIATIONS

DAG, diacylglycerols; DHA, docosahexaenoic acid; EPA, eicosapentaenoic acid; FA, fatty acids; FAME, fatty acid methyl esters; GLC, gas liquid chromatography; HDL, high density lipoproteins; HPLC, high performance liquid chromatography; HPSEC, high performance size exclusion chromatography; LDL, low density lipoproteins; MAG, monoacylglycerols; MW, molecular weight; SEC, size exclusion chromatography; SPE, sucrose polyesters; TAG, triacylglycerols; THF, tetrahydrofuran; TLC-FID, thin layer chromatography-flame ionization detector; VHDL, very high density lipoproteins; VLDL, very low density lipoproteins.

REFERENCES

1. Aitzetmüller,K., *J. Chromatogr.*, **71**, 355-360 (1972).
2. Aitzetmüller,K., *Prog. Lipid Res.*, **21**, 171-193 (1982).
3. Aitzetmüller,K., *Chem. Ind. (London)*, 452-464 (1988).
4. Arvidson,G.A.E., *J. Chromatogr.*, **103**, 201-204 (1975).
5. Ault,J.A., Schofield,C.M., Johnson,L.D. and Waltz,R.H., *J. Agric. Food Chem.*, **27**, 825-832 (1979).
6. Barth,H.G. and Boyes,B.E., *Anal. Chem.*, **62**, 381R-394R (1990).
7. Barth,H.G. and Boyes,B.E., *Anal. Chem.*, **64**, 428R-442R (1992).
8. Bartosiewicz,R.L., *J. Paint Technol.*, **39**, 28-39 (1967).
9. Birch,C.G. and Crowe,F.E., *J. Am. Oil Chem. Soc.*, **53**, 581-583 (1976).

10. Bombaugh,K.J., Dark,W.A. and Levangie,R.F., *Z. Anal. Chem.*, **236**, 443-451 (1968).
11. Borgstrom,B., *Biochem. Biophys. Acta*, **106**, 171-183 (1965).
12. Brooks,C.J.W. and Keates,R.A.B., *J. Chromatogr.*, **44**, 509-521 (1969).
13. Burkow,I.C. and Henderson,R.J., *Lipids*, **26**, 227-231 (1991).
14. Burkow,I.C. and Henderson,R.J., *J. Chromatogr.*, **552**, 501-506 (1991).
15. Busbee,D.L. Payne,D.M., Jasheway,D.W., Carlisle,S. and Lacko,A.G., *Clin. Chem.*, **27**, 2052-2058 (1981).
16. Calderon,M. and Bauman,W.J., *J. Lipid Res.*, **11**, 167-169 (1970).
17. Carrol,R.M. and Rudel,L.L., *J. Lipid Res.*, **24**, 200-207 (1983).
18. Chamberlain,S.J., *Analyst (London)*, **115**, 1161-1165 (1990).
19. Chandrasekhara Rao,T., Kale,V., Vijayalakshmi,P., Gangadhar,A., Subbarao,R. and Lakshminarayama,G., *J. Chromatogr.*, **466**, 403-406 (1989).
20. Chang,T.L., *Anal. Chem.*, **40**, 989-992 (1968).
21. Cho,S.Y., Miyashita,K., Miyazawa,T., Fujimoto,K. and Kaneda,T., *J. Am. Oil Chem. Soc.*, **64**, 876-879 (1987).
22. Christie,W.W., *High-performance Liquid Chromatography and Lipids. A Practical Guide* (1987) (Pergamon Press, Oxford).
23. Christopoulou,C.N. and Perkins,E.G., *J. Am. Oil Chem. Soc.*, **63**, 679-684 (1986).
24. Christopoulou,C.N. and Perkins,E.G., *J. Am. Oil Chem. Soc.*, **66**, 1338-1343 (1989).
25. Clifton,P.M., Mackinnon,A.M. and Barter,P.J., *J. Chromatogr.*, **414**, 25-34 (1987).
26. Cohen,D.E. and Carey,M.C., *J. Lipid Res.*, **31**, 2103-2112 (1990).
27. Cooper,B.S., *J. Chromatogr.*, **46**, 112-115 (1970)
28. Daft,J., Hopper,M., Hensley,D. and Sisk,R., *J. Assoc. Off. Anal. Chem.*, **73**, 992-994 (1990).
29. Dobarganes,M.C., Pérez-Camino,M.C. and Márquez-Ruiz,G., *Fat Sci. Technol.*, **90**, 308-311 (1988).
30. Dobarganes,M.C., Pérez-Camino,M.C. and Márquez-Ruiz,G., in *Actes du Congr/s International "Chevreul" pour l'étude des corps gras*, Vol 2, pp. 578-584 (1989) (edited by ETIG, Paris).
31. Dobarganes,M.C., Pérez-Camino, .C., Márquez-Ruiz,G. and Ruiz-Méndez, M.V., in *Edible Fats and Oils Processing: Basic Principles and Modern Practices*, pp. 427-429 (1990) (editedby D.R. Erickson, AOCS, Champaign).
32. Doeden,W.G., Bowers,R.H. and Ingala,A.C., *J.Am. Oil Chem. Soc.*, **56**, 12-14 (1979).
33. Dubin,P.L., in *Advances in Chromatography*, Vol 31, pp. 119-151 (1992) (edited by J.C. Giddings, E. Grushka & P.R. Brown, Marcel Dekker Inc., New York).
34. Fallon, A., Booth, R.F.G. and Bell, L.D., in *Laboratory Techniques in Biochemistry and Molecular Biology, Vol 17: Applications of HPLC in Biochemistry*, pp. 56-64 (1987) (edited by R.H. Burdon & P.H. Knippenberg, Elsevier, Amsterdam).
35. Ferreri,L.F., *J.Dairy Sci.*, **65**, 1912-1920 (1982).
36. Ferreri, L.F., in *Lipid Research Methodology*, pp. 133-156 (1984) (edited by J.A. Story, Alan R. Liss, Inc., New York).
37. Gardner,D.R. and Sanders,R.A., *J. Am. Oil Chem. Soc.*, **67**, 788-795 (1990).
38. Gere,A., *Fette Seifen Anstrichm.*, **85**, 111-117 (1983).
39. Gomes,T., *Riv. Ital. Sostanze Grasse*, **65**, 433-438 (1988).
40. Gomes,T. and Catalano, M., *Riv. Ital. Sostanze Grasse*, **65**, 125-127 (1988).
41. Gomes,T., in *Actes du Congr/s International "Chevreul" pour l'étude des corps gras*, Vol 3, pp. 1169-1175 (1989) (edited by ETIG, Paris).
42. Gomes,T., *J. Am. Oil Chem. Soc.*, **69**, 1219-1223 (1992).
43. Grob,K. and Klin,I., *J. High Resolut. Chromatogr.*, **14**, 451-454 (1991).
44. Grob,K. and Klin,I., *J. Agric. Food Chem.*, **39**, 1950-1953 (1991).
45. Grummer,R.R., Davis,C.L. and Hegarty,H.M., *Lipids*, **18**, 795-802 (1983).
46. Ha,Y.C. and Barter,P.J., *J. Chromatogr.*, **341**, 154-159 (1985).
47. Hara,I., Okazaki,M. and Ohno,Y., *J. Biochem. (Tokyo)*, **87**, 1863-1865 (1980).
48. Hara,I., Shiraishi,K. and Okazaki,M., *J. Chromatogr.*, **239**, 549-557 (1982).
49. Hara,I. and Okazaki,M., *Progress in HPLC*, **1**, 95-103 (1985).
50. Hara,I. and Okazaki,M., *Methods Enzymol.*, **129**, 57-78 (1986).
51. Harris,W.C., Crowell,E.P. and Burnett,B.B., *J. Am. Oil Chem. Soc.*, **50**, 537-539 (1973).
52. Harvey,P.R.C., Somjen,G., Lichtenberg,M.S., Petrunka,C., Gilat,T. and Strasberg,S.M., *Biochim. Biophys. Acta*, **921**, 198-204 (1987).

53. Hillman, D.E., *Anal. Chem.*, 43, 1007-1013 (1971).
54. Holasová, M. and Blattná, J., *J. Chromatogr.*, 123, 225-230 (1976).
55. Holmquist,L. and Carlson,L.A., *Lipids*, **20**, 378-388 (1985).
56. Hopia,A.I., Piironen,V.I. and Lampi,A.M., *INFORM*, **3**, 493 (1992).
57. Hopia,A.I., Piironen,V.I., Koivistoinen,P.E. and Hyvönen,L.E.T., *J. Am. Oil Chem. Soc.*, **69**, 772-776 (1992).
58. Horler,D.F., *J. Sci. Food Agr.*, **19**, 229-231 (1968).
59. Huang,C.H., *Biochemistry*, **8**, 344-352 (1969).
60. Husain,S., Sastry,G.S.R., Raju,N.P. and Narasimha,R., *J. Chromatogr.*, **454**, 317-326 (1988).
61. Inoue,H., Konishi,K. and Taniguchi,N., *J. Chromatogr.*, **47**, 348-354 (1970).
62. Jensen,O.N. and Moller,J., *Fette Seifen Anstrichm.*, **88**, 352-357 (1986).
63. Kanner,J. and Rosenthal,I., *Pure & Appl. Chem.*, **64**, 1959-1964 (1992).
64. Kato,Y., in *Handbook of HPLC for the Separation of Amino Acids, Peptides and Proteins*, Vol 2, pp. 363-369 (1984) (edited by W.S. Hancock, CRC Press, Boca Raton).
65. Kieft,K.A., Bocan,T.M.A. and Krause,B.R., *J. Lipid Res.*, **32**, 859-866 (1991).
66. King,R.J. and Clements,J.A., *J. Lipid Res.*, **11**, 381-385 (1970).
67. Knobler, H., Fainaru, M. and Sklan, D., *J. Chromatogr.*, **421**, 136-140 (1987).
68. Kodama,T., Akanuma,Y., Okazaki,M., Aburatani,H., Itakura,H., Takahashi,K., Sakuma,M., Takaku,F. and Hara,I., *Biochim. Biophys. Acta* , **752**, 407-415 (1983).
69. Kuehl,D.W. and Leonard,E.N., *Anal. Chem.*, **50**, 182-185 (1978).
70. Kulin,L.I., Flodin,P., Ellingsen,T. and Ugelstad,J., *J. Chromatogr.*, **514**, 1-9 (1990).
71. Kupranycz,D.B., Amer,M.A. and Baker,B.E., *J. Am. Oil Chem. Soc.*, **63**, 332-337 (1986).
72. Lee,T.J. and Smith,B.F., *J. Lipid Res.*, **30**, 491-498 (1989).
73. Lesieur,S., Grabielle-Madelmont,C., Paternostre,M.T. and Ollivon,M., *Anal. Biochem.*, **192**, 334-343 (1991).
74. Márquez-Ruiz,G., Pérez-Camino,M.C. and Dobarganes,M.C., *J. Chromatogr.*, **514**, 37-44 (1990).
75. Márquez-Ruiz,G., Pérez-Camino,M.C. and Dobarganes,M.C., *Fat Sci. Technol.*, **94**, 307-311 (1992).
76. Márquez-Ruiz,G., Pérez-Camino,M.C. and Dobarganes,M.C., *J. Am. Oil Chem. Soc.*, **69**, 930-934 (1992).
77. Mittelbach,M. and Trathnigg,B. in *Edible Fats and Oils Processing: Basic Principles and Modern Practices*, pp. 441-442 (1990) (edited by D.R. Erickson, AOCS, Champaign).
78. Mills, G.L., Lane, P.A. and Weech, P.K., in *Laboratory Techniques in Biochemistry and Molecular Biology, Vol 14: A Guidebook to Lipoprotein Technique*, pp. 18-116 (1984) (edited by R.H. Burdon & P.H. Knippenberg, Elsevier, Amsterdam).
79. Mulder,J.L. and Buytenhuys,F.A., *J. Chromatogr.*, **51**, 459-477 (1970).
80. Norman,A., *Proc. Soc. Exp. Biol. Med.*, **116**, 902-905 (1964).
81. Nozaky,Y., Lasic,D.D., Tanford,C. and Reynolds,J.A., *Science*, **217**, 366-367 (1982).
82. Nyyssönen,K. and Salonen,J.T., *J. Chromatogr.*, **570**, 382-389 (1991).
83. Okazaki,M., Ohno,Y. and Hara,I., *J. Biochem. (Tokyo)*, **89**, 879-887 (1981).
84. Okazaki,M., Shiraishi,K., Ohno,Y. and Hara,I., *J. Chromatogr.*, **223**, 285-293 (1981).
85. Okazaki,M., Hara,I., Tanaka,A., Kodama,T. and Yokoyama,S., *New England J. Med.*, **304**, 1068 (1981).
86. Okazaki,M., Hagiwara,N. and Hara,I., *J. Biochem. (Tokyo)*, **91**, 1381-1389 (1982).
87. Okazaki,M., Hagiwara,N. and Hara,I., *J. Chromatogr.*, **231**, 13-23 (1982).
88. Okazaki,M., Itakura,H., Shiraishi,K. and Hara,I., *Clin. Chem.*, **29**, 768-773 (1983).
89. Okazaki,M. and Hara,I., in *Handbook of HPLC for the Separation of Amino Acids, Peptides and Proteins*, Vol 2, pp. 393-403 (1984) (edited by W.S. Hancock, CRC Press, Boca Raton).
90. Ollivon,M., Walter,A. and Blumenthal,R., *Anal. Biochem.*, **152**, 262-274 (1986).
91. Parizcova,H., Pokorny,S. and Pokorny,J., *J. Chromatogr.*, **170**, 259-263 (1979).
92. Pattinson,N.R., Willis,K.E. and Frampton,C.M., *J. Lipid Res.*, **32**, 205-214 (1991).
93. Pérez-Camino,M.C., Márquez-Ruiz,G., Ruiz-Méndez,M.V. and Dobarganes,M.C., *Grasas y Aceites*, **41**, 366-370 (1990).
94. Pérez-Camino,M.C., Márquez-Ruiz,G., Ruiz-Méndez,M.V. and Dobarganes,M.C., in

Proceedings of Euro Food Chem VI, Hamburg, Vol 2, pp. 569-574 (1991) (edited by W. Baltes, T. Eklund, R. Fenwick, W. Pfannhauser, A. Ruiter, & H.P. Thier, Lebensmittelchemische Gesellschaft, Frankfurt)
95. Perkins,E.G., Taubold,R. and Hsieh,A., *J. Am. Oil Chem. Soc.*, **50**, 223-225 (1973).
96. Perrin,J.L., Redero,F. and Prevot,A., *Rev. Franc. Corps Gras*, **31**, 131-133 (1984).
97. Pfaffinger,D., Edelstein,C. and Scanu,A.M., *J. Lipid Res.*, **24**, 796-800 (1983).
98. Pokorny,J., Tai,P.T. and Janicek,G., *Riv. Ital. Sostanze Grasse*, **53**, 255-258 (1976).
99. Pokorny,S., Coupek,J., Luan,N.T. and Pokorny,J., *J. Chromatogr.*, **84**, 319-328 (1973).
100. Quarfordt,S.H., Nathans,A., Dowdee,M. and Hilderman,H.L., *J. Lipid Res.*, **13**, 435-444 (1972).
101. Reers,M., Elbracht,R., Rudel,H. and Spener,F., *Chem. Phys. Lipids*, **36**, 15-28 (1984).
102. Reynolds,J.A., Nozaki,Y. and Tanford,C., *Anal. Biochem.*, **130**, 471-474 (1983).
103. Reuben,A., Howell,K.E. and Boyer,J.L., *J. Lipid Res.*, **23**, 1039-1052 (1982).
104. Rhoden,V. and Goldin,S., *Biochemistry*, **18**, 4137-4176 (1979).
105. Ríos,J.J., Pérez-Camino,M.C., Márquez-Ruiz,G. and Dobarganes,M.C., *Food Chem.*, **44**, 357-362 (1992).
106. Ríos,J.J., Pérez-Camino,M.C., Márquez-Ruiz,G. and Dobarganes,M.C., in *Proceedings of the Symposium on Chemical Reactions in Foods*, pp. 238-243 (1992) (edited by J. Velísek, Prague).
107. Roos,A.H., van Munsteren,A.J., Nab,F.M. and Tuinstra, L.G.M.Th., *Anal. Chim. Acta*, **196**, 95-102 (1987).
108. Rudel,L.L., Lee,J.A., Morris,M.D. and Felts,J.M., *Biochem. J.*, **139**, 89-95 (1974).
109. Rudel,L.L., Pitts,L.L., and Nelson,C.A., *J. Lipid Res.*, **18**, 211-222 (1977).
110. Rudel,L.L., Marzetta,C.A. and Johnson,F.L., *Methods Enzymol.*, **129**, 45-57 (1986).
111. Sata,T., Estrick,D.L., Wood,P.D.S. and Kinsell,L.W., *J. Lipid Res.*, **11**, 331-340 (1970).
112. Sata,T., Havel,R.J. and Jones, A.L. *J. Lipid Res.*, **13**, 757-768 (1972).
113. Schulte,E., *Fette Seifen Anstrichm.*, **84**, 178-180 (1982).
114. Schurtenberger,P. and Hauser,H., *Biochim. Biophys. Acta*, **778**, 470-480 (1984).
115. Sebedio,J.L., Septier,Ch. and Grandgirard,A., *J. Am. Oil Chem. Soc.*, **63**, 1541-1543 (1986).
116. Shansky,R.E. and Kane,R.E., *J. Chromatogr.*, **589**, 165-170 (1992).
117. Shukla,V.K.S. and Perkins,E.G., *Lipids*, **26**, 23-26 (1991).
118. Sömjen,G.J. and Gilat,T., *FEBS Lett.*, **156**, 265-268 (1983).
119. Sömjen,G.J. and Gilat,T., *J. Lipid Res.*, **26**, 699-704 (1985).
120. Stalling,D.L., Tindle,R.C. and Johnson,J.L., *J. Assoc. Off. Anal. Chem.*, **55**, 32-38 (1972).
121. Stellwagen,E., *Methods Enzymol.*, **182**, 317-328 (1990).
122. Thureborn,E., *Nature (London)*, **197**, 1301-1302 (1963).
123. Ugelstad,J., Moerk,P.C., Herder Kaggerud,K., Ellingsen,T. and Berge,A., *Adv. Colloid Interface Sci.*, **13**, 101-140 (1980).
124. Unbehend,V.M., Scharmann,H., Strauss,H.J. and Billek,G., *Fette Seifen Anstrichm.*, **75**, 689-696 (1973).
125. van Gent,T. and van Tol,A., *J. Chromatogr.*, **525**, 433-441 (1990).
126. Vercaemst,R. Rosseneu,M. and Van Biervliet,J.P., *J. Chromatogr.*, **276**, 174-181 (1983).
127. Waltking, A.E. and Wessels, *J. Assoc. Off. Anal. Chem.*, **64**, 1329-1330 (1981).
128. Wehr,C.T., Cunico, .L., Ott,G.S. and Shore,V.G., *Anal. Biochem.*, **125**, 386-394 (1982).
129. White,P.J. and Wang,Y., *J. Am. Oil Chem. Soc.*, **63**, 914-920 (1986).
130. Williams,M.C., Kelley,J.L. and Kushwaha,R.S., *J. Chromatogr.*, **308**, 101-109 (1984).
131. Williams,M.C., Kushwaha,R.S. and McGill,H.C., *Lipids*, **22**, 366-374 (1987).
132. Williams,M.C. and Kushwaha,R.S., *J. Chromatogr.*, **433**, 257-263 (1988).
133. Wolff,J.P., Mordret,F.X. and Dieffenbacher,A., *Pure Appl. Chem.*, **63**, 1163-1171 (1991).
134. Young,P.M., Boehm,T.M. and Brown,J.E., *J. Chromatogr.*, **311**, 79-92 (1984).

Chapter 4

MERCURY ADDUCT FORMATION IN THE ANALYSIS OF LIPIDS

Jean Louis Sébédio

INRA, Station de Recherches sur la Qualité des Aliments de l'Homme, Unité de Nutrition Lipidique, 17 rue Sully 21034 Dijon Cédex.

A. INTRODUCTION

For a number of years, gas-liquid chromatography (GLC) has been the method of choice to analyse fatty acids from oils or biological samples [5,11]. The introduction of fused silica columns as well as new stationary phases now allows the separation and quantification of numerous naturally occurring fatty acids, including geometrical and positional isomers of mono-, di- and triethylenic fatty acids, mostly of C_{18} and C_{20} chain length [49,55,56].

However, the retention time of compounds on a given stationary phase, under the same conditions (carrier gas, temperature, *etc.*) depends on the chain length, the degree of unsaturation and the position and geometry of the ethylenic bond(s). The introduction of new phases and new types of columns can lead to excellent separations especially in the field of fish oil fatty acids where as many as 60 different fatty acids can be separated and quantified [4]. This can at the same time be a major drawback of GLC when complex mixtures of *cis* and *trans*, methylene and non-methylene interrupted, dienes and trienes have to be analysed. Partially hydrogenated oils, especially those of marine origin, come into this category [35,37,43]. The presence of hundreds of peaks results in large overlaps of isomers and gives unreliable quantitative analyses even on capillary GLC columns [43].

It is then necessary to use prefractionation techniques prior to GLC analyses in order to quantify the different fatty acid families (monoene, diene, triene and so forth). Silver-nitrate thin layer chromatography (TLC) has always been a powerful tool, as fractionation on the TLC plate depends upon the degree of unsaturation, the number of carbon atoms, and also the geometry and position of the ethylenic bond(s) [18]. Another possibility is to use silver-ion high performance liquid chromatography [14]. When highly complex mixtures have to be analysed using these techniques, it is not possible to obtain in one fraction molecules which only differ by the degree of unsaturation.

This goal can however be achieved by derivatization of the unsaturated fatty acids into mercury adducts. The mercury adducts are then further separated by TLC to give fractions which can be analysed by GLC. One must not forget that mercury adducts are not only used as a prefractionation step; they have been used as a preparative tool as well as derivatives for mass spectral identification. This chapter will describe their major applications in the lipid field.

B. PREPARATION OF UNSATURATED FATTY ACID CONCENTRATES

The reaction between mercuric acetate and olefinic bonds in methanol giving acetoxymercuri-methoxy derivatives was first applied to lipids in the early 1950's.

$$R{-}CH{=}CH{-}R \xrightarrow[CH_3OH]{Hg(OAc)_2} R{-}\underset{\displaystyle |}{\overset{CH_3O}{CH}}{-}\underset{}{\overset{HgOAc}{CH}}{-}R$$

The original double bond can be regenerated by reaction with aqueous acid with no double bond migration or *cis-trans* isomerization.

At present, only a little use has been made of mercuric derivatives for the isolation of specific unsaturated fatty acids from complex mixtures [32,51,54]. However, Stearn *et al.* [51] described a procedure to isolate methyl linoleate of 95% purity and linolenate of 90% purity from natural oils (sunflower and linseed oils) after reacting the fatty acid methyl esters with mercuric acetate in methanol. As an extension of this work, White and Quackenbush [54] obtained methyl linolenate (99% purity) by partitioning the mercuric adducts between 10% methanol in water and diethyl ether.

The possibility of applying this reaction to the highly unsaturated fatty acids present in fish oils such as menhaden oil was also studied. However the method looked tedious and the yields were low (about 15%). Since this work and in consideration of the increased interest in long-chain polyunsaturated fatty acids, methods have been developed to prepare fish oil concentrates containing high amounts of eicosapentaenoic and docosahexaenoic acids [3,50].

C. ANALYSIS OF OILS, FATTY ACIDS AND RELATED MOLECULES

1. Unsaturated fatty acids

Mercury adduct formation has been more widely applied as an analytical than a preparative tool, especially to complex fatty acid mixtures, such as marine oil fatty acid methyl esters. For example, this technique permitted concentration of minor polyunsaturated fatty acids which would not be detected directly or quantified by GLC analysis *per se.* Different derivatives have been proposed (acetoxy-mercuri- and methoxyhalogenomercuri-adducts) [15,22,26,31,52,53]. A disadvantage in the use of methoxyacetoxymercury adducts is that they are very polar compared to the parent esters [26]. Conversion of methoxyacetoxymercury adducts to methoxybromomercury derivatives reduces the polarity of the molecule and gives an improvement in the chromatographic properties [52]. Consequently, the methoxybromomercury derivatives have been the more widely utilized.

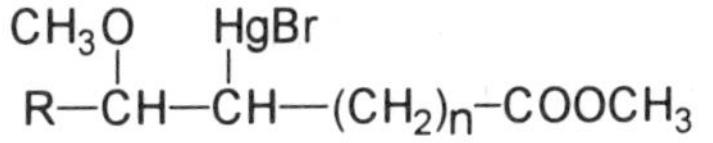

$$R\text{—}\underset{\displaystyle |}{\overset{CH_3O}{}}\!\!CH\text{—}\overset{HgBr}{CH}\text{—}(CH_2)_n\text{—}COOCH_3$$

The procedure in quite simple [52]. It consists of first reacting the fatty acid methyl esters with mercuric acetate in methanol. The resultant acetoxymercuri-adducts are then converted to the bromo-derivatives by reacting the mixture with an excess of sodium bromide in methanol. Then the derivatives can be fractionated by TLC, and after extraction of the bands, the parent esters can be recovered by reaction of the adducts with hydrochloric acid. The intention in this type of fractionation is generally to find a suitable elution solvent where the migration of the adducts depends only on the degree of unsaturation, and not on the geometry or the position of the ethylenic bond.

The extent of mercuration has been tested on fatty acids differing in chain length, degree of unsaturation and geometry of the ethylenic bond (Table 4.1). High mercuration yields were obtained with oleate, linoleate and linolenate, compared to a little lower yield when increasing the degree of unsaturation (about 95%). However, one must be careful as under different reaction conditions to those described by White [52], the *cis*-18:1 isomer reacted faster than the *trans* one. Under typical experimental

Table 4.1
Extent of mercuration (from H.B. White [52]).

Fatty acid	% Reaction
18:0	0.0
cis-9-18:1	100.0
trans-9-18:1	99.6
13-22:1	97.8
9,12-18:2	99.0
9,12,15-18:3	97.0
5,8,11,14-20:4	94.7
5,8,11,14,17-20:5	95.8
4,7,10,13,16,19-22:6	95.1

conditions [52], most unsaturated fatty acids can be separated as a function of the degree of unsaturation (18:0 (R_f = 0.73), 18:1*c* (R_f = 0.63), 18:2(*n*-6) (R_f = 0.47), 18:3(*n*-3) (R_f = 0.30), 20:4(*n*-6) (R_f = 0.19), 20:5(*n*-3) (Rf=0.07), and 22:6(*n*-3) (R_f = 0.01)). In fact, the separation between 5 and 6 ethylenic bonds is usually very tedious when mixtures of isomers are present.

This type of fractionation coupled with GLC analysis of the fractions has been used to study fish oil methyl esters [53], animal tallows [15], human blood plasma [53] and C_{20} polyunsaturated fatty acids [40], for example.

As outlined in Figure 4.1 and in Table 4.2, this technique was used to identify all the isomers formed after a hydrazine reduction of 20:5(*n*-3) [40]. A total of 32 fatty acids were separated into six fractions; one saturated component, a monoenoic faction having 5 isomers, a dienoic with 10 isomers, a trienoic with 10 isomers, a tetraenoic with 5 isomers and the unreacted eicosapentaenoic acid. In this instance, the combination of mercury adduct fractionation and GLC analysis of the resulting fractions was the only way to study such a complex mixture considering the overlap between the monoenes, dienes, trienes and tetraenes on GLC analysis.

2. *Partially hydrogenated oils*

Partial hydrogenation has been the research field where the fractionation of methoxymercuri-adducts has been most widely used [20,23,41-43,45]. One effect of partial hydrogenation is to transform the natural *cis* unsaturated acids into *trans* isomers with simultaneous formation of positional isomers. As far as vegetable oils are concerned, where the most unsaturated fatty acid is linolenic acid, partially hydrogenated oils can be studied by a combination of GLC and silver-nitrate TLC. However, this is not the case for marine oils, which are rich in long-chain polyunsaturated fatty acids and give complex mixtures of C_{20} and C_{22} positional and geometrical isomers upon partial

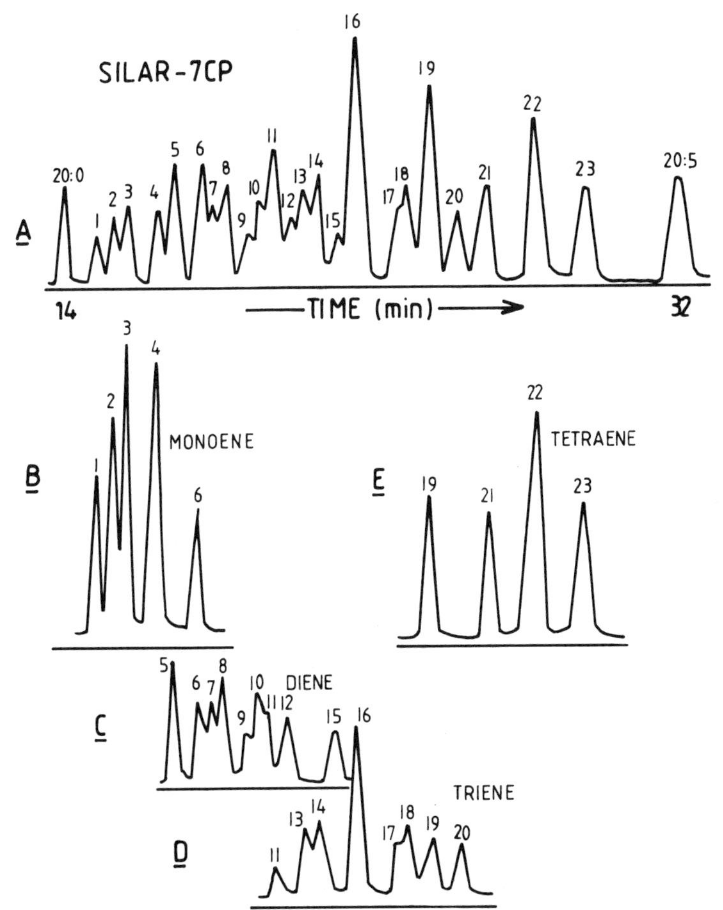

Fig. 4.1. Complete C_{20} section of GLC chart for methyl esters of products from partial reduction of 5,8,11,14,17- eicosapentaenoic acid (A), with corresponding isolates of monoenes (B), dienes (C), trienes (D) and tetraenes (E); Column: Wall-coated stainless steel, 47 m x 0.25 mm i.d., liquid phase SILAR-7CP, operated at 160°C. (Reproduced by kind permission of the *Journal of Chromatographic Science* and redrawn from the original paper [40]).

hydrogenation [42,43]. Marine oils can be divided into two groups, those such as herring oil which contain large quantities of long-chain monounsaturated fatty acids, and those such as menhaden oil which are rich in the C_{20} and C_{22} polyunsaturated fatty acids with five and six ethylenic bonds [42,43]. As for vegetable oils, the low-iodine value partially hydrogenated marine oils (Figure 4.2) can be studied using silver nitrate-TLC and GLC analysis, since partial hydrogenation results in the formation of large quantities of positional and geometrical monoethylenic isomers only [47]. In any case, it is not possible to quantify precisely the amount of dienoic and trienoic fatty acids, as shown in Figure 4.2.

This is not true for the hydrogenation of high-iodine value marine oil [42,43]. For menhaden oil (Figure 4.3), as the iodine value decreased from

Table 4.2

Comparison of experimental and calculated equivalent chain-length (ECL) values and fractional chain-length (FCL) values for the hydrazine reaction products of all-*cis* 5,8,11,14,17-20:5 on Silar-7CP at 160°C (from J.L. Sebedio and R.G. Ackman [40]).

	Diene n	Peak No. (Fig. 4.1)	Experimental ECL	FCL	Calc. ECL - Exp. ECL
Monoenes					
Δ 5	-	1	20.23	0.23	-
Δ 8	-	2	20.32	0.32	-
Δ 11	-	3	20.37	0.37	-
Δ 14	-	4	20.53	0.53	-
Δ 17	-	6	20.74	0.74	-
					Calc. ECL
Dienes					
Δ 5,8	1	5	20.60	20.55	-0.05*
Δ 5,11	4	5	20.60	20.60	0.00
Δ 5,14	7	6	20.74	20.76	0.02
Δ 8,11	1	7	20.80	20.69	-0.11*
Δ 8,14	4	8	20.85	20.85	0.00
Δ 5,17	10	9	20.96	20.97	0.01
Δ 11,14	1	10	21.00	20.90	-0.10*
Δ 8,17	7	11	21.06	21.06	0.00
Δ 11,17	4	12	21.12	21.11	-0.01*
Δ 14,17	1	15	21.32	21.27	-0.05*
Trienes					
Δ 5,8,11	-	11	21.06	21.08	0.02
Δ 5,8,14	-	13	21.19	21.13	-0.06
Δ 5,11,14	-	14	21.25	21.23	-0.02
Δ 5,8,17	-	16	21.36	21.43	0.07
Δ 5,11,17	-	16	21.36	21.34	-0.02
Δ 8,11,14	-	16	21.36	21.35	-0.01
Δ 5,14,17	-	17	21.53	21.54	0.01
Δ 8,11,17	-	18	21.57	21.55	-0.02
Δ 8,14,17	-	19	21.65	21.64	-0.01
Δ 11,14,17	-	20	21.77	21.79	-0.02
Tetraenes					
Δ 5,8,11,14	-	19	21.65	21.69	0.04
Δ 5,8,11,17	-	21	21.83	21.81	-0.02
Δ 5,8,14,17	-	22	21.98	21.98	0.00
Δ 5,11,14,17	-	22	21.9	22.04	0.06
Δ 8,11,14,17	-	23	22.14	22.15	0.01

*Diene adjustment for Δ 5,8, Δ 8,11, Δ 11,14 and Δ 14,17 respectively

159 in the starting oil to 84.5 in the hydrogenated oil, the overlap between the different 20:1, 20:2 and 20:3 isomers became very important. A substantial change was obtained when lowering the iodine value from 131.5 to 96.5. Only the utilization of the methoxybromomercuri-adduct permitted quantification of the different fatty acid classes. A small overlap only occurred between the monoenes and dienes (Figure 4.4), while the overlap between the dienes and trienes was quite appreciable. The latter could easily be explained from a consideration of the number of *cis* and *trans* dienoic and trienoic isomers. Starting with the major C_{20} polyunsaturated fatty acids (20:2, 20:3, 20:4 and 20:5), the positions of the

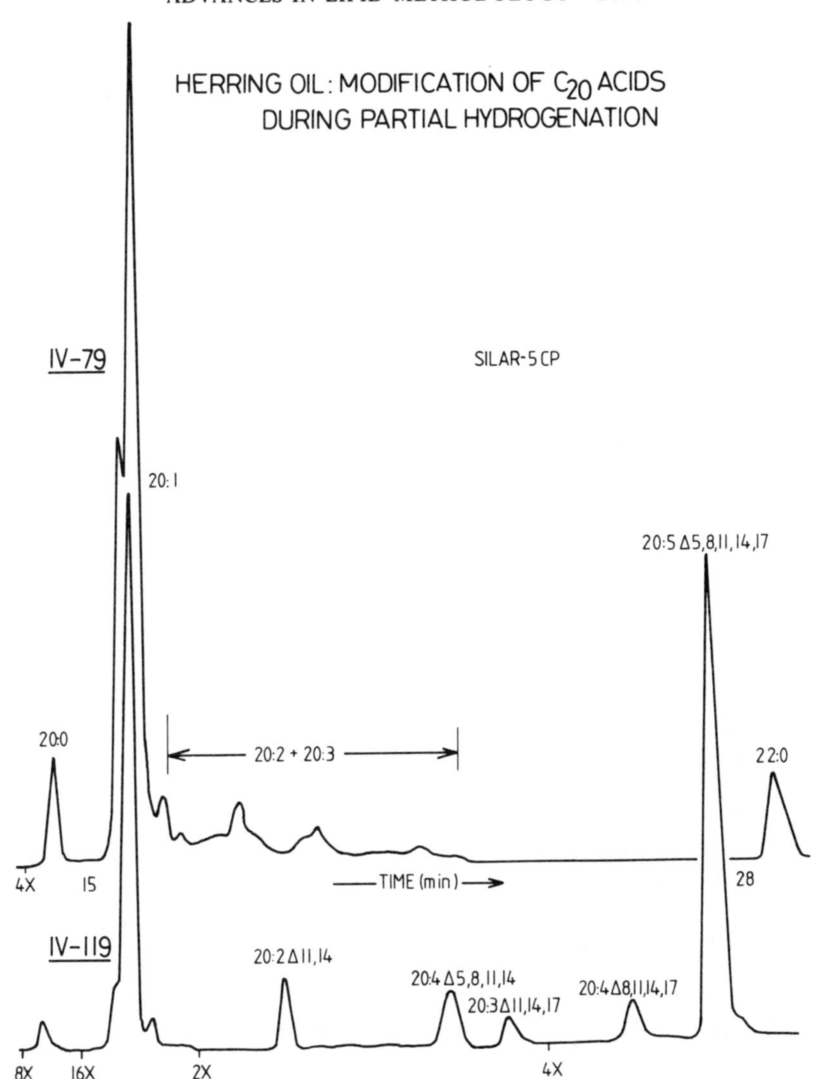

Fig. 4.2. Comparison of parts of GLC charts for fatty acid methyl esters from refined herring oil (below) and from partially hydrogenated herring oil of iodine value 79 (above). Column, open-tubular, 47 m x 0.25 mm i.d., 180°C, SILAR-5CP. (Reproduced by kind permission of the *Journal of the American Oil Chemists' Society* and redrawn from the original paper [47]).

ethylenic bond of the *cis* and *trans* configuration in the dienes and trienes ranged from the Δ3 to the Δ18 positions [43]. These studies on partially hydrogenated marine oils were carried out using the method described by White [52] after a few modifications [42], necessary to obtain a complete reaction of mercuric acetate with the isomeric polyunsaturated fatty acids found in fish oils. Table 4.3 represents the comparison between the direct GLC analysis of a refined menhaden oil with that reconstituted from the methoxybromomercuri-adduct (MBM) fractionation. A fairly good

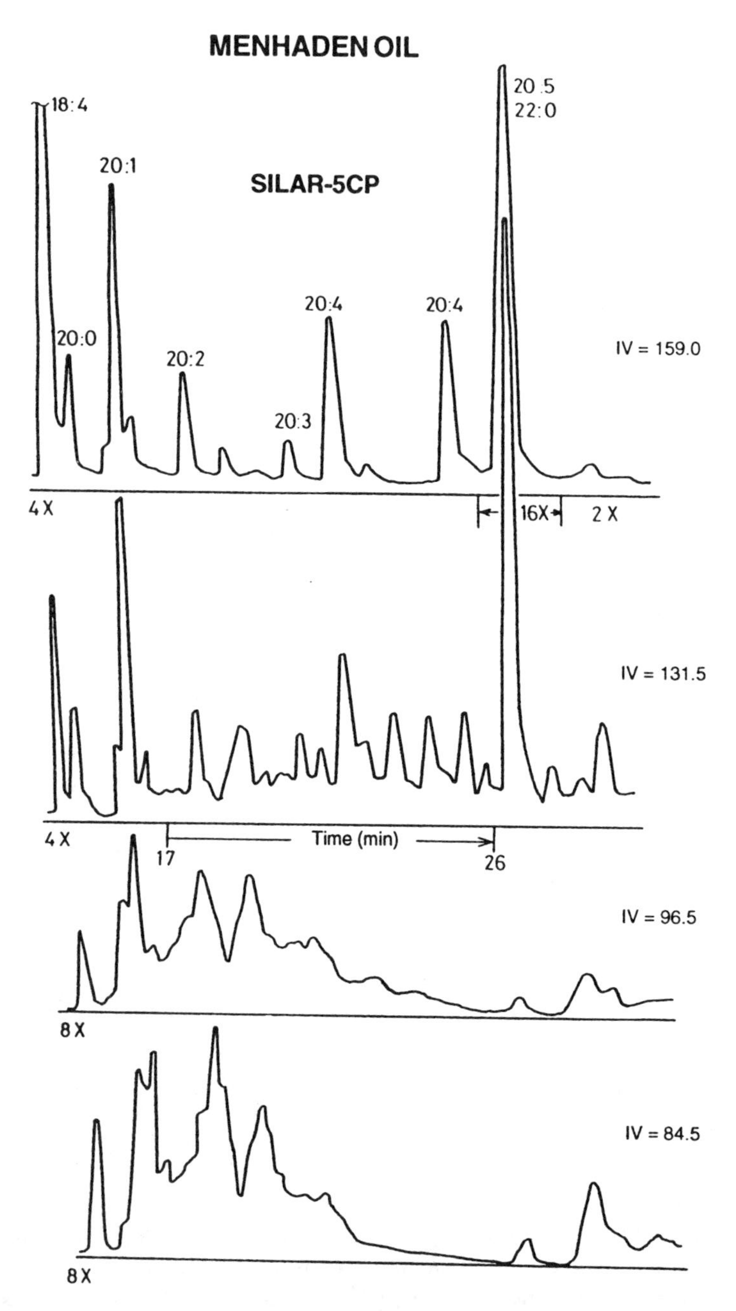

Fig. 4.3. Comparison of parts of GLC charts for fatty acid methyl esters from refined and bleached menhaden oil (IV = 159.0) and from three partially hydrogenated menhaden oils with iodine values of 131.5, 96.5 and 84.5. Column: open-tubular, 47 m x 0.25 mm i.d., 175°C, SILAR-5CP. (Reproduced by kind permission of the *Journal of the American Oil Chemists' Society* and redrawn from the original paper [43]).

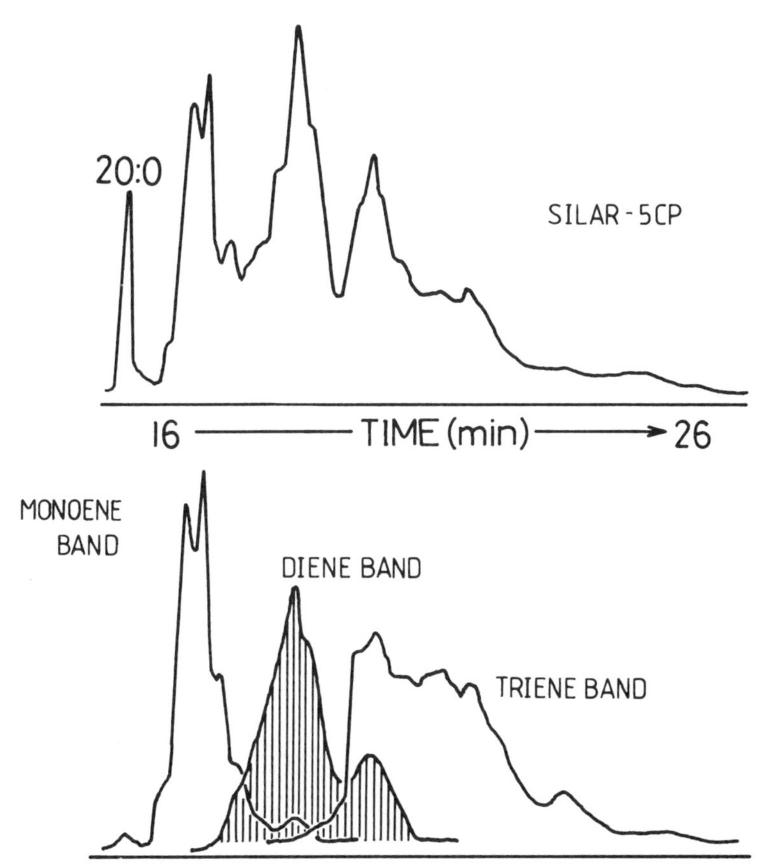

Fig. 4.4. Gas-liquid chromatograms of methyl esters of total C_{20} fatty acids (above) from partially hydrogenated menhaden oil (IV = 84.5), and of respective individual bands resulting from MBM-adduct fractionation (below), on SILAR-5CP. (Reproduced by kind permission of *Lipids* and redrawn from the original paper [42]).

agreement was obtained. However, care is required when carrying out the experiments, as a minor modification of the procedure could lead to incomplete reaction and erroneous results.

The major drawback of this method is that it can only be applied to samples which contain minor quantities of conjugated fatty acids. Only about 70% the original conjugated acids were recovered after MBM-adduct formation compared to oleic or linoleic acids. Furthermore, this loss was nonselective as the GLC profiles of the conjugated acids before and after MBM fractionation were different (Figure 4.5) [27]. The same phenomenon was also observed by Plank *et al.* [29] when reacting mercuric acetate with methyl α-eleostearate. It was concluded that the addition of mercuric acetate to the conjugated system was followed by

Table 4.3
Comparison of the fatty acid compositions of refined and of partially hydrogenated menhaden oils (PHMO) by direct GLC analysis and following MBM-adduct fractionation. Adapted from J.L. Sebedio and R.G. Ackman [42].

Fatty acid	Refined		PHMO[d]
	GLC	After MBM[d]	
14:0	10.8	9,8	10.5
16:0	23.2	21.6	24.1
18:0	4.2	4.0	5.2
20:0	0.4	0.3	0.7
22:0	trace	0.1	0.3
others[a]	3.0	3.2	3.0
Σ saturates	41.6	39.0	43.8
16:1	11.4	13.0	15.0
18:1	10.5	11.3	12.5
20.1	1.3	1.5	4.9
22:1	0.2	0.3	1.7
others[b]	0.5	0.2	0.1
Σ monoenes	23.9	26.3	34.2
C16 dienes	1.5	1.6	0.9
C18 dienes	1.8	2.3	2.4
C20 dienes	0.6	0.3	6.6
C22 dienes	-	-	3.3
Σ dienes	3.9	4.2	13.2
C16 trienes	2.2	2.1	-
C18 trienes	1.7	1.7	0.2
C20 trienes	0.3	0.2	3.8
C22 trienes	-	-	4.3
Σ trienes	4.2	4.0	8.3
C16 tetraenes	1.0	1.2	-
C18 tetraenes	2.1	2.4	-
C20 tetraenes	2.3	2.1	0.1
C22 tetraenes	0.2	0.1	0.3
Σ tetraenes	5.6	5.8	0.4
5,8,11,14,17-20:5	11.9	12.2	-
6,9,12,15,18-21:5	0.5	0.4	-
C22 pentaenes + hexaenes	8.8	8.6	trace
Σ conjugated acids[c]	0.7	0.7	0.5

[a]Includes 12:0, 19:0, *iso*- and *anteiso*-15:0, *iso*-16:0 and *iso*-18:0
[b]Includes 7-methyl-16:1, 17:1, 19:1 and 24:1
[c]AOCS official method Cd-7-58
[d]Average of two analyses

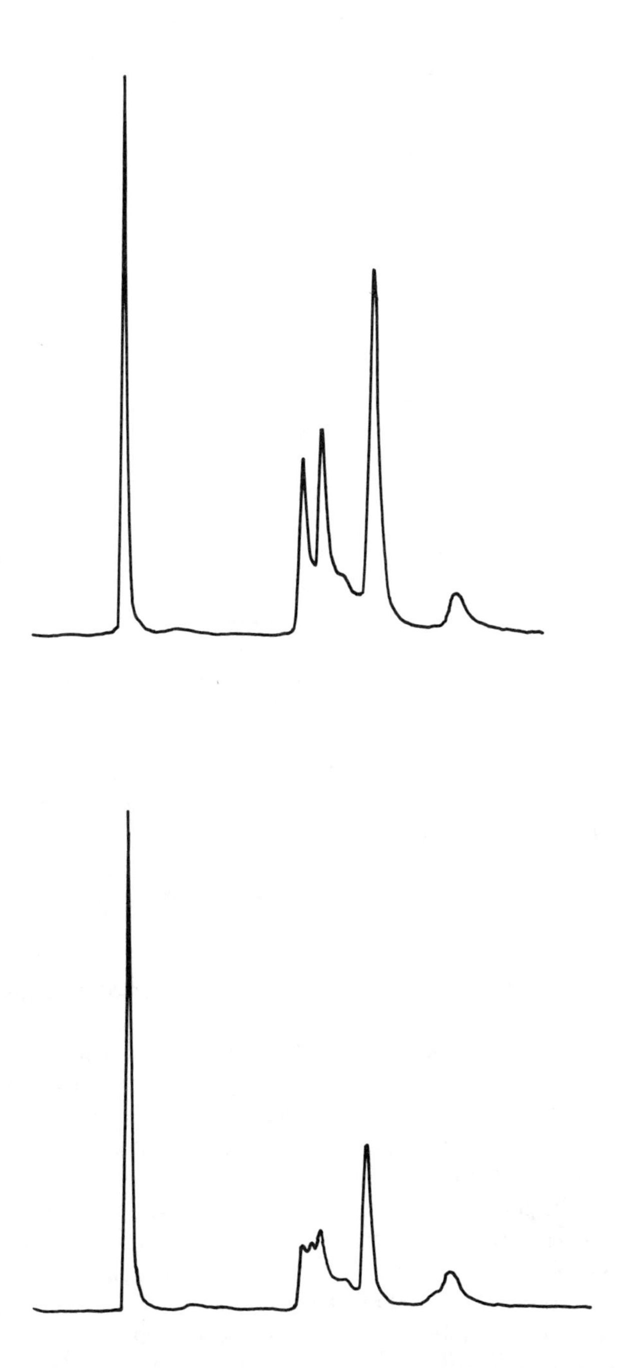

Fig. 4.5. GLC analysis of a mixture of 18:2(*n*-6) and of C_{18} synthetic conjugated acids before (above) and after mercury adduct fractionation.

decomposition, giving a compound of lower mercury content than expected. Despite this loss, bromomercuri-adduct fractionation is applicable without restriction to oils containing low proportions of conjugated acids.

3. Other components

Mercury adduct fractionation has been applied to other types of components such as triglycerides [9,28], lecithins [10], saturated and branched chain fatty acids [16,21] and cyclic fatty acids from heated fats and oils [38].

Some investigations were also carried out in order to improve the original method using thin-layer chromatography to fractionate mercury adducts of the esters. For example, in order to isolate saturated fatty acids, Johnson [21] used an ion exchange resin which could readily separate the saturates and the monoethylenic fatty acids, while diene and triene adducts were removed from the column by passing methanolic hydrogen chloride through it, followed by methanol. This technique is of limited use to fractionate fatty acid methyl ester mixtures, but it seemed to be suitable for the analysis of large numbers of samples and for isolation of saturated and monoethylenic fatty acids.

Another interesting application was the utilization of the TLC/flame ionization detection technique (TLC/FID) to the analysis of triglycerides according to the degree of unsaturation [28]. This instrumental method combines the efficiency of thin-layer chromatography with the ease of detection by the FID, as used for GLC analyses [2,6]. In this instance, the fractionation of the mercury acetate adducts is carried out on silica gel sintered rods. Analyses were performed on palm oil, shea butter and mango kernel oil. When more detailed compositions of triglyceride mixtures are needed, one would need to use HPLC, although this TLC/FID technique is quite useful for fractionation of triglycerides as a function of the degree of unsaturation. However, great care must be taken for the quantification, as the response of the detector is not linear over a wide range of sample loads. It was therefore necessary to find the range of linearity of different model substances, as the response factor was found to depend on the degree of unsaturation of the molecule. A further conclusion was that fats and oils with a high content of polyunsaturated triglycerides would not be suitable for analysis by this technique because of a relatively poor separation. However, this study was effected in 1982 with an older TLC/FID Iatroscan model and the SII rods. The new generation of instruments as well as the new chromarods might permit better separations to be obtained [48].

Mercury adduct fractionation has also been used to study unusual fatty acids, such as cyclic fatty acids [38], which are formed from linoleic and linolenic acids mainly during heat treatment of fats and oils [46]. Those formed from linoleic acid are C_{18} disubstituted fatty acids having one ethylenic bond while those formed from linolenic acid are diunsaturated isomers [46]. TLC fractionation of the mercuric adducts of total fatty acid methyl esters from a heated linseed oil gave 5 major bands (Figure 4.6), a triene band of R_f = 0.58, a diene band of R_f = 0.82, a monoene band of R_f = 0.87, a saturated band of R_f = 0.90, and a band of R_f = 0.78 which contained a complex mixture of unknown components. The analyses of each band after total hydrogenation showed that the cyclic fatty acids were mainly distributed in three bands, the triene band, the diene band and one between the diene and triene bands (Table 4.4). The cyclic fatty acids are now known to be a mixture of *cis,trans* diethylenic isomers [46], so that this study indicates that the migration of the adducts of cyclic fatty acids is influenced by more than the degree of unsaturation of the molecule. However, the knowledge of the structures of cyclic fatty acids was not yet sufficient to elucidate which factors influence the formation and migration of the adducts. Some further experiments may permit the use of this fractionation technique in combination with other methods such as silver ion HPLC [14] to obtain simple fractions for the structure elucidation of complex cyclic fatty acid mixtures present in heated fats and oils.

D. STRUCTURAL ELUCIDATION OF UNSATURATED FATTY ACIDS

The location of an ethylenic bond in a molecule can be determined by different techniques. For monoethylenic fatty acids, the position of the ethylenic bond is obtained after ozonolysis [1] and analysis of the fragments either by GLC or by GLC coupled with mass spectrometry (GC-MS). This technique can also be applied to diethylenic fatty acids [7,33,39]. The analysis of polyunsaturated fatty acids is much more time-consuming and requires the utilization of a combination of chemical methods such as hydrazine reduction, mercury adduct fractionation and silver-nitrate TLC prior to ozonolysis [36,44]. An alternative is the use of GLC coupled to mass spectrometry. Numerous derivatives have been proposed to elucidate the structures of the unsaturated fatty acids [8,12,13,19,34].

A rapid procedure using mercuration-demercuration reactions has also been developed to locate double bonds [17,24,25,30]. This involved the reaction of the ester with an equi-molar amount of mercuric acetate in methanol followed by a sodium borohydride reduction of the mercury adducts as outlined in Figure 4.7. This yields methoxylated fatty acid methyl esters, readily isolable by TLC. The ethylenic bonds are then located from the mass spectra of the derivatives (Figure 4.8), which are characterized by intense peaks due to a cleavage adjacent to methoxyl groups.

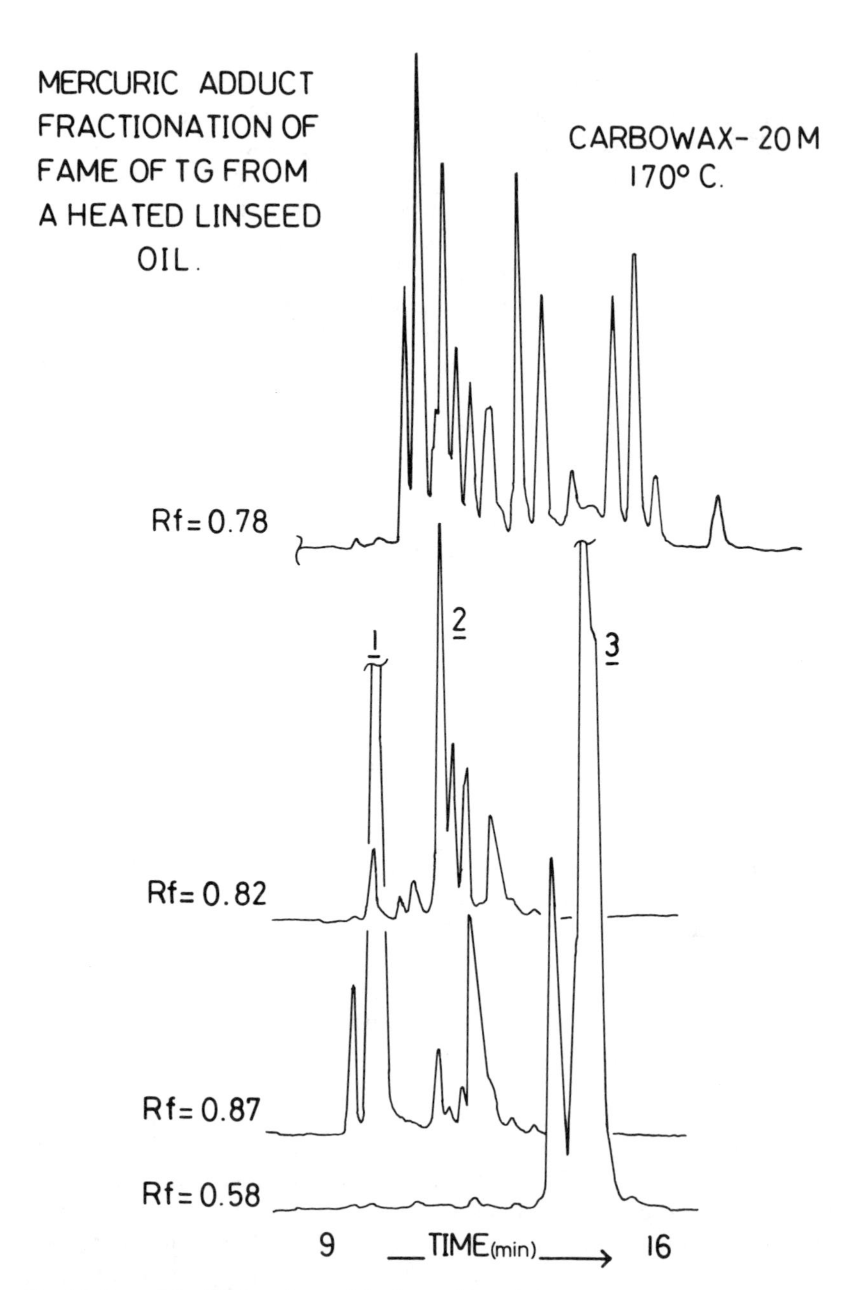

Fig. 4.6. Part of the GLC analyses of the fatty acid methyl esters of four MBM adduct bands obtained from the triglycerides isolated from a heated linseed oil (275°C for 12 hours under N_2). Column: Carbowax 20-TA, 25 m x 0.30 mm i.d., 170°C. (Reproduced by kind permission of *Fette Seifen Anstrichmittel* and redrawn from the original paper [38]).

Table 4.4
Percentage of cyclic fatty acid methyl esters (CFAM) (as of total CFAM) in the MBM adduct fractions obtained from a heated linseed oil. Adapted from J.L. Sébédio [38].

MBM bands R_f (Fig 4.6)	%CFAM
0.87	11.2
0.82	38.9
0.78	40.6
0.63	4.8
0.58	4.5

$$R_1-CH=CH-R_2 \xrightarrow[MeOH]{Hg(OAc)_2} \begin{matrix} R_1-CH(OMe)-CH(HgOAc)-R_2 \\ R_1-CH(HgOAc)-CH(OMe)-R_2 \end{matrix} \xrightarrow{NaBH_4} \begin{matrix} R_1-CH(OMe)-CH_2-R_2 \\ R_1-CH_2-CH(OMe)-R_2 \end{matrix}$$

Fig. 4.7. Formation of methoxylated esters. (Reproduced by kind permission of the authors and of *Lipids* and redrawn from the original paper [25]).

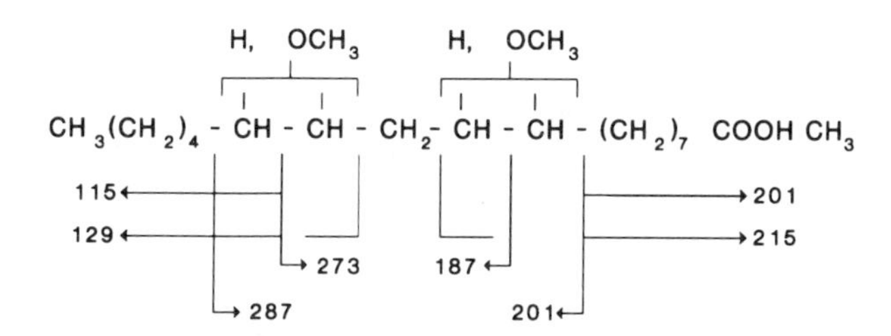

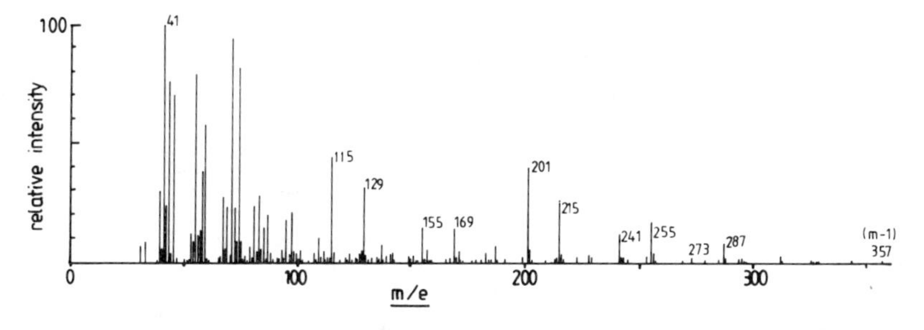

Fig. 4.8. Mass spectrum of methoxylated esters derived from methyl linoleate by methoxy-mercuration/demercuration. (Reproduced by kind permission of the authors and of *Lipids* and redrawn from the original paper [25]).

E. CONCLUSIONS

Mercury adduct formation has been a very useful technique for the analysis of lipids. It can be used as a preparative tool and to make derivatives for GC-MS analysis but its major application is still to fractionate complex mixtures containing positional and geometrical fatty acid isomers. Of particular value is the property that the separation can be largely independent of the geometry of the double bonds. The technique also has some potential for analysis of unusual fatty acids such as those formed during heat treatment of fats and oils. Although many new and interesting methods have been developed in recent years, mercury adducts still have appreciable value for lipid analysts.

Abbreviations

CFAM, cyclic fatty acids methyl esters; ECL, equivalent chain-length values; FCL, fractional chain-length values; GLC, gas-liquid chromatography; IV, iodine value; MBM, methoxybromomercury; TLC/FID, thin-layer chromatography/flame ionization detection.

REFERENCES

1. Ackman,R.G., *Lipids*, **12**, 293-294 (1977).
2. Ackman,R.G., *Methods Enzymol.*, **72**, 205-252 (1981).
3. Ackman,R.G., *Chem. Ind.* (*London*), 139-145 (1988).
4. Ackman,R.G., in *Marine Biogenic Lipids, Fats and Oils, Vol. I*, pp. 103-137 (1989) (edited by R.G. Ackman, CRC Press, Boca Raton, Florida).
5. Ackman,R.G., in *Analyses of Fats, Oils and Lipoproteins*, pp. 270-300 (1991) (edited by E.G. Perkins, American Oil Chemists'Soc., Champaign, IL).
6. Ackman,R.G., McLeod,C.A. and Banerjee,A.K., *J. Planar Chromatogr.*, **3**, 450-490 (1990).
7. Ackman,R.G., Sebedio,J.L. and Ratnayake,W.N., *Methods Enzymol.*, **12**, 253-276 (1981).
8. Andersson,B.A. and Holman,R.T., *Lipids*, **9**, 185-190 (1974).
9. Banks,W., Christie,W.W., Clapperton,J.L. and Girdler,A.K., *J. Sci. Food Agric.*, **39**, 303-316 (1987).
10. Blank,M.L., Nutter,L.J. and Privett,O.S., *Lipids*, **1**, 132-135 (1966).
11. Christie,W.W., *Gas Chromatography and Lipids* (1989) (The Oily Press, Ayr, Scotland).
12. Christie,W.W., Brechany,E.Y., Gunstone,F.D., Lie Ken Jie,M.S.F. and Holman,R.T., *Lipids*, **22**, 664-666 (1987).
13. Christie,W.W., Brechany,E.Y. and Holman,R.T., *Lipids*, **22**, 224-228 (1987).
14. Christie,W.W., Brechany,E.Y. and Stefanov,K., *Chem. Phys. Lipids*, **46**, 127-135 (1988).
15. Craske,J.D. and Edwards,R.A., *J. Chromatogr.*, **53**, 253-261 (1970).
16. Duncan,W.R.H., Lough,A.K., Garton,G.A. and Brooks,P., *Lipids*, **9**, 669-673 (1974).
17. Gunstone,F.D. and Inglis,R.P., *Chem. Phys. Lipids*, **10**, 73-88 (1973).
18. Gunstone,F.D., Ismail,I.A. and Lie Ken Jie,M., *Chem. Phys. Lipids*, **1**, 376-385 (1967).
19. Harvey,D.J., *Biomed. Mass Spectrom.*, **11**, 340-347 (1984).
20. Holmer,G. and Aaes-Jorgensen,E., *Lipids*, **4**, 507-514 (1969).
21. Johnson,C.B., *J. Am. Oil Chem. Soc.*, **66**, 935-937 (1989).
22. Kuemmel,D.F., *Anal. Chem.*, **34**, 1003-1007 (1962).
23. Lambertsen,G., Myklestad,H. and Braekkan,O.R., *J. Food Sci.*, **31**, 48-52 (1966).
24. Minnikin,D.E., *Lipids*, **10**, 55-56 (1975).
25. Minnikin,D.E., Abley,P., Mcquillin,F.J., Kusamran,K., Masken,K. and Polgar,N., *Lipids*, **9**, 135-140 (1974).

26. Minnikin,D.E. and Smith,S., *J. Chromatogr.*, **103**, 205-207 (1975).
27. Mounts,T.L., Dutton,H.J. and Glover,D., *Lipids*, **5**, 997-1005 (1970).
28. Petersson,B., *J. Chromatogr.*, **242**, 313-322 (1982).
29. Planck,R.W., O'Connor,R.T. and Goldblatt,L.A., *J. Am. Oil Chem. Soc.*, **33**, 350-353 (1956).
30. Plattner,R.D., Spencer,G.F. and Kleinman,R., *Lipids*, **11**, 222-227 (1976).
31. Pohl,P., Glasl,H. and Wagner,H., *J. Chromatogr.*, **42**, 75-82 (1969).
32. Privett,O.S., *Prog. Chem. Fats Other Lipids*, **9**, 409-452 (1971).
33. Ratnayake,W.M.N. and Ackman,R.G., *Lipids*, **14**, 580-584 (1979).
34. Ratnayake,W.M.N. and Ackman,R.G., in *The Role of Fats in Human Nutrition*, pp. 515-565 (1989) (edited by A.J. Vergroesen & M. Crawford, Academic Press, New York).
35. Ratnayake,W.M.N. and Beare-Rogers,J.L., *J. Chromatogr. Sci.*, **28**, 633-639 (1990).
36. Ratnayake,W.M.N., Grossert,J.S. and Ackman,R.G., *J. Am. Oil Chem. Soc.*, **67**, 940-946 (1990).
37. Ratnayake,W.M.N. and Pelletier,G., *J. Am. Oil Chem. Soc.*, **69**, 95-105 (1992).
38. Sebedio,J.L., *Fette Seifen Anstrichm.*, **87**, 267-273 (1985).
39. Sebedio,J.L. and Ackman,R.G., *Can. J. Chem.*, **56**, 2480-2485 (1978).
40. Sebedio,J.L. and Ackman,R.G., *J. Chromatogr. Sci.*, **20**, 231-234 (1982).
41. Sebedio,J.L. and Ackman,R.G., *Fette Seifen Anstrichm.*, **85**, 339-346 (1983).
42. Sebedio,J.L. and Ackman,R.G., *Lipids*, **16**, 461-467 (1981).
43. Sebedio,J.L. and Ackman,R.G., *J. Am. Oil Chem. Soc.*, **60**, 1986-1991 and 1992-1996 (1983).
44. Sebedio,J.L., Farquharson,T.E. and Ackman,R.G., *Lipids*, **17**, 469-475 (1982).
45. Sebedio,J.L., Finke,G. and Ackman,R.G., *Chem. Phys. Lipids*, **34**, 215-225 (1984).
46. Sebedio,J.L. and Grandgirard,A., *Prog. Lipid Res.*, **28**, 303-336 (1989).
47. Sebedio,J.L., Langman,M.F., Eaton,C.A. and Ackman,R.G., *J. Am. Oil Chem. Soc.*, **58**, 41-48 (1981).
48. Sebedio,J.L. and Juaneda,P., *J. Planar Chromatogr.*, **1**, 35-41 (1991).
49. Shantha,N.C. and Ackman,R.G., *Lipids*, **26**, 237-239 (1991).
50. Stansby,M.E. in *Fish oils in Nutrition*,pp. 73-119 (1990) (edited by M.E. Stansby, Van Nostrand Reinhold, New York).
51. Stearns,E.M., White,H.B. and Quackenbush,F.W., *J. Am. Oil Chem. Soc.*, **39**, 61-62 (1962).
52. White,H.B., *J. Chromatogr.*, **21**, 213-222 (1966).
53. White,H.B. and Powell,S.S., *J. Chromatogr.*, **32**, 451-457 (1968).
54. White,H.B. and Quackenbush,F.W., *J. Am. Oil Chem. Soc.*, **39**, 517-519 (1962).
55. Wijesundera,R.C., Ratnayake,W.M.N. and Ackman,R.G., *J. Am. Oil Chem. Soc.*, **66,** 1822-1830 (1989).
56. Wolff,R.L., *J. Chromatogr. Sci.*, **30**, 17-22 (1992).

Chapter 5

CAPILLARY ISOTACHOPHORESIS IN THE ANALYSIS OF LIPOPROTEINS

Gerd Schmitz, Grazyna Nowicka* and Christoph Möllers

*Institute for Clinical Chemistry and Laboratory Medicine, University of Regensburg (FRG); *National Food and Nutrition Institute, Warsaw (Poland)*

Address for correspondence: Prof. Dr. Gerd Schmitz, Institute for Clinical Chemistry and Laboratory Medicine, University of Regensburg, 93042 Regensburg, FRG.

A. INTRODUCTION

Electrophoretic techniques are widely used in the analysis of biological macromolecules. These techniques are simple, reliable, inexpensive, easy to automate and commonly used in clinical laboratories.

Many compounds exist in aqueous solvents as more or less dissociated species, either with a positive (cation) or a negative charge (anion). The net charge of ions depends on particle composition, properties of the solvent, and on interactions with other particles. In an electric field ions begin to migrate to the electrodes of opposite charge. In steady state conditions the velocity is proportional to the strength of the electric field. The proportionality constant in this case defined as the electrophoretic mobility depends on the average charge, the size and shape of the species and on the properties of the solvent, especially the pH, polarity and viscosity. This relationship determines the basis of electrophoretic separations.

The mobility of charged particles can be influenced by a number of different parameters and in protein analysis this is predominantly the pH of the solvent.

1. Principles of isotachophoresis

Isotachophoresis (ITP) is carried out in a discontinuous electrolyte system formed by a leading and a terminating electrolyte. The sample is applied between the two electrolytes (Figure 5.1). In a particular separation either cations or anions can be evaluated, however, it is not possible to measure both ion species at the same time [3, 10, 12, 14, 17]. If an anionic sample is analyzed, the leading electrolyte contains an anion of higher mobility (such as chloride$^-$) than any of the sample ions of interest, whereas the terminating electrolyte contains an anion of lower mobility.

When an electric field is applied the negatively charged ions begin to migrate towards the anode with a velocity depending on the effective mobility of the leading ion, according to the equation:

$$v_i = m_i \times E_i \qquad (1)$$

where v is the velocity, E the field strength, m the effective mobility and the subscribed i stands for the different ions.

The anions with the highest effective mobility will move first followed

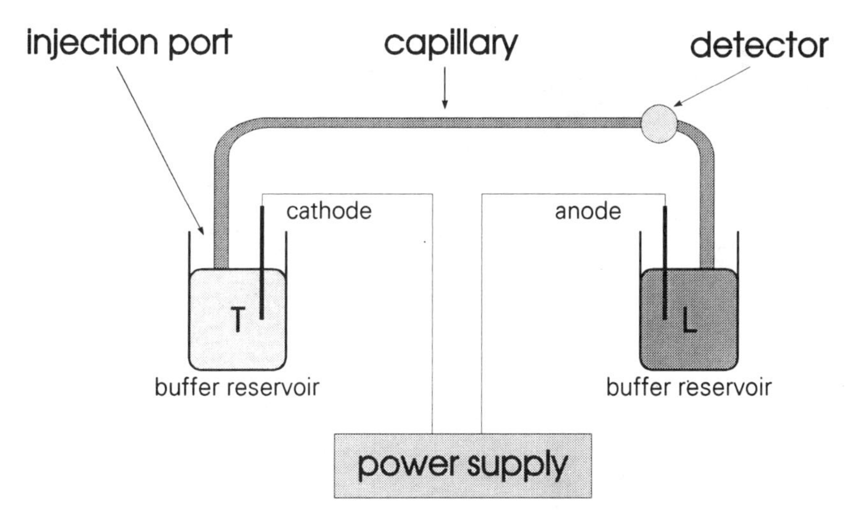

Fig. 5.1. Principle of an analytical capillary isotachophoresis system. Sample ions to be separated are introduced between leading and terminating electrolyte. The analyte is separated into a number of zones, which according to ITP principle migrate with the same velocity. The separated zones are monitored with a detector system.

by those with lower effective mobility in decreasing order of magnitude. The current density in such a system must be homogeneous and therefore the zones are forced to follow in direct contact with each other with equal velocity [12, 14].

To fulfil equation (1), the electric field strength E is therefore increased when the mobility decreases and at equilibrium individual anionic sample components migrate as distinct bands located between the leading and terminating electrolyte (Figure 5.2).

If a sample ion diffuses forward into a preceding band, it will experience a lower field strength and it slows its rate of migration according to equation (1). Therefore the sample ion will be readily caught up by its own zone. Similarly, if a sample ion diffuses back to a succeeding band, because of the higher field strength its velocity will be increased and the ion will rejoin its original band. This phenomenon, known as zone-sharpening effect, is one of the basic effects in isotachophoresis responsible for the high resolution of this technique.

The second effect is the concentration effect, which is based on the Kohlrausch equation

$$\frac{c_a}{c_b} = \frac{m_a}{m_a + m_r} \times \frac{m_b + m_r}{m_b} \quad (2)$$

where c = concentration, m = mobility and the subscribed a, b and r refer to ions a^-, b^- and r^+. Since the effective mobility of each participating

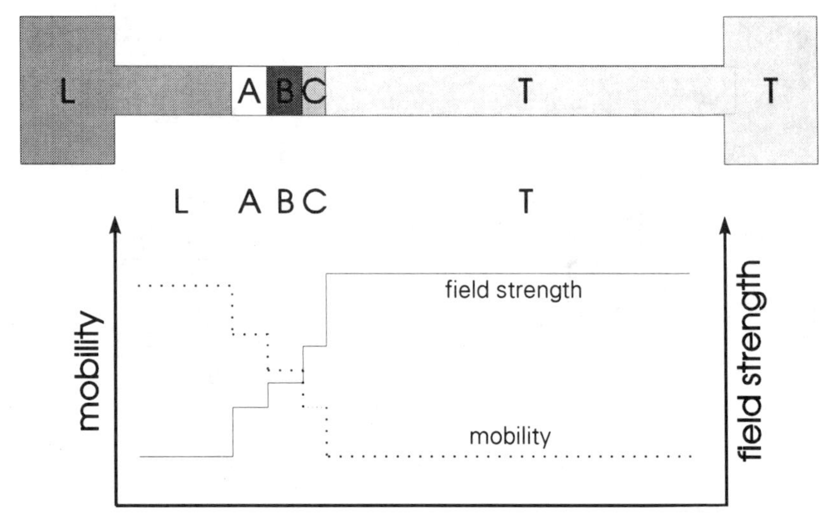

Fig. 5.2. When the sample is completely resolved into a zone of A-, B- and C- ions, we find that there is a field strength step at each zone boundary. Since each unit length must conduct the same amount of current at any given time, the electrical field strength over each zone will be different, due to the differences in net mobility. The fall in net mobility at a zone boundary is therefore accompanied by a proportionately-large rise in field strength. This stepwise increase results in an important beneficial effect, the zone sharpening effect.

ion is constant under defined conditions, the Kohlrausch law can be rewritten as

$$c_a = c_b \times \text{constant} \tag{3}$$

This indicates that at equilibrium the sample ion concentrations are directly proportional to the leading ion concentration [3, 12, 14]. If a component A^- (Figure 5.3I) is introduced at very low concentration, the concentration of A^- at equilibrium will be increased until it reaches the theoretically defined level (Figure 5.3II). Consequently a concentrated sample will be diluted until the correct equilibrium value is adjusted (Figure 5.3III).

Since in isotachophoresis the resolved sample zones are forced to run in immediate contact with each other, the UV-detection of separated bands is difficult. To achieve an optical separation of consecutive sample zones the addition of compounds with intermediate mobility was elaborated (Figure 5.4). These compounds, called spacers, force the consecutive zones apart and, because of their non UV-absorbing properties, permit the optical detection of UV-absorbing samples. In contrast UV-absorbing spacers are used when the sample does not absorb UV light.

Isotachophoretic separation can be performed analytically and preparatively. Analytical capillary ITP is carried out on commercially available capillary electrophoresis systems, which consist of a capillary ending in electrolyte reservoirs which also contain electrodes connected to a power supply (Figure 5.1) Advantages of analytical capillary ITP are the

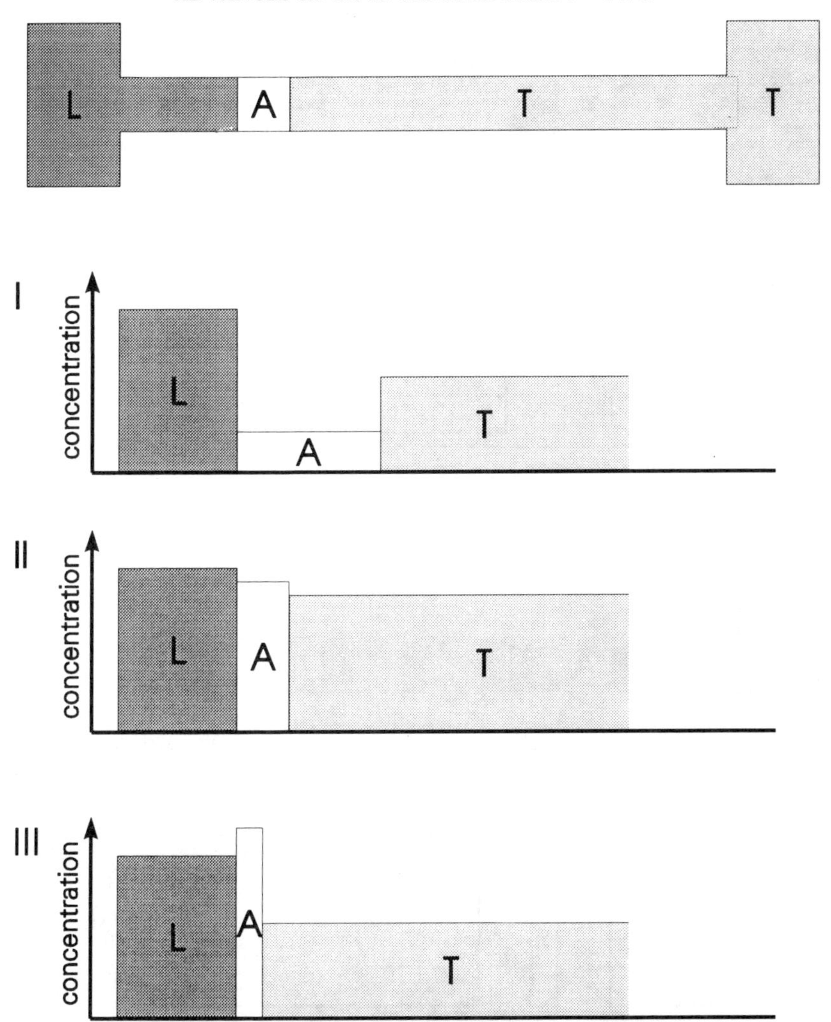

Fig. 5.3. The concentration of ions in each of the separated zones is defined by an equation derived by Kohlrausch as long ago as 1897. Kohlrauschs Law states that, at equilibrium, the concentration of ions in each sample zone is defined by the concentration of ions in the preceding zone and therefore ultimately on the concentration of the leading electrolyte. The equilibrium concentration (panel II) of sample ions can thus be obtained in two ways: a) the sample is initially dilute (panel I), in order to reach the equilibrium concentration, the sample must be concentrated. b) the sample is initially concentrated (panel III) and the sample must be diluted to reach the equilibrium concentration.

high resolution and the small sample volume. The technique is simple to handle, fast, reliable and can be fully automated.

2. *Lipoproteins*

Lipoproteins represent the transport forms of lipids in the circulation [36]. They are composed of free and esterified cholesterol, triglycerides,

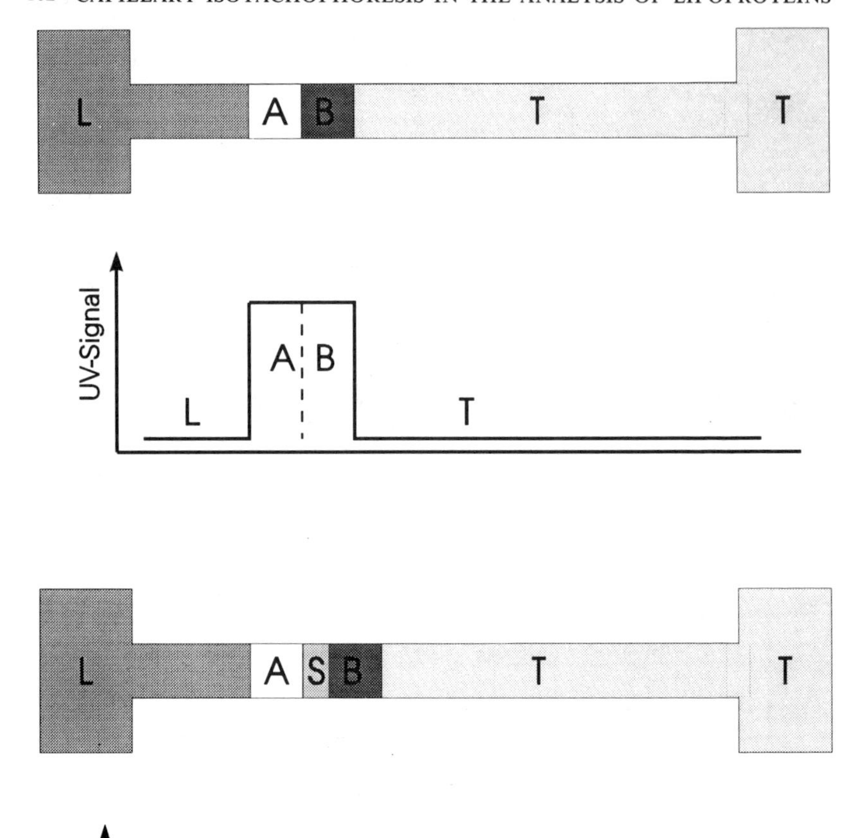

Fig. 5.4. Since the zones migrate in isotachophoresis in immediate contact with each other, distinct zones giving the same UV absorbance are not resolved by the UV detector (upper two panels). Those zones can be resolved by adding an ion which has a net mobility in-between the net mobilities of A and B and which also has little or no UV absorbance. The ion S is called a spacer ion. As a result of adding S, the zones of A and B can now be clearly resolved (lower two panels).

phospholipids and proteins called apolipoproteins. Several methods have been used for lipoprotein separation and characterisation e.g. sequential and gradient ultracentrifugation [18,21,34], gradient gel electrophoresis [7,27], isoelectric focusing [44], and chromatographic [20,46] and immunological [1,4,9,22] methods. Lipoproteins can be divided into various categories on the basis of their compositional and functional properties (see Table 5.1). Two major lipoprotein families can be recognised: apo B-containing lipoproteins consisting of chylomicrons, very

Table 5.1. Size relations of lipoprotein subclasses separated by ultracentrifugation and electrophoresis, electroimmunodiffusion and isotachophoresis. (+ + +, Particle detected). Data are from Alaupovic *et al.*, *Biochim. Biophys Acta*, **260**, 689-707 (1972) and the references therein and ref. 41, respectively.

Gradient Gel Electrophoresis		Ultra-centrifugation	Electroimmunodiffusion					Isotachophoresis		
Subfraction	Diameter [nm]	Density A-II	LpA-I:	LpA-I	LpB	LpC-III:B	LpB:E	Fast	Inter-mediate	Slow
HDL_{2b}	12.93-9.71	1.063-1.1		+ + +				30	21	32
HDL_{2a}		9.71-8.77	1.1-1.125	+ + +				10	29	9
HDL_{3a}	8.77-8.17			+ + +				41	25	39
HDL_{3b}	8.17-7.76	1.125-1.21	+ + +					12	22	8
HDL_{c}	7.76-7.21							7	3	12
VLDL	30-80	<1.006				+ + +		two main subfractions		
IDL	25-35	1.006-1.019					+ + +	one major subfraction		
LDL	18-25	1.019-1.063			+ + +			two main subfractions		

low density lipoproteins (VLDL), intermediate density lipoproteins (IDL) and low density lipoproteins (LDL), and apo A-containing lipoproteins which represent high density lipoproteins (HDL).

It was shown that each of the major apo B-containing fraction is structurally and functionally heterogeneous. Large triglyceride-rich VLDL are cleared rapidly from the circulation and only small portions are converted into low density lipoproteins, while smaller cholesteryl ester-enriched VLDL have a longer plasma circulation time and form large amounts of LDL [26,33]. Among intermediate density lipoproteins two subclasses of overlapping buoyant density and with mean particle diameters of 29.1 nm and 28.3 nm have been isolated: IDL_1, which appear to form a continuous spectrum with small VLDL, and the smaller more dense IDL_2 [25]. IDL_1 and IDL_2 may be precursors of two functionally different LDL subclasses [26]. LDL is constituted of particle species that differ from each other in terms of size, buoyant density, lipid composition, apolipoprotein content, and metabolic properties [8,23]. It was shown that the turnover of lighter LDL ($d < 1.04$ g/ml) subspecies is faster than that of heavier LDL ($d > 1.04$ g/ml) subfractions [45]. The heterogeneity of LDL has been characterised in normal and hyperlipidemic individuals. Large, cholesterol-rich LDL particles accumulate in many patients with familial hypercholesterolemia, while in hypertriglyceridemia or hyper-apoB-lipoproteinemia the accumulation of smaller, more cholesterol depleted LDL were observed when compared to normal subjects.

The catabolism of apo B-containing lipoproteins is partly mediated by specific cell receptors [16]. Extrahepatic cells possess specific binding sites (apo B,E receptors) that recognise apo B-100 and apo E. These apo B,E receptors are responsible for uptake and catabolism of LDL, but they also recognise IDL and VLDL. However, VLDL binding is significantly correlated with the particle composition and apo B and apo E conformation [24]. In normal subjects only the smallest VLDL are able to interact specifically with extrahepatic receptors.

Liver cells, which play a central role in lipoprotein metabolism, contain classical apo B,E receptors as well as receptors suggested to recognise apo E (LRP = LDL-Receptor related protein = α_2 Makroglobulin receptor) and thereby are critically involved in the clearance of remnants originating from the metabolism of triglyceride-rich lipoproteins. VLDL seem to interact also with these receptors in addition to the specific binding to hepatic apo B,E receptors.

High density lipoproteins represent a series of dynamically changing HDL populations differing in density, size, and composition. HDL_2 ($d = 1.063\text{-}1.125$ g/ml) and HDL_3 ($d = 1.125\text{-}1.210$ g/ml) are the major subpopulations [11]. In addition, several other HDL subspecies have been identified by gradient ultracentrifugation [18], gradient gel electrophoresis [7], and immunological [20,22] and chromatographic [9,46] methods.

Among the several apolipoproteins of HDL are apo AI, apo AII, apo AIV, C apolipoproteins, apo D, E apolipoproteins, apo H, apo J, and cholesteryl ester transfer protein [41]. Two major HDL particle groups have been isolated according to their content of apo AI with and without apo AII as the major apolipoproteins [9,20]. They occur within the HDL_2 and HDL_3 density classes and consist of particles that are heterogeneous with respect to their size and apolipoprotein content.

It has been proposed that HDL play an important role in the transport of cholesterol from peripheral cells to the liver for bile acid formation and excretion [11,41]. Hepatic and extrahepatic cells such as fibroblasts, endothelial cells, smooth muscle cells, adipocytes, and macrophages possess specific high-affinity binding sites (receptors) for HDL. HDL binding increases when cells are loaded with cholesterol or incubated with acyl-CoA:cholesterol acyltransferase (ACAT) inhibitors, suggesting that these receptors are regulated in response to changes in cellular cholesterol metabolism [40,41]. Apo AI-containing HDL particles bind to high affinity receptor sites on cells and it has been suggested that they have a high capacity to promote cholesterol efflux. Also HDL particles containing apo AIV have been found to be very effective in removing cellular cholesterol, while particles rich in apo AII and C apolipoproteins reveal very low capacity or are even ineffective in this process [29,41,43].

B. ISOTACHOPHORESIS OF LIPOPROTEINS

1. Analytical capillary isotachophoresis of lipoproteins

Initial studies were performed on the LKB Tachophor 2127-capillary electrophoresis system equipped with a 500 μm i.d. Teflon capillary. Therefore we developed a procedure specifically designed for these physical properties [29,30,32]. This procedure can be easily adapted to other capillary electrophoresis systems, which became available more recently.

Isotachophoretic lipoprotein patterns have been obtained from sera/plasma that were preincubated for 30 min at 4°C with the nonpolar dye Sudan Black B (1% solution in ethylene glycol) and mixed with spacers (2:1, v/v, final concentration of each spacer 0.2 mg/ml). The following compounds were used as spacers: glycyl-glycine, alanyl-glycine, valyl-glycine, glycyl-histidine, serine, glutamine, methionine, histidine, glycine, 3-methyl-histidine, and pseudouridine. For separation, 2 μl of the mixture was injected into the capillary system between the leading and terminating electrolyte. The leading electrolyte consisted of 5 mM H_3PO_4, 0.25% HPMC, and 20 mM ammediol, pH 9.2. The terminating electrolyte contained 100 mM valine and 20 mM ammediol, pH 9.4. Separation was performed at a current of 150 μA (6 kV). During the run the current can be changed to 100 μA and before the detection to 50 μA. The separated zones have been monitored at 570 nm.

Hydroxypropylmethylcellulose has the function to decrease the electroendosmotic flow and to increase the buffer viscosity which leads to better zone sharpening and to a more constant flow profile. The second function is to stabilise the proteins by increasing the osmolarity of the buffer which influences positively the reproducibility.

The Sudan black B solution in 1% ethylene glycol (w/v) can be used for more than 6 month when stored under light exclusion at 4°C. The stability of the dye-lipid complex during electrophoresis or isotachophoresis is not critical. Once the dye is bound to the lipoprotein it sticks very firmly under the described conditions. Firstly, all separations should be carried out at 4-8°C if possible, but not above the transition temperature for phospholipids which is at about 23°C. Secondly, a stained serum should be analysed during the next 1-2 hours because the lipid-dye complexes are stable at room temperature for nearly 3 hours and for about 6 hours at 4°C. Because lipoproteins may be metabolized on the bench at room temperature, they are more critical for sample stability than the dye-lipid complex. Excessive dye is also not a problem because it does not interfere with the separation profile of the serum lipoproteins. Nevertheless the excessive amount of dye should be as small as possible.

We consider serum to be the most convenient specimen in isotachophoretic analysis, although plasma can be used as well (see Section C.2.).

According to our experiments the best results are achieved by analysing the sera within half a day. This includes the time for blood collection and blood clotting, followed by centrifugation and aliquotation. Upon centrifugation (15 min, 20.000 g, 4°C) the sample should be stored at 4°C and analysed within 3 to 4 hours. Aliquots which are not used have to be stored at -20 to -70°C. For the separation a 100 μl serum aliquot is stained for 30 min with 50 μl staining solution (1 % Sudan Black B (w/v) in ethylene glycol) at 4°C. These precautions are necessary since the samples contain active enzymes (lecithin:cholesterol acyltransferase (LCAT) and some phospholipases and hydrolases) which are attached to the lipoprotein system in serum or plasma and can alter the lipoprotein subclass profile, even at the bench level, when samples are stored at room temperature.

Fresh serum provides good and reliable results in a routine laboratory. However, for patient and drug studies one might not have always fresh material available. For these studies immediate freezing of the fresh serum samples is the best way for handling. A sample that was thawed only once gives almost identical results. In normolipidemic sera, a second thawing would also provide quite stable results, but in the moment when triglyceride levels are above 240 mg/dl, reproducibility decreases significantly with the number of thawings.

In conclusion, for good and reliable results fresh samples analysed within 4 hours or samples that have been thawed only once are recommended.

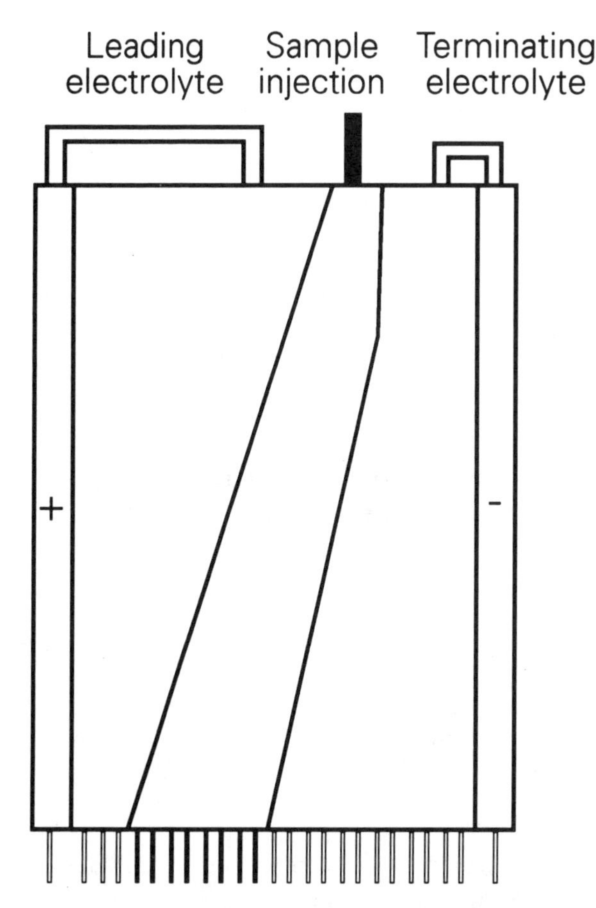

Fig. 5.5. Principle of a preparative isotachophoresis system. Samples to be separated are injected continuously between the laminar streaming leading- and terminating buffer which flow perpendicular to an electric field. Sample ions with different electrophoretic mobility are separated, move in an isotachophoretic "steady state' configuration and can be collected at the bottom of the chamber for preparative isolation of the components.

In addition to the storage conditions, frozen samples are stable for six months at -40°C, for about five years at - 70°C or on liquid nitrogen.

2. *Preparative isotachophoresis of lipoproteins*

Preparative subfractionation of lipoproteins was carried out alternatively in a flat bed electrophoresis cell Elphor VAP 22 (Bender & Hobein, Munich) (Figure 5.5) or in a new free-flow electrophoresis instrument, which has been developed recently in our laboratory. The separation principle and separation conditions are the same in both instruments and fractionations on both instruments are directly comparable. However, to achieve a better resolution of isolated subfractions we do not use a continuous sample flow, and instead the chamber is filled once with whole sample. The separation distance of the chamber in our instrument is longer than that of Elphor VAP 22. The chamber has a size of 20 cm/12 cm/1 mm and the cooling allows that separations can be performed at 4°C. Because of these modifications the electrodynamic disturbances are significantly diminished and reproducible separations with good resolution have been achieved. For separation, leading and terminating electrolytes were used as for analytical ITP. The serum (1 ml) was mixed with terminating buffer (2 ml) containing spacers in a final concentration of 1 mg/ml each. The separation was carried out at 10 mA, voltage of 800 V, at 4°C for 3 hours. The fractions that have been collected were subject to further analysis.

C. ISOTACHOPHORETIC ANALYSIS OF WHOLE SERUM LIPOPROTEINS AND LIPOPROTEIN SUBCLASSES UNDER DIFFERENT METABOLIC CONDITIONS

1. Normal serum ITP lipoprotein profile

A normal serum ITP-lipoprotein profile obtained from whole serum by analytical capillary isotachophoresis consists of fourteen subfractions [29,30]. Six of them (peaks 1 to 6 in Figure 5.6) describe subpopulations of high density lipoproteins and eight (peaks 7 to 14 in Figure 5.6) represent subpopulations of apo B-containing lipoproteins: chylomicron derived particles, very low density lipoproteins and intermediate and low density lipoproteins. The separated lipoprotein subpopulations have been characterised according to their composition and biological properties (Figures 5.7 and 5.8).

The fast-migrating HDL subpopulations (Figure 5.7, peaks 1 and 2) represent apo AI and phosphatidylcholine-rich HDL subclasses which possess high affinity to HDL binding sites on macrophages and are effective in removal of cholesterol from preloaded cells. The HDL subpopulations of intermediate mobility (peak 3) contain particles rich in apo AII, apo E and C apolipoproteins. The lipid moiety of these particles consists mainly of cholesteryl esters and sphingomyelin. Their affinity to HDL receptors on macrophages is significantly lower than that of the HDL particles in the fast-migrating subpopulations. The slow-migrating HDL subpopulations (peaks 4 to 6) correspond to HDL subclasses containing apo AI, apo AIV and LCAT activity. They express the highest non-specific binding to macrophages, as compared to other subpopulations, and contain also a small amount of particles which interact with HDL receptors by high affinity binding. The slow-migrating HDL are good acceptors of cellular cholesterol. They show similarities to earlier defined pre-beta-migrating HDL [13]. Both of these subpopulations are rich in apo AI, contain a high LCAT activity and are

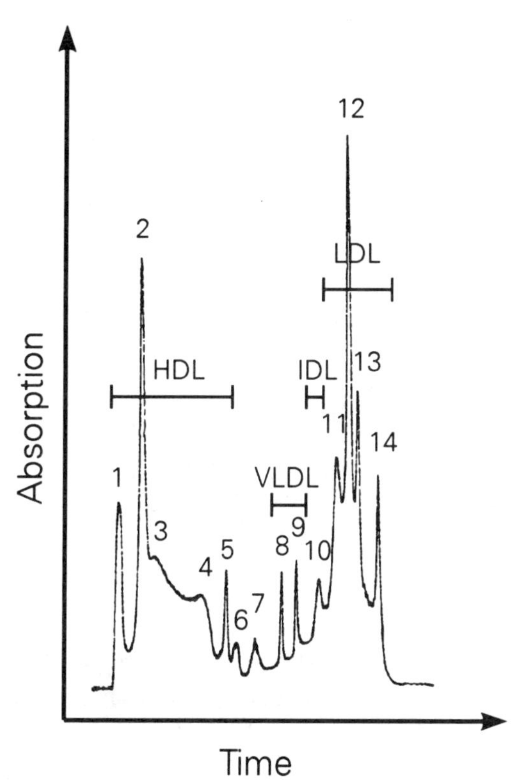

Fig. 5.6. Normal serum lipoprotein profile determined by analytical capillary isotachophoresis. Subfractions of four major lipoprotein classes have been separated: HDL, high density lipoproteins; VLDL, very low density lipoproteins; IDL, intermediate density lipoproteins; LDL, low density lipoproteins.

good acceptors for cell derived cholesterol, indicating that pre-beta-migrating HDL could contribute to slow-migrating HDL separated by isotachophoresis [13,29].

Among apo B-containing lipoproteins four major functional populations were recognised (Figure 5.8). Particles migrating in peak 7, which represent chylomicron derived particles, and particles migrating in peak 8, which represent large triglyceride-rich VLDL, do not bind to apo B,E-receptors on fibroblasts but do bind with high affinity to hepatic apo E-receptors. Particles creating peaks 9 and 10 represent small VLDL and IDL, respectively. They bind to apo B,E-receptors on fibroblasts and apo B,E- and apo E-receptors on hepatocytes. Low density lipoproteins are separated by analytical ITP in four subpopulations (peaks 11 to 14) forming two functional groups. LDL particles creating peaks 11 and 12 bind with lower affinity to apo B,E-receptors on fibroblasts compared to LDL particles migrating in peaks 13 and 14 [30].

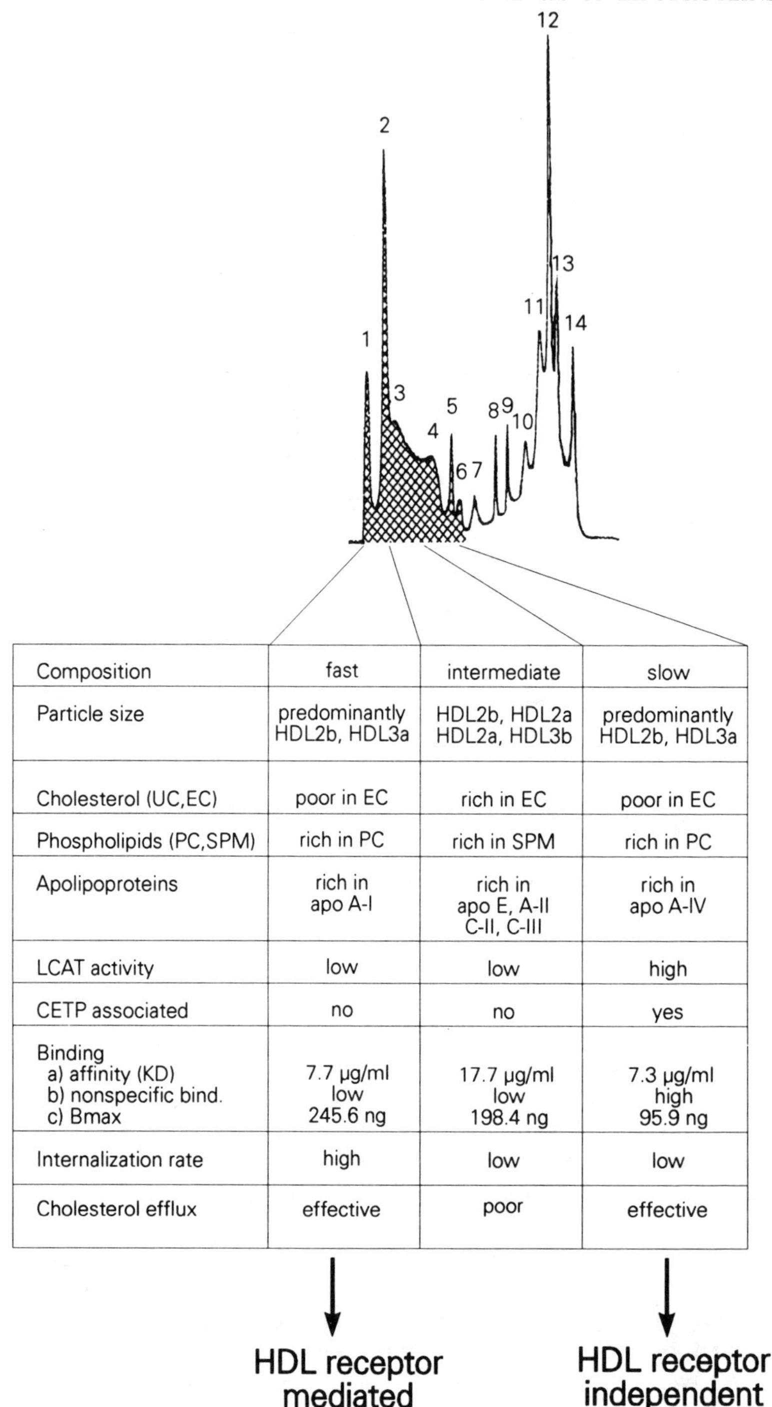

Composition	fast	intermediate	slow
Particle size	predominantly HDL2b, HDL3a	HDL2b, HDL2a HDL2a, HDL3b	predominantly HDL2b, HDL3a
Cholesterol (UC,EC)	poor in EC	rich in EC	poor in EC
Phospholipids (PC,SPM)	rich in PC	rich in SPM	rich in PC
Apolipoproteins	rich in apo A-I	rich in apo E, A-II C-II, C-III	rich in apo A-IV
LCAT activity	low	low	high
CETP associated	no	no	yes
Binding a) affinity (KD) b) nonspecific bind. c) Bmax	 7.7 µg/ml low 245.6 ng	 17.7 µg/ml low 198.4 ng	 7.3 µg/ml high 95.9 ng
Internalization rate	high	low	low
Cholesterol efflux	effective	poor	effective

Fig. 5.7. Characterisation of subpopulations of high density lipoproteins separated from normal serum by analytical and preparative isotachophoresis.

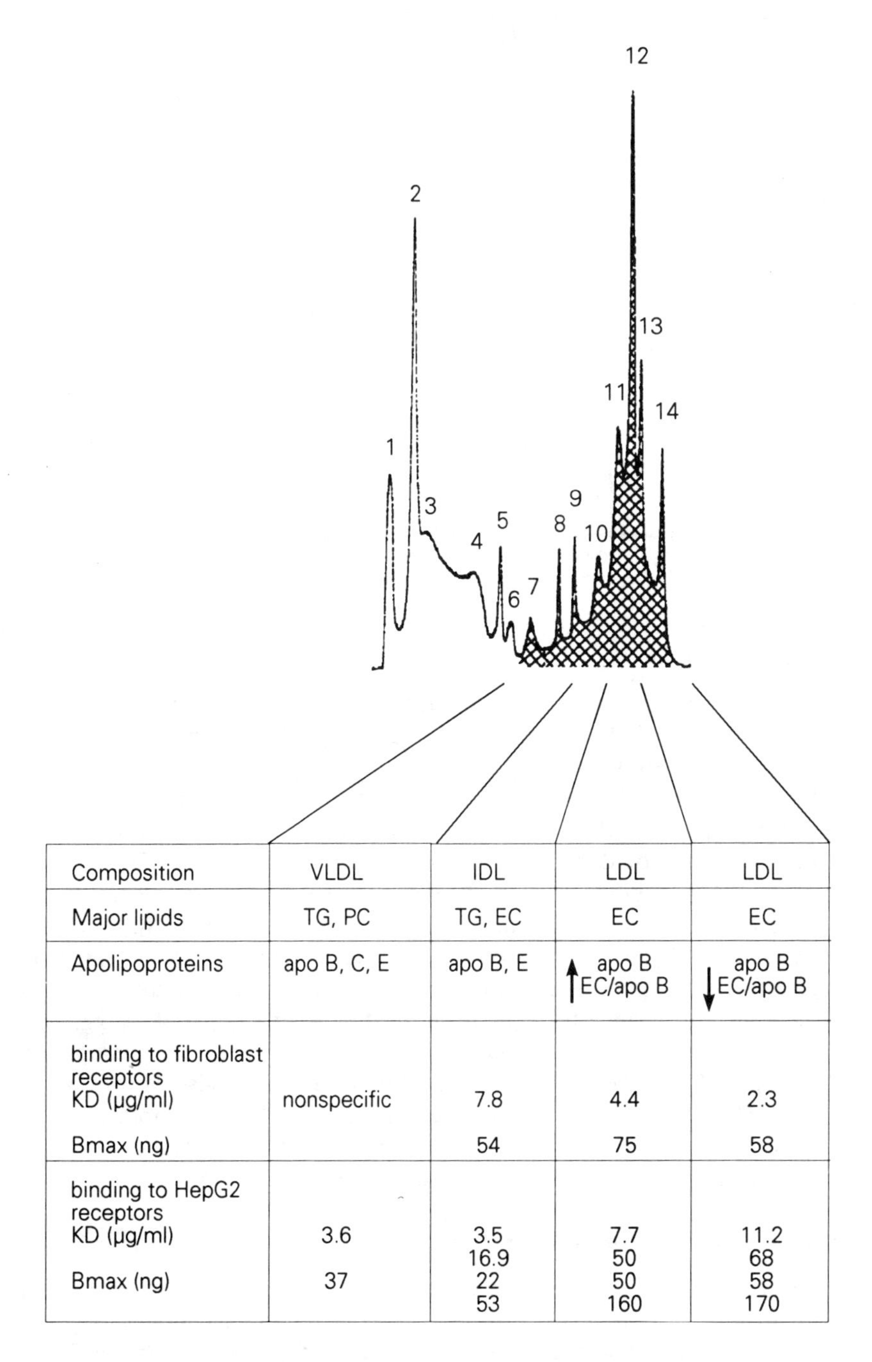

Composition	VLDL	IDL	LDL	LDL
Major lipids	TG, PC	TG, EC	EC	EC
Apolipoproteins	apo B, C, E	apo B, E	↑ apo B EC/apo B	↓ apo B EC/apo B
binding to fibroblast receptors KD (μg/ml) Bmax (ng)	 nonspecific	 7.8 54	 4.4 75	 2.3 58
binding to HepG2 receptors KD (μg/ml) Bmax (ng)	 3.6 37	 3.5 16.9 22 53	 7.7 50 50 160	 11.2 68 58 170

Fig. 5.8. Characterisation of subpopulations of apo B-containing lipoproteins (VLDL, IDL, LDL) separated from normal sera by analytical and preparative isotachophoresis.

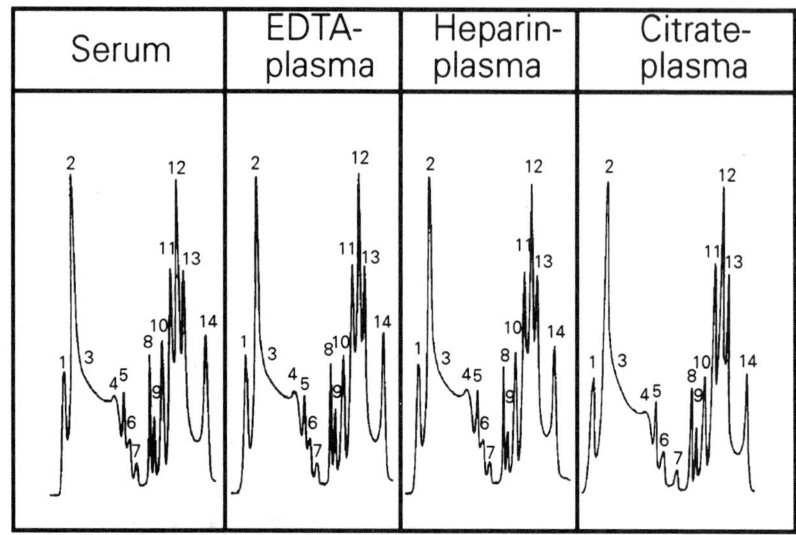

Fig. 5.9. Comparison of serum and plasma lipoprotein profiles of the same donor. A, serum; B, EDTA plasma; C, heparin plasma; D, citrate plasma

2. *Comparison of ITP lipoprotein patterns obtained from serum and plasma samples*

The ITP lipoprotein patterns obtained from serum and plasma samples from the same donor were compared [30]. The profiles typical for normolipidemic subjects are presented in Figure 5.9.

We did not find significant differences between serum and plasma lipoprotein profiles of normal donors. However, it is well accepted that citrate plasma should not be used for lipoprotein analysis. Heparin can bind to heparin binding sites of apolipoproteins and may change the separation pattern, especially in hypertriglyceridemic patients. In sera from patients with hypertriglyceridemia the apo E content is increased in triglyceride-rich lipoproteins and therefore apo E, which is known to bind to heparin, can change the separation profile. Therefore we do suggest to refrain from using heparin-plasma for lipoprotein analysis. In our opinion EDTA-plasma is better suited for lipoprotein analysis. However, in routine clinical practice serum is the material that is used more often in ITP analysis because of its stability. Therefore, we decided to use and recommend serum samples for the isotachophoretic separation.

3. *ITP pattern of lymph lipoproteins*

We analysed lymph lipoprotein profiles by capillary ITP and compared it with the serum lipoprotein pattern of the same donor (Figure 5.10). We found significant differences between both profiles. In the lymph HDL pattern only one subpopulation corresponding to serum fast-migrating apo AI-HDL (peaks 1 to 2) and significantly enhanced concentrations (as

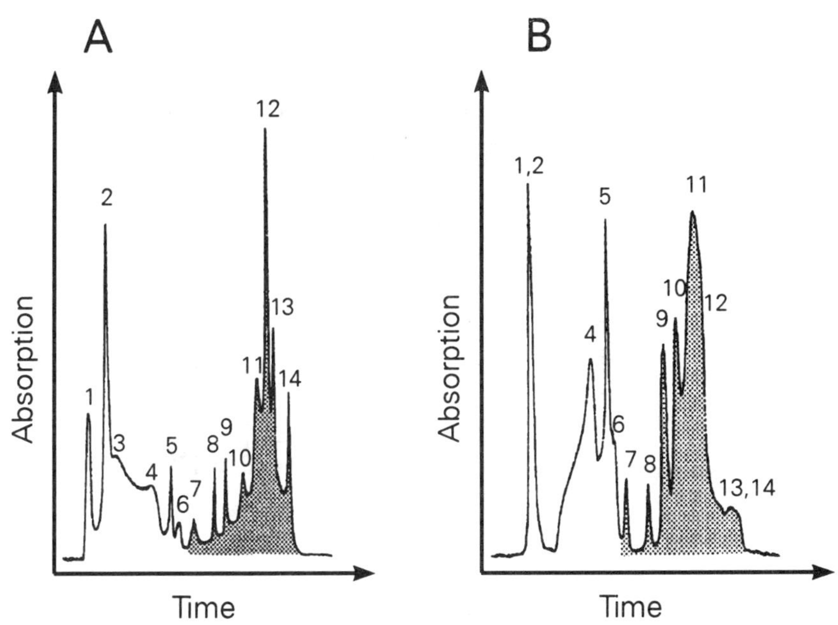

Fig. 5.10. Lymph lipoprotein profile of normal subject. A, serum lipoprotein pattern; B, lymph lipoprotein pattern.

compared to serum) of slow-migrating HDL (peaks 4 to 6) have been recognised. The lymph particles of IDL mobility are the major components of apo B-containing lipoproteins, whereas slow-migrating LDL particles are almost undetectable. This is in good agreement with other studies of lymph lipoprotein composition [42]. Lymph HDL compared with plasma HDL contain more phospholipids and free cholesterol. Their apolipoprotein compositions also differ significantly. Lymph HDL are rich in apo AI, apo AIV and apo E. Particle groups containing apo AI/AIV and apo AI/AIV/E were found. These particles correspond to slow-migrating HDL. Low levels of LCAT activity in lymph have also been reported. Lymph HDL are enriched in discoidal particles, which correspond to fast-migrating HDL. Lymph VLDL are small, contain less triglycerides and more cholesterol, and show similarities to small plasma VLDL and IDL. Also lymph LDL are enriched in free cholesterol and phospholipids, and resemble plasma IDL and light LDL (peak 11).

4. ITP pattern of isolated lipoprotein subclasses

The relationship between the hydrated densities of lipoprotein subclasses and their corresponding isotachophoretic mobility was studied. The ITP analysis of lipoprotein subclasses isolated by ultracentrifugation from sera of normal, hypercholesterolemic and hypertriglyceridemic subjects presented in Figure 5.11 and Figure 5.12 and our results

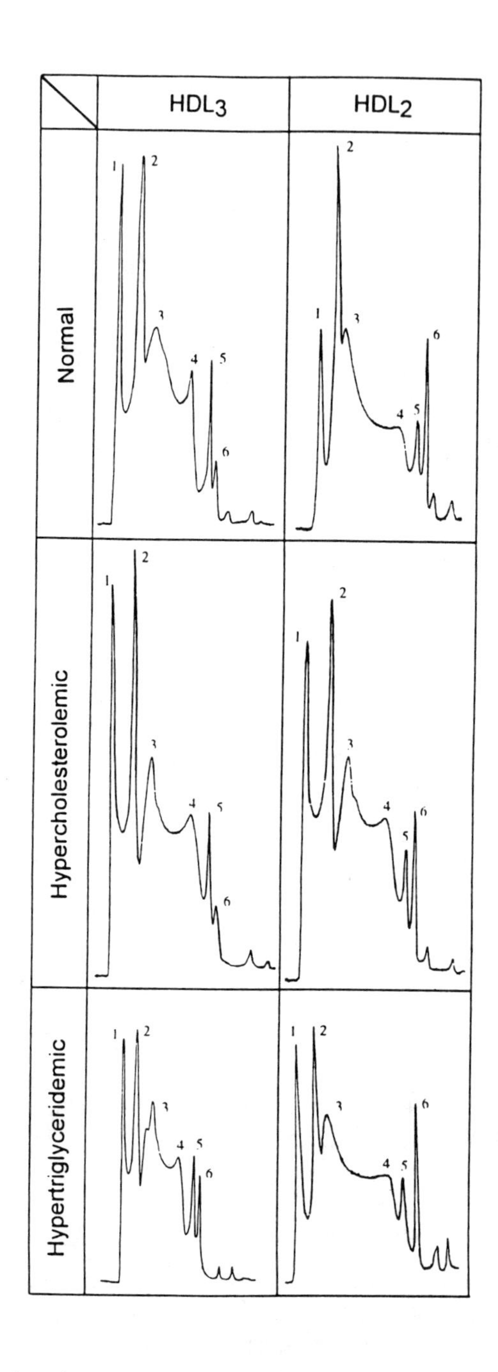

Fig. 5.11. Isotachophoretic analysis of ultracentrifugally isolated HDL subclasses.

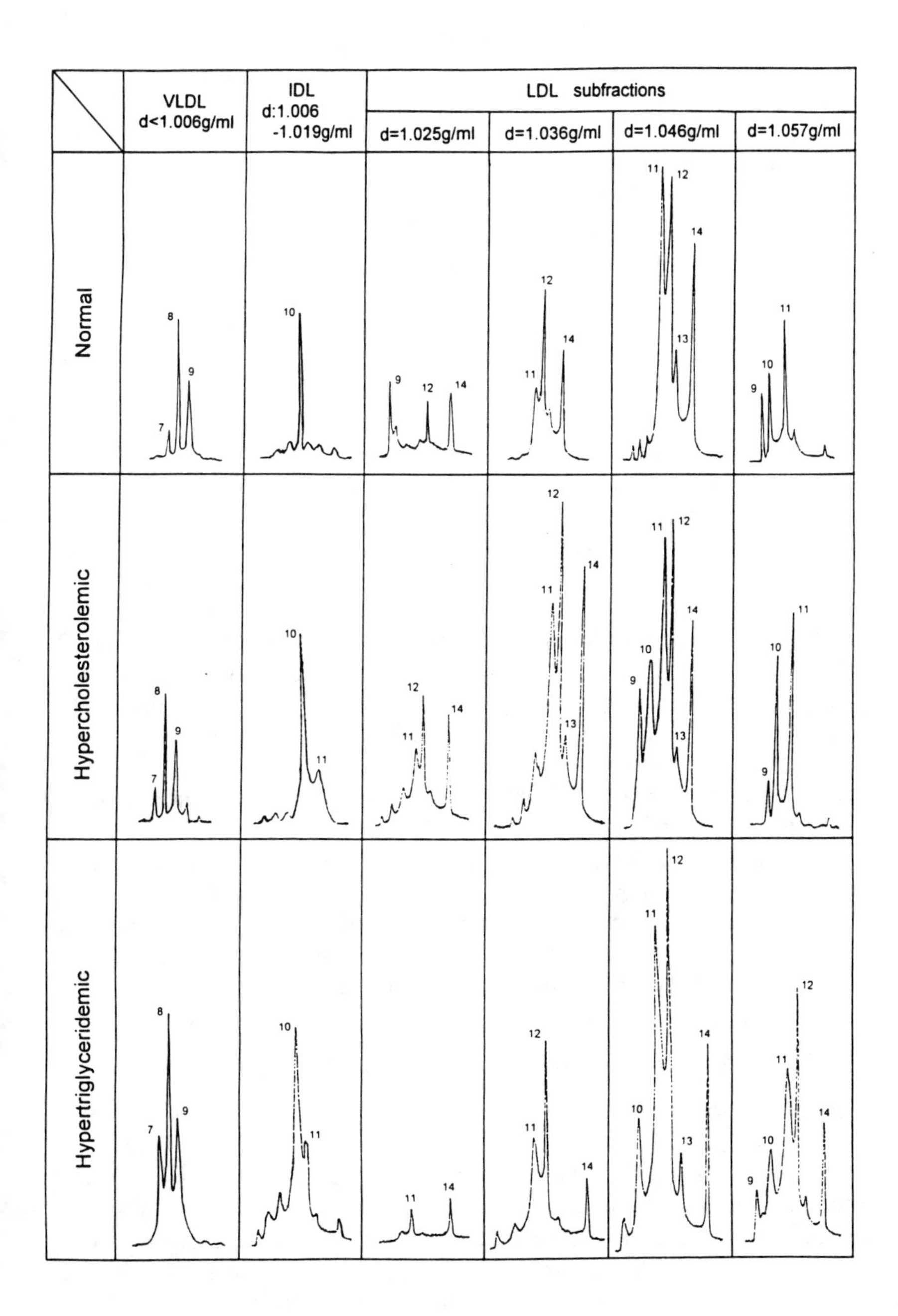

Fig. 5.12. Isotachophoretic analysis of ultracentrifugally isolated subpopulations of apo B-containing lipoproteins.

published elsewhere show that particle subpopulations of defined mobility can be found in subfractions of different densities. In VLDL of normal and hyperlipidemic sera three subpopulations dominate: peaks 7 to 9. In normal sera IDL particles with the mobility of peak 10 dominate, while in hyperlipidemic sera IDL particles creating peaks 10 and 11 prevail. Differences between normal and hyperlipidemic sera occur also in ITP profiles of isolated LDL subfractions. Peaks 9, 12 and 14 were detected in the lightest LDL subfraction (d=1.025 g/ml) isolated from sera of normal subjects. In contrast, in hypertriglyceridemic and hypercholesterolemic sera this LDL-subfraction was represented by peaks 11 and 14 and 11, 12 and 14, respectively. In the LDL subfraction of d=1.036 g/ml particles with the mobility of peaks 11, 12 and 14 are pre-eminent, whereas in the subfraction of d=1.046 g/ml particles with the mobility of peaks 11 to 14 predominate. In both subfractions isolated from hyperlipidemic sera particles with faster mobility (peaks 10 and 9) appear. In all HDL subfractions analysed (HDL_2 and HDL_3 - Figure 5.10) particle groups of different mobility occur, but their concentrations differ significantly between subfractions isolated from different donors. Therefore, the net electric mobility of the individual particle groups does not always correspond to the flotation properties of these lipoproteins, and the individual peak groups between different subjects and different density classes reflect a variability in individual peak concentration rather than changes in peak mobility. In order to recognise possible artefacts that could be formed during ultracentrifugation, each subfraction was mixed with both normal and hyperlipidemic sera and changes in the serum ITP profile were monitored (data not shown). In all experiments no additional peaks were observed. The mobility of the particle groups detected in the isolated subfractions strictly corresponds to the mobility of the individual peak groups detectable in whole serum. Therefore, lipoprotein particle groups of defined mobility have been recognised in both normolipidemic and hyperlipidemic sera, but in different subjects they exist in different concentrations and create individual peak patterns.

5. *ITP lipoprotein patterns in different types of hyperlipoproteinemia*

Using analytical capillary isotachophoresis we analysed serum lipoprotein patterns of patients with different types of hyperlipoproteinemia (HLP) classified according to WHO recommendations (Fredrickson phenotypes). The WHO classification system [5] provides the classical way of describing the lipoprotein profile of commonly occurring species of hyperlipoproteinemia. There is essentially a classification of biochemical abnormalities and it does not specify whether the disorder is genetically determined or is secondary to underlying disease or environmental factors. Type IIa HLP represents the hypercholesterolemia due to an increase in LDL cholesterol concentration.

In type IIb HLP an increase of LDL cholesterol level is accompanied by an increase in VLDL concentration, especially small VLDL, and also IDL. Type III HLP is characterised by moderate or massive accumulation of chylomicron- and VLDL-remnants and IDL, with low LDL level. Type IV HLP represents pure hypertriglyceridemia due to an increase in VLDL, among which large triglyceride-rich VLDL dominate. Type I and V HLP describe chylomicronemia syndromes. In type V HLP appearance of chylomicrons is accompanied by an increase in VLDL. In both phenotypes LDL levels are very low.

ITP lipoprotein profiles typical for patients with defined HLP types (IIa, IIb, III, IV, I and V) are presented in Figure 5.13. In the ITP lipoprotein pattern of hypercholesterolemic patients (type IIa) and patients with mixed hyperlipidemia (type IIb), the LDL subpopulations dominate, but in type IIb significantly enhanced concentrations of VLDL and IDL particles (peaks 8 to 10) occur.

In remnant hyperlipoproteinemia (type III), lipoprotein particles with the mobility of peaks 9, 10 and 11 dominate and almost no slow-migrating LDL subpopulations (peaks 13, 14) can be found. In hypertriglyceridemia (type IV) the ITP lipoprotein profile is characterised by the accumulation of large VLDL particles migrating in peak 8 and the concentration of VLDL remnants and IDL is significantly lower as compared to type III (remnant HLP). The lipoprotein patterns of type I and type V patients are determined by the accumulation of unhydrolysed chylomicrons and large VLDL particles. In both patterns a particle group with the mobility of peak 8a appears, that might represent a subgroup of large triglyceride-rich particles. Other subgroups of these particle population formed one subfraction of the mobility of peaks 8 and 9. The concentration of small particles of apo B-containing lipoproteins (IDL, LDL) is significantly diminished.

The analysis of the lipoprotein pattern from patients with different types of hyperlipoproteinemia together with the characterisation of the composition and biological properties of isolated lipoprotein subpopulations permits to discriminate between the apo B-containing lipoproteins particle groups, which are substrates for lipoprotein lipase and hepatic lipase, and particle groups which are ligands for the remnant receptor and the LDL receptor. In addition, it provides an insight into subpopulations of high density lipoproteins: apo AI-rich subpopulations *versus* apo AII-rich subpopulations and HDL receptor active subfractions.

6. Selected genetic disorders of lipoprotein synthesis, secretion and metabolism

We analysed the serum lipoproteins of patients with rare genetic disorders of synthesis, secretion and metabolism of high density lipoproteins and apo B-containing lipoproteins.

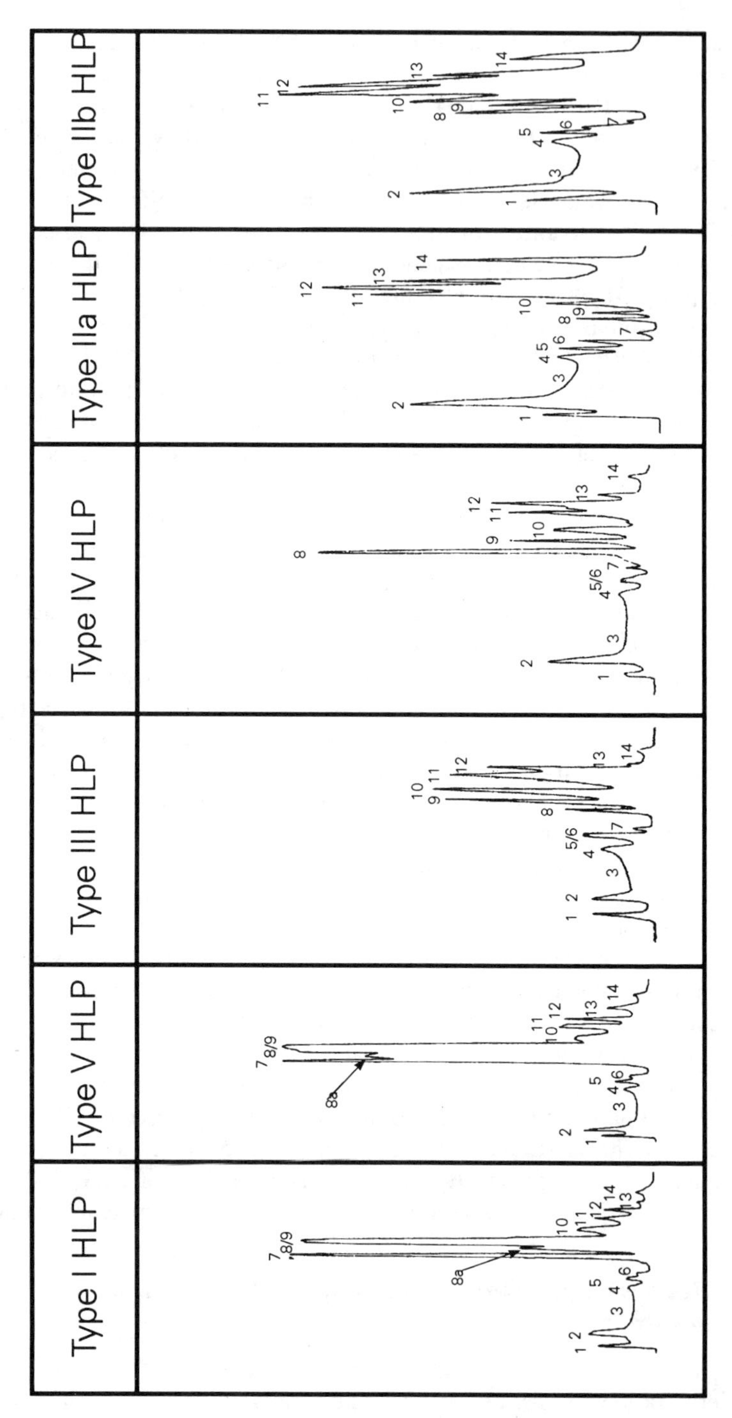

Fig. 5.13. ITP-profiles typical for patients with defined phenotypes of hyperlipoproteinemia.

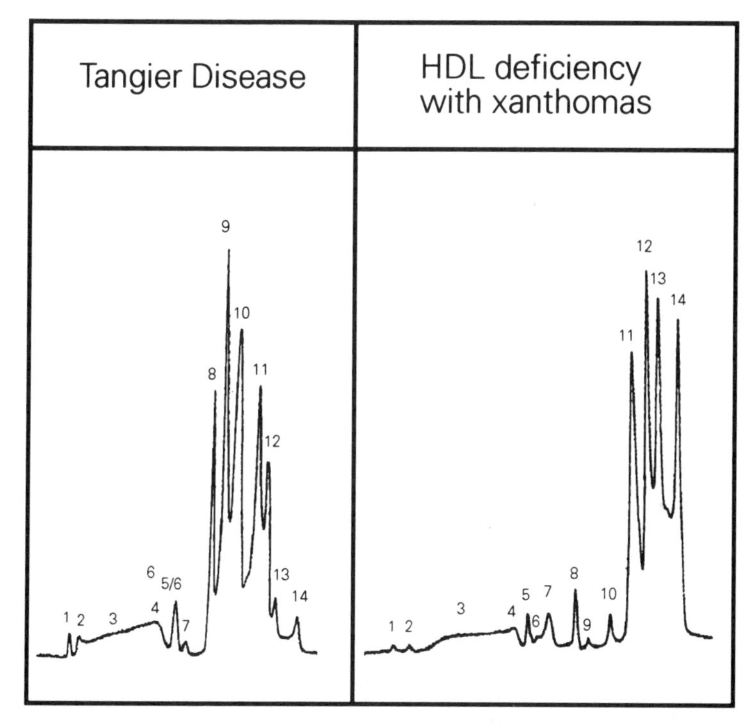

Fig. 5.14. ITP-lipoprotein profiles obtained from serum of patients with Tangier disease and HDL deficiency with xanthomas.

Tangier disease, HDL deficiency with plane xanthomas, familial LCAT deficiency and Fish eye disease represent HDL deficiency syndromes. Tangier disease is associated with disorders of intracellular HDL catabolism, and in HDL deficiency with plane xanthomas the molecular defect in apo AI synthesis was identified [2,37,38]. The ITP lipoprotein profile of Tangier disease (Figure 5.14) is determined by very low concentrations of HDL subpopulations, significantly enhanced concentrations of larger and smaller particles of apo B-containing lipoproteins (peaks 8, 9 and 10) and very low concentrations of slow-migrating LDL subpopulations (peaks 13 and 14). In HDL deficiency with plane xanthomas HDL subpopulations are almost absent and in the profile LDL subpopulations (peaks 11 to 14) dominate. In comparison with Tangier disease and normal subjects very low concentrations of particles creating peaks no 8 to 10 appear.

Familial LCAT deficiency and Fish eye disease are related to defects in LCAT synthesis and function, and characterised by a deficiency in cholesteryl esters [28]. In both diseases (Figure 5.15) within the HDL range the fast-migrating particles creating peaks 1 and 2 are almost not detectable, while the slow-migrating particles corresponding to peaks 5 and 6 are still present. Significant differences between these disorders are

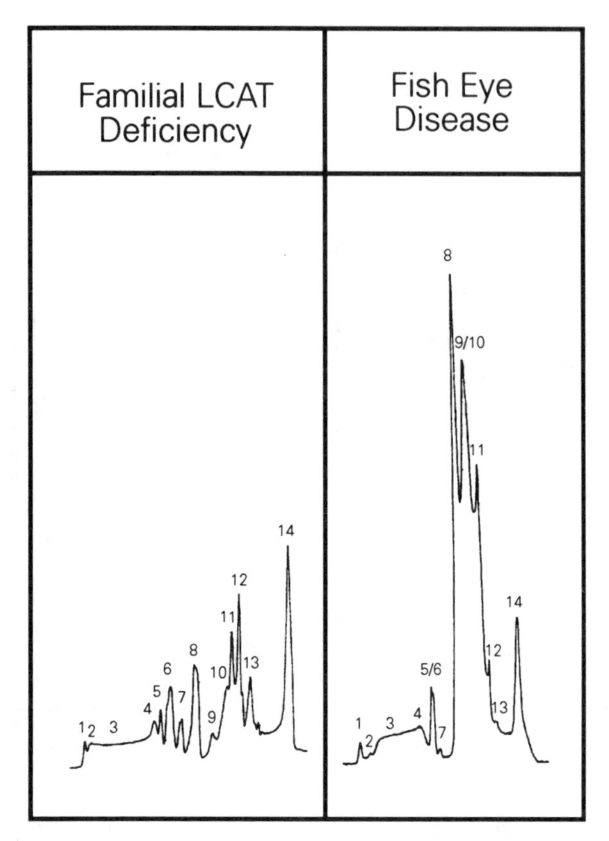

Fig. 5.15. ITP-lipoprotein profiles of patients with familial LCAT deficiency and Fish eye disease.

observed in subpopulations of apo B-containing lipoproteins. In familial LCAT deficiency in the VLDL/IDL range peak 8 dominates, while peak 9 is almost absent and peak 10 is closely connected with LDL peak 11. Among LDL subpopulations (peaks 11 to 14) relatively high concentrations of particles creating the last peak (no 14) appear. This subpopulation contains LpX or LpX like particles as seen in patients with liver disorders. In Fish eye disease among apo B-containing lipoproteins particles dominate which have the mobility of peaks 8, 9, 10 and 11. The LDL peaks 12 and 13 are almost absent, while peak 14 is still present.

Secondly, deficiencies in cholesterol esterification, but not so severe as in genetic LCAT deficiency, have been observed in patients with liver diseases and especially with biliary obstruction as well as in patients with adult respiratory distress syndrome (ARDS). We compared the lipoprotein profile determined by analytical capillary ITP in sera from patients with adult respiratory distress syndrome, fetal respiratory distress and cholestasis (Figure 5.16). In adult respiratory distress syndrome and in fetal distress the fast-migrating particles, creating peaks 1 and 2 in the

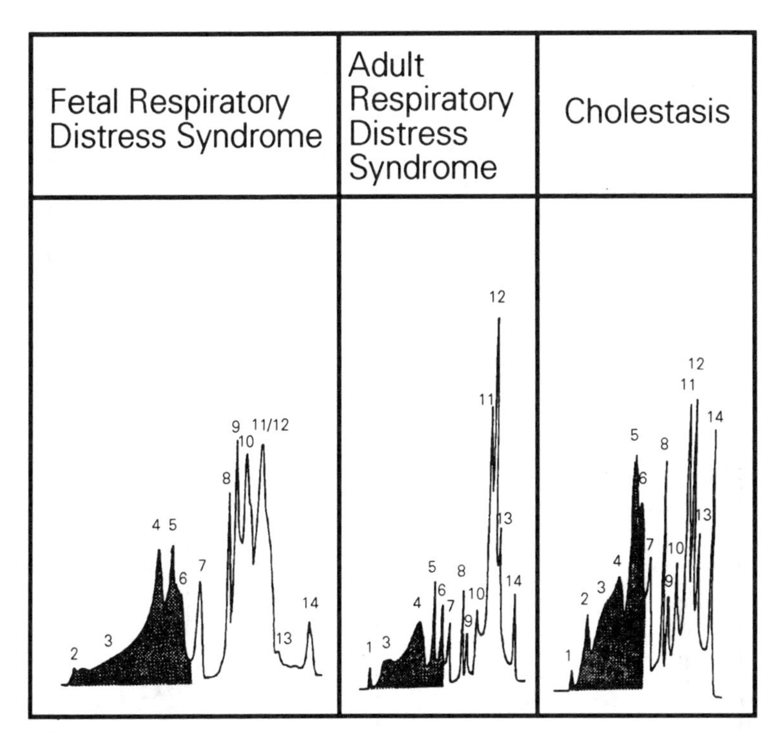

Fig. 5.16. ITP-lipoprotein profiles typical for patients with respiratory distress syndrome and cholestasis.

normal pattern are almost not recognised. The concentration of particles migrating in peak 3 is also significantly reduced compared to profiles obtained from control probands. In both disorders slow-migrating particles creating peaks 4, 5 and 6 dominate in the HDL pattern. In the range of apo B-containing lipoproteins of patients with ARDS subpopulations with the mobility of peaks 11 and 12 prevail, but all subpopulations occurring in the normal lipoprotein pattern are present. In fetal distress particles with the mobility of peaks 11 and 12 dominate and in this case form only one peak. In patients with cholestasis HDL particles creating peaks 1 and 2 appear in a very low concentration. The amount of particles migrating in peak 3 is also significantly reduced, while peak 4 is recognised distinctly. The most abundant particles show the mobility of peaks 5 and 6. Among apo B-containing lipoproteins high concentrations of VLDL particles which form peak 8 and LDL particles with the mobility of peaks 11, 12 and 14 occur. It was recognised (data not shown here) that in peak 14 not only LDL particles but also LpX particles, typical for cholestasis, migrate.

Abetalipoproteinemia and hypobetalipoproteinemia represent disorders characterised by deficiency of apo B-containing lipoproteins. In abetalipoproteinemia no secretion of apo B-containing lipoproteins occurs

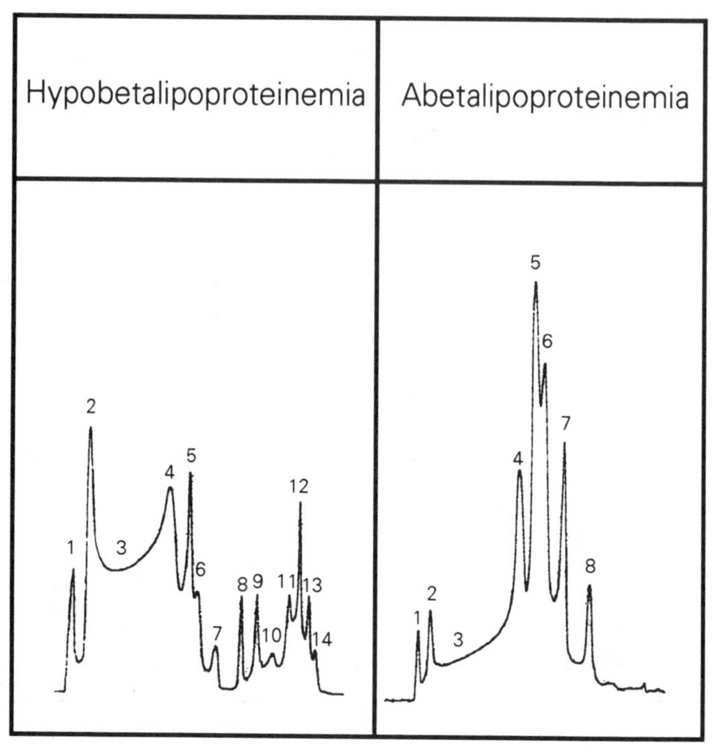

Fig. 5.17. ITP-lipoprotein profiles of patients with abetalipoproteinemia and hypobetalipoproteinemia.

and therefore in the ITP lipoprotein pattern no apo B-containing lipoprotein subpopulations are detectable. Among the HDL subpopulations high concentrations of slow-migrating HDL appear (Figure 5.17). Hypobetalipoproteinemia can be caused by different mutations of the apo B gene resulting in secretion of truncated apo B forms and deficiency of subpopulations of apo B-containing lipoproteins [15]. In the ITP lipoprotein profile HDL subpopulations dominate and a low concentration of apo B-containing lipoproteins is observed.

7. *Postprandial studies*

Postprandial lipemia provides a daily state of challenge to triglyceride transport. A poor ability to cope with this challenge, *i.e.* a low triglyceride metabolic capacity, is likely to increase risk of cardiovascular disease [19,36]. The diminished triglyceride metabolic capacity associated with accumulation of triglyceride-rich lipoproteins of both intestinal and hepatic origin enhances cholesterol ester transfer into triglyceride-rich lipoproteins and converts them into potentially atherogenic agents. These disturbances lead to enhanced concentrations of small, cholesterol ester-enriched particles of apo B-containing lipoproteins, and a low

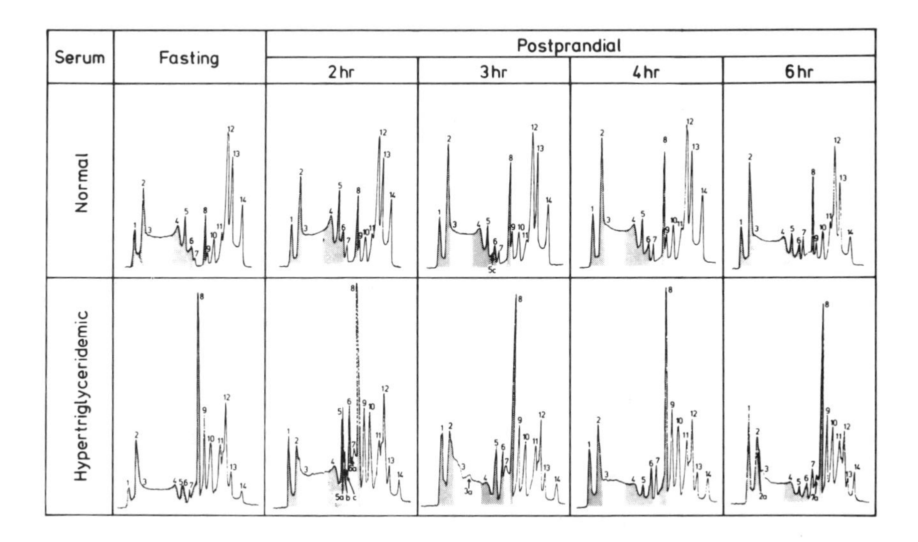

Fig. 5.18. Changes in serum lipoprotein profiles occurring under postprandial conditions.

concentration of HDL, which is believed to be associated with increased frequency of coronary artery disease. It was reported that the analysis of postprandial lipemia can help to recognise patients with enhanced risk of CHD. We analysed changes in the serum lipoprotein pattern occurring after an oral fat load test (postprandial state) in normal subjects and patients with lipoprotein disorders. The ITP lipoprotein pattern of normal and hypertriglyceridemic subjects obtained before (fasting) and 2, 3, 4 and 6 h after the fat load (postprandial) are presented in Figure 5.18. Significant differences and time dependent changes in the concentration of chylomicron derived particles (peak 7), large triglyceride-rich VLDL (peak 8), VLDL remnants and IDL particles (peaks 9 to 11) have been recognised between subjects analysed. Appreciable changes occur also among HDL subpopulations (especially slow and fast-migrating subpopulations). It is generally accepted [19,35] that an individual triglyceride metabolic capacity observed during postprandial lipemia is a highly reproducible pattern. Our results indicate that analytical ITP is a helpful tool to analyse time-dependent changes occurring during postprandial lipemia in the lipoprotein pattern, and can recognise patients with accumulation of large triglyceride-rich VLDL and patients in whom accumulation of smaller cholesterol ester-enriched remnant particles and IDL occur.

8. Newborn studies

It is well known that after birth significant changes in lipid, lipoprotein and apolipoprotein concentrations occur [6]. Estimations of changes in

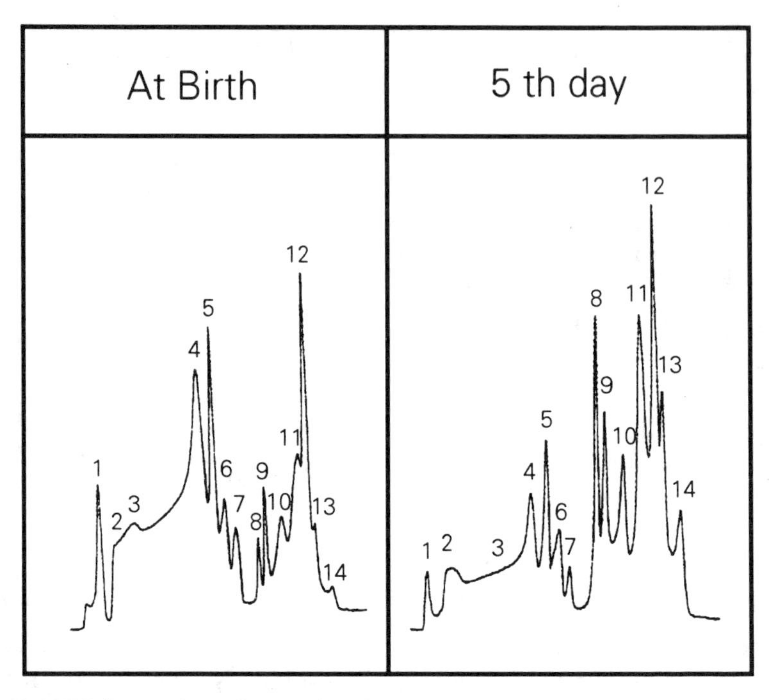

Fig. 5.19. ITP-lipoprotein profiles obtained from healthy new-born directly after birth and at 5th day of life.

lipoprotein pattern from the fetal state up to the period of infancy are of importance for understanding the development of lipid disorders as well as for their early diagnosis. Analysis of newborn lipoprotein subclasses by conventional methods is limited by difficulties in obtaining blood samples. Therefore, we decided to use analytical ITP for newborn lipoprotein analysis because this technique is able to give lipoprotein profile with high resolution and precision starting with extreme small sample volume. Analysis was carried out at birth and at 5th day of live. At birth umbilical vein blood was obtained, but no differences between lipoprotein profiles obtained from umbilical vein and artery blood occur. Representative lipoprotein patterns are presented in Figure 5.19. At birth low lipoprotein levels are observed and HDL represent the main lipoprotein class. In the HDL range slow particles-rich in apo AI, AIV, E and phospholipids dominate while within the first days of life this group is significantly reduced. After birth rapid changes in the lipoprotein transport occur, that can be related to the introduction of energy *via* feeding and higher liver synthesis of very low density lipoproteins. Therefore, in children and adults there is a significant increase in VLDL, IDL and LDL levels, and LDL becomes the predominant lipoprotein class.

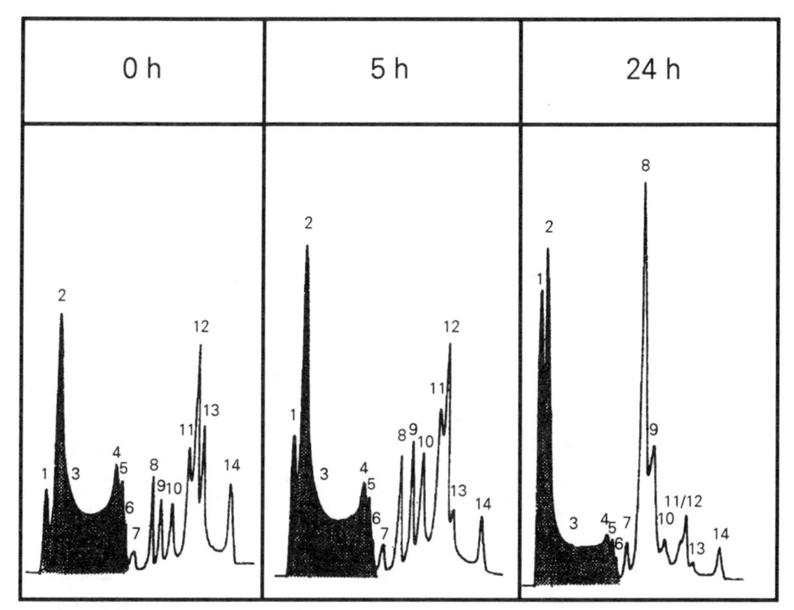

Fig. 5.20. Changes in lipoprotein profiles occurring under serum incubation at 37 °C *in vitro* .

9. *Interconversion of HDL subpopulations* in vitro

HDL contain numerous structurally and functionally different particle populations. The composition of HDL particles and their metabolic behaviour are continuously modified as a result of a dynamic exchange and net transfer of lipids between different HDL particles, between HDL and other lipoproteins, and between HDL and cells. Incubation of serum at 37°C *in vitro* results in changes of the size and composition of HDL particles. The transformation of HDL particles occurs in the presence of LCAT activity, apo B-containing lipoproteins and lipid transfer proteins. The absence or defect of some of these components can minimise or abolish those changes. It was reported that upon incubation larger and smaller HDL particles appear. We used isotachophoresis to characterise changes occurring among HDL subfractions upon serum incubation *in vitro* [31]. Serum incubation at 37°C leads to a time-dependent increase in the concentration of the fast-migrating apo AI-HDL particles (with a high affinity HDL binding site), while the slow-migrating LCAT-rich HDL particles (apo AI/A IV/LCAT-HDL) disappear (Figure 5.20). If an LCAT inhibitor is added to the sera before incubation *in vitro*, no significant changes are visible. These results indicate that LCAT plays a major role in the formation of high affinity binding apo AI-rich HDL particles by converting apo AI/AIV-rich particles into apo AI-rich particles. These results are underlined by the isotachophoretic analysis of the HDL pattern in patients with primary and secondary LCAT deficiency syndromes. In

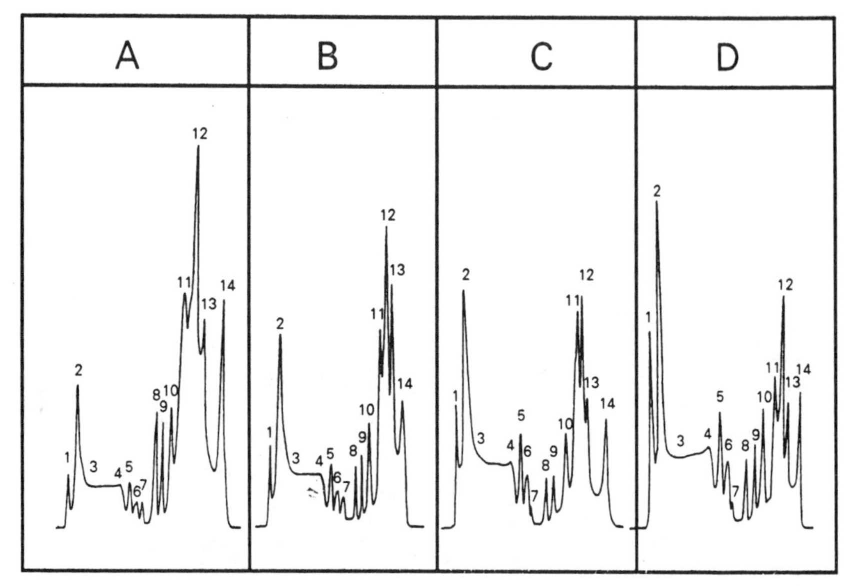

Fig. 5.21. Changes in serum ITP-lipoprotein profile of hyperlipidemic patients occurring under treatment with hyperlipidemic drugs. A before therapy; B after 1 month of treatment with 20 mg of Simvastatin (HMG-CoA reductase inhibitor); C after 1 month of treatment with 40 mg of Simvastatin; D after 1 month of treatment with combined therapy: 40 mg Simvastatin plus 8 g of cholestyramine.

these disorders the fast-migrating apo AI-HDL particles are almost absent and no interconversion of the HDL subpopulations is observed upon serum incubation.

Upon incubation the HDL subfractions were depleted in cholesterol esters and enriched in triglycerides, while opposite effects were observed among apo B-containing lipoprotein subpopulations indicating the action of cholesterol ester transfer proteins (CETP).

10. Therapeutic monitoring of lipoprotein subclasses

The rationale of the treatment with lipid lowering agents in hyperlipoproteinemia is based on the assumption that progression of the atherosclerotic process will be delayed or prevented. It is clear that atherosclerosis is a multifactorial process, and the levels and composition of individual lipoprotein subpopulations, among other factors, are of importance for the development of the disease. Therefore, effective lipid-lowering agents should correct abnormalities in lipoprotein composition and metabolism. We have determined the effect on HMG-CoA reductase inhibitor alone and in combination with a bile acid sequestrant on individual lipoprotein subpopulations separated by analytical capillary isotachophoresis. The results presented in Figure 5.21 have shown that HMG-CoA reductase inhibitors significantly reduce all subpopulations of apo B-containing lipoproteins. The addition of bile acid sequestrant further reduce LDL subpopulations (peaks 11 to 14) but enhance VLDL and IDL subpopulations (peaks 8 to 10). Under the therapy significant changes were observed among HDL subpopulations. The HMG-CoA reductase inhibitor increased fast-migrating apo AI-HDL subpopulations with high affinity to HDL receptors, and slow-migrating AI/AIV/LCAT-HDL subpopulations, which are good acceptors of cellular cholesterol.

The results indicate that ITP can be used for monitoring changes in lipoprotein subpopulations occurring upon treatment.

D. DEVELOPMENTAL CONCEPT OF LIPOPROTEIN ANALYSIS WITH ANALYTICAL CAPILLARY ISOTACHOPHORESIS - DIAGNOSTIC SUPPORT FOR DISORDERS OF LIPID AND LIPOPROTEIN METABOLISM

In clinical practice analysis of lipoproteins is needed to recognise and classify disorders of lipoprotein metabolism and to describe lipoprotein profiles, and to relate them to the risk of coronary heart disease.

1. Diagnostic strategy and interpretation of the data

Current lipid and lipoprotein analysis in the clinical laboratory induces the following parameter panel:

Step I:

- triglycerides
- total-cholesterol
- HDL-cholesterol (precipitation method.)
- LDL-cholesterol (Friedewald formula)
- alternatively semiquantitative lipoprotein electrophoresis
- 4°C storage of serum for chylomicrons

Step II:

- lipoprotein electrophoresis if β-VLDL has to be excluded
- apolipoprotein A-I
- apolipoprotein B
- lipoprotein(a)
- fibrinogen, *etc.*

Step III:

- apolipoprotein E polymorphism
- activity of lipoprotein lipase (LPL) and hepatic triglyceride lipase (HTGL)
- apolipoprotein C II and C III polymorphism
- LCAT activity
- LDL-receptor analysis
- apo B-3500 analysis (dysfunctional apo B)

- homocysteine
- lipid peroxides and modified lipoproteins
- molecular genetics of proven families

The major limitations of today's clinical practice are related to the fact, that beyond step I, there is an enormous increase in analytical efforts and costs necessary to answer additional clinical questions for further diagnostic and therapeutic discussions. Here, with a single isotachophoretic analysis (using lipophilic dyes) associated with a simple result interpretation algorithm, the following results could be generated:

i. Precursor/product relationships for the major lipolytic enzymes

a) the ratio of peaks 4+5+6 = slow-migrating HDL, resembling preβ-migrating HDL *versus* peaks 1+2 = fast-migrating HDL gives an estimate of LCAT-activity since slow-migrating HDL are converted by LCAT to fast-migrating HDL. At the same time cholesterol is transferred preferentially to peak 11 resembling IDL.
b) the ratio of peaks 1+2+4+5+6 = fast + slow-migrating HDL *versus* peak 3 = intermediate migrating HDL gives an estimate for the relationship between apo AI-rich HDL *versus* apo E and apo AII-rich HDL.
c) the ratio of peaks 7+8 (unhydrolysed triglyceride lipoproteins) *versus* peaks 9+10 provides a rough estimate of LPL activity in the metabolic overload syndrome and in the genetic defects of triglyceride metabolism.
d) the ratio of peaks 10+11 *versus* peaks 12+13+14 provides a rough estimate of HTGL activity in mixed hyperlipidemias.

ii. Presence of certain lipoprotein subclasses

a) remnants of triglyceride-rich lipoproteins = peaks 9 and 10
b) preβ-migrating $HDL_{1,2,3}$ = peaks 4,5,6 in ITP
c) Lp(x) (migrates behind peak 14)

iii. Estimation of receptor active particle populations

Within the lipoprotein subclasses there is no distribution of high affinity particle for the different lipoprotein receptor sites. Based on our results the following estimates can be made:

a) LDL receptor active particles (peaks 12 to 14)
b) Remnant receptor active particles (peaks 9 to 11)
c) HDL-receptor active particles (peaks 1+2)

iv. Postprandial changes of lipoprotein subclasses

Lipolysis studies can be done by oral fat load subjection and repetitive ITP at 0 h, 3 h and 5 h in order to monitor whether patients with borderline hyperlipidemia can still effectively metabolise oral fat.

From a theoretical point of view, especially from an academic perspective this approach might be criticised as scientifically imprecise. However, the guidelines for practical clinical decisions, which can be derived from this single laboratory analysis, would be enormously helpful.

2. Diagnostic support for disorders of lipid and lipoprotein metabolism

The results published before and described in the present review clearly indicate that ITP analysis of serum lipoproteins together with total triglycerides, total cholesterol and HDL cholesterol measurements can significantly improve the routine diagnosis for disorders of lipoprotein metabolism. ITP can provide the information about precursor/product relationships for the key enzymes of lipoprotein metabolism - LPL, HTGL, LCAT and estimate their activity; can give the information of the concentration of particles which are ligands for HDL receptors, remnant receptors and LDL receptors; can recognise remnant particles during one run and significantly improves the diagnosis of dysbetalipoproteinemia. Based on ITP analysis we can estimate the proportion between large triglyceride-rich VLDL and small cholesterol ester-enriched VLDL remnants and IDL and improve the diagnosis of familial combined hyperlipidemia, the most frequent disorder of lipoprotein metabolism associated with increased risk of CHD. ITP analysis is a very helpful tool in the estimation of the concentration of apo AI and apo AI/AII-containing HDL particles. In addition, apo AI-HDL particles with a high affinity to HDL receptors can be recognised. It can significantly improve the recognition of lipoprotein patterns and help to relate it to CHD risk. ITP can also be used for the diagnosis of new disorders of lipoprotein metabolism.

As described before disturbances in postprandial lipoprotein metabolism associated with increased accumulation of cholesterol ester-enriched VLDL and IDL particles and low HDL levels can be associated with increased risk of CHD. Therefore, there is a growing interest in studies describing triglyceride metabolic capacity.

ITP can be a method of choice to study changes in lipoprotein profile after a fat load and to relate these changes to CHD risk.

ITP can be also used for indirect lipoprotein turnover studies. We have analysed the lipoprotein profile in patients who were subjected to either selective immunoabsorption of LDL, lipoprotein(a), apo E-containing particles or plasmapheresis. The reappearance rate of individual lipoprotein subclasses can be related to secretion and plasma elimination rates.

We have realised that capillary ITP is an interesting tool to monitor lipoprotein subclass changes in new-borns. By comparison of normal babies with babies with fetal distress syndromes we found out that the pattern can be related to intestinal and liver development. Moreover, parental nutritional control in paediatric intensive care units is an important problem and here ITP with the use of capillary blood provides an ideal tool for patient monitoring.

In intensive care medicine under the conditions of long term high calorie parental nutrition capillary ITP can provide information which helps the clinician to avoid fat-embolism.

3. Advantages of analytical capillary isotachophoresis

We tested various methods for the analysis of lipoproteins in biological fluids, including free flow isoelectric focusing, gradient gel electrophoresis, carrier free zone electrophoresis, field step electrophoresis, immunoabsorption and free flow isotachophoresis. The major conclusions drawn from these evaluations are as follows:

i) A technique using carrier free electrophoresis is superior to gel or membrane systems.

ii) Net charge of lipoprotein particles as a separation principle is nearer to functional properties of individual lipoproteins than size. If particle size is used as the separation principle triglyceride and cholesterol ester-containing particles are separated which may have an identical size but completely different biological function.

iii) Zone electrophoresis has the general disadvantage of diffusion which is also present in small capillaries. Moreover, if the separation mixture contains components with large concentration differences it is difficult to reliably quantify "small" and 'large" peaks in the same run.

iv) Free flow isoelectric focusing is an interesting alternative, however, since focusing appears at the isoelectric point, particles rapidly tend to precipitate and then disintegrate. Therefore, despite these technique provides a high discrimination, it is not suitable for stable separation and quantification of macromolecules in whole body fluids. Moreover, isoelectric focusing is more difficult to automate. The same limitations can be related to field step electrophoresis.

v) Immunoabsorption and separation of lipoprotein particles according to the Alaupovic concept is an interesting tool to generate isolated lipoprotein particles for structural and functional analysis and therefore very helpful in lipoprotein research. However, the quantification of individual particles directly in whole serum has great limitations. A major disadvantage is that in order to generate reliable pathophysiologically relevant data numerous particle populations have to be analysed in sequential runs with different antibodies. Moreover, immunological properties of particles in whole serum are not stable and therefore numerous artefacts appear when individual apolipoproteins dissociate from the particles. This phenomenon appears, as we have analysed, even at the bench level. Another measuring problem is related to the antibody specificity itself. If Lp-AI is measured it contains also Lp-AI/AIV or other Lp-AI particle mixtures. Since the Lp-AI *versus* the Lp-AI/AIV particles have different functions it is unclear what the results really reflect and how many variables are included.

E. CONCLUSION

Analytical capillary isotachophoresis is an interesting new technique for lipoprotein analysis for the following reasons: a) In isotachophoresis the concentration- and zone sharpening-effect allows a simultaneous separation of major and minor lipoprotein subpopulations in the same run. Moreover, with the selective addition of spacer mixtures the separation conditions can be improved. b) In isotachophoresis lipoprotein analysis can be done directly from whole serum, plasma lymph and other biological fluids in a very short time. It can be carried out with only a few nanomols of sample and with the new generation of capillary electrophoresis systems an automation of the technique is not problematic. c) UV-detection and consequently the staining of the lipoproteins using Sudan black B is almost quantitative when different calibration curves are used for the individual lipoprotein classes (three different slopes: slope 1. for HDL subclasses slope 2. for triglyceride-rich particles and slope 3. for cholesterol-rich IDL and LDL). However, we think that the specificity of lipoprotein staining can be further improved in a new capillary electrophoresis system. The new laser-induced fluorescence detection allows new staining procedures with lipophilic fluorescent dyes for whole lipoproteins as well as fluorescence labelled monoclonal antibodies against apolipoproteins. Moreover, with the use of different fluorescent dyes the determination of multiparametric apolipoprotein profiles directly from whole serum will be possible. If either 2- or 3-parameter fluorescence detectors are developed, the fluorescence biochemistry, which is nowadays used in fluorescence-flow cytometry can be transferred to the capillary electrophoresis system technology.

The following parameter combinations would be of interest:

i) apo AI, apo B-48 and apo B-100
ii) apo B-100 and lipoprotein(a)
iii) apo E, cholesterol ester transfer protein and other transfer proteins
iv) apo C III_0 and apo C $III_{2 \text{ or } 1}$

With all these characteristics analytical capillary isotachophoresis can become the method of choice for a fast and reliable automated analysis of lipoprotein subpopulations in biological fluids in the clinical laboratory.

ABBREVIATIONS

ACAT, acyl-CoA:cholesterol acyltransferase; apo, apolipoprotein; ARDS, adult respiratory distress syndrome; CETP, cholesterol ester transfer protein; CHD, coronary heart disease; EDTA, ethylenediaminetetraacetic acid; HDL, high density lipoprotein; HLP, hyperlipoproteinemia; HMG-CoA, 3-hydroxy-3-methyl-glutaryl-coenzyme A; HPMC, hydroxypropylmethylcellulose; HTGL, hepatic triglyceride lipase; IDL, intermediate density lipoprotein; ITP, isotachophoresis; LCAT, lecithin:cholesterol acyltransferase; LDL, low density lipoprotein; Lp, lipoprotein; LPL, lipoprotein lipase; LRP, LDL-receptor related protein; UV, ultraviolet; VLDL, very low density lipoprotein.

REFERENCES

1. Alaupovic,P., Beckaert,E.D. and Koren,E., In *Biotechnology of Dyslipoproteinemias: Clinical Applications in Diagnosis and Control, Atherosclerosis Reviews*, **69**, 179-188 (1990).
2. Assmann,G., Schmitz,G. and Brewer,H.B., In *The Metabolic Basis of Inherited Disease*, 6th ed., pp. 1267-82 (1989) (edited by C.R. Scriver, A.L. Beandet, W.S. Sly and D. Valle, McGraw-Hill, New York).
3. Baldesten,A., *Sci. Tools*, **27**, 2-7 (1980).
4. Bard,J.M., Candelier,L., Agnani,G., Clavey,V., Torpier,G., Steinmatz,A. and Fruchart,J.C., *Biochim. Biophys. Acta*, **1082**, 170-176 (1991).
5. Beaumont,J.L., Carlson,L.A., Cooper,G.R., Fejfar,Z., Fredrickson,D.S. and Strasser,T., *Bull. WHO*, **43**, 891-908 (1970).
6. Biervliet,J.P., Rossenau,M., Bury,J., Caster,H., Stul,M.S. and Lamote,R., *Pediat. Res.*, **20**, 324-331 (1986).
7. Blanche,P.J., Gong,E.L., Forte,T.M. and Nichols,A.V., *Biochim. Biophys. Acta*, **665**, 408-419 (1981).
8. Chapman,M.J., Lapland,P.M., Luc,G., Forgez,P., Bruckert,E., Goulient,S. and Lagrange,D., *J. Lipid Res.*, 29, 442-458 (1988).
9. Cheung,M.C. and Albers,J.J., *J. Biol. Chem.*, **259**, 12201-12209 (1984).
10. Delmotte,P., *Sep. Purif. Methods*, **10**, 29-45 (1981).
11. Eisenberg,S., *J. Lipid Res.*, **25**, 1017-1058 (1984).
12. Everaets,F.M., Beckers,J.L. and Verheggen,T., *Isotachophoresis - Theory, Instrumentation and Applications.* (Elsevier, Amsterdam) (1976).
13. Francone,O.L., Guraker,A. and Fielding,C., *J. Biol. Chem.*, **264**, 7066-7072 (1989).
14. Gaal,O., Medgyesi,G.A. and Vereczkery,L., *Electrophoresis in the Separation of Biological Macromolecules.* (Akademiai Kiado, Budapest, and John Wiley & Sons, Chichester, New York) (1980).
15. Gabelli,C., *Curr. Opin. Lipidol.*, **3**, 208-214 (1992)
16. Gianturco,S.H., Via,D.P., Gotto,A.M. and Bradley,W.A., In *The Role of Receptors in Biology and Medicine*, pp. 167-180 (1989) (edited by A.M. Gotto and B.W. O'Malley, Raven Press, New York).
17. Gordon,M.J., Huang,X., Pentoney,S.L. and Zare,R.N., *Science*, **242**, 224-228 (1988).
18. Groot,P.H.E., Scheek,L.M., Havekes,L., van Noort,W.L. and van't Hoof,F.M., *J. Lipid Res.*, **23**, 1342-1353 (1982).
19. Groot,P.H.E., van Stiphout,W.A.H.J., Krauss,X.H., Jansen,H., van Toll,A., van Ramshorst,E., Chin-On,S., Hofman,A., Cresswell,S.R. and Havekes,L., *Arteriosclerosis*, **11**, 653-662 (1991).
20. James,R.W., Hochstrasser,D., Tissot,J.D., Funk,M., Appel,R., Barja,F., Pellagrini,C., Muller,A.M. and Pometta,D., *J. Lipid Res.*, **29**, 1557-1571 (1988).
21. Kohlon,T.S., Glines,L.A. and Lindgren,F.T., *Methods Enzymol.*, **129**, 26-45 (1986).
22. Koren,E., Puchois,P., Alaupovic,P., Fesmire,J., Kandoussi,A. and Fruchart,J.C., *Clin. Chem.*, **33**, 38-43 (1987).
23. Krauss,R.M. and Bruke,D.J., *J. Lipid Res.*, **23**, 97-104 (1982).
24. Krul,E.S., Tikkanen,M.J., Cole,T.G., Davie,J.M. and Schonfeld,G., *J. Clin. Invest.*, **75**, 361-369 (1985).
25. Musliner,T.A., Giotas,C. and Krauss,R.M., *Arteriosclerosis*, **6**, 79-87 (1986).

26. Musliner,T.A., McVicker,K.M., Josefa,J.F. and Krauss,R.M., *Arteriosclerosis*, **7**, 408-420 (1987).
27. Nichols,A.V., Krauss,R.M. and Musliner,T.A., *Methods Enzymol.*, **128**, 417-431 (1986).
28. Norum,K.R., Gjone,E. and Glomset,J.A., In *The Metabolic Basis of Inherited Disease*, 6th ed., pp. 1181-1194 (1989) (edited by C.R. Scriver, A.L. Beandet, W.S. Sly and D. Valle, McGraw-Hill, New York).
29. Nowicka,G., Brüning,T., Böttcher,A., Kahl,G. and Schmitz,G., *J. Lipid Res.*, **31**, 1947-1963 (1990).
30. Nowicka,G., Brüning,T., Grothaus,B., Kahl,G. and Schmitz,G., *J. Lipid Res.*, **31**, 1173-1186 (1990).
31. Nowicka,G., Dieplinger,H., Williamson,E., Böttcher,A., Stender,S. and Schmitz,G., (submitted 1993)
32. Nowicka,G., Gheeraert,P. and Schmitz,G., In *Intestinal Lipid and Lipoprotein Metabolism*, pp. 151-161 (1989) (edited by E. Windler and B. Greten, W.Zuchschwerdt, München).
33. Packard,C.J., Munro,A., Lorimer,A.R., Gotto,A.M. and Shepherd,J., *J. Clin. Invest.*, **74**, 2178-2192 (1984).
34. Patsch,J.R. and Patsch,W., *Methods Enzymol.*, **129**, 3-21 (1986).
35. Patsch,J.R., Miesenbock,G., Hopferwieser,T., Muhlberger,V., Knapp,E., Dunn,J.K., Gotto,A.M. and Patsch,W., *Arteriosclerosis*, 12, 1336-1345 (1992).
36. Scanu,A.M. and Spector,A., *Biochemistry and Biology of Plasma Lipoproteins.* (Dekker, New York) (1986).
37. Schmitz,G., Brüning,T., Williamson,E. and Nowicka,G., *Eur. Heart J.*, **11** (Suppl. E), 197-211 (1990).
38. Schmitz,G., Brüning,T., Williamson,E. and Nowicka,G., In *Treatment of Severe Dyslipoproteinemia in the Prevention of Coronary Heart Disease.* Vol.III., pp. 17-34 (1992) (edited by A.M. Gotto, M. Mancini, W.O. Richter and P. Schwandt, Karger GmbH, Munchen).
39. Schmitz,G., Nowicka,G., Giesel,J. and Oette,K., (submitted 1993).
40. Schmitz,G., Robenek,H., Lohmann,U. and Assmann,G., *EMBO J.*, **4**, 613-622 (1985).
41. Schmitz,G. and Williamson,E., *Curr. Opin. Lipidol.*, **2**, 177-189 (1991).
42. Sloop,Ch.H., Dory,L. and Roheim,P.S., *J. Lipid Res.*, **28**, 225-237 (1987).
43. Stein,O., Stein,Y., Lefevre,M. and Roheim,P.S., *Biochim. Biophys. Acta*, **878**, 7-13 (1986).
44. Sundaram,S.G., Shakir,K.M. and Margolis,S., *Anal. Biochem.*, **88**, 425-433 (1978).
45. Thompson,G.R., Teny,B. and Sniderman,A.D., *Am. Heart J.*, **113**, 514-519 (1987).
46. Weisgraber,K.M. and Mahley,R.W., *J. Lipid Res.*, **21**, 316-325 (1980).

Chapter 6

PREPARATION OF LIPID EXTRACTS FROM TISSUES

William W. Christie
The Scottish Crop Research Institute, Invergowrie, Dundee (DD2 5DA), Scotland

A. INTRODUCTION

Quantitative isolation of lipids from tissues in their native state and free of non-lipid contaminants must be accomplished before analysis is attempted. Although this first step can be tedious, time-consuming and relatively uninteresting, any slackness will almost certainly result in the loss of specific components and in the production of artefacts. When appreciable amounts of free fatty acids, diacylglycerols, phosphatidic acid and certain other lipids are detected in extracts, for example, faulty storage of the tissues or of lipid isolates or inappropriate extraction conditions are generally indicated. Many solvents or solvent combinations can be used to extract lipids from tissues, but care must be taken to ensure that lipolytic and other enzymes are deactivated and that the recovery is complete. Any non-lipid contaminants that are co-extracted must then be eliminated from the recovered lipids by washing or other solvent partition procedures, before the sample can be subjected to detailed analysis. At every stage, precautions must be taken to minimise the risk of hydrolysis of lipids or of autoxidation of unsaturated fatty acids.

The choice of extraction procedure will depend on the nature of the tissue matrix, and for example, whether the sample is of animal, plant or microbial origin. Another factor is the amount of information required from the sample; many simple extraction procedures can be used for triacylglycerol-rich tissues such as adipose tissue or oil seeds if the main lipid class only is required for analysis. On the other hand, if a detailed knowledge of every minor lipid class is required, no short cuts are possible. Particular care is necessary for the more polar near-water-soluble lipids such as gangliosides or polyphosphoinositides. This review describes the principles and good practice of tissue handling, lipid extraction and elimination of non-lipid contaminants and the many potential pitfalls. The topic has been reviewed in general terms elsewhere [17,47,73-76,85,123,124], and in addition there exist many specialist reviews that deal with the extraction of specific lipid classes such as gangliosides or inositides (see the appropriate sections below).

B. STORAGE OF TISSUES AND PRELIMINARY TREATMENTS PRIOR TO EXTRACTION

All tissues, whatever their origin, should ideally be extracted immediately after removal from the living organism, so that there is little opportunity for changes to occur to the lipid components. It is of course essential that plasma or tissue samples be taken with the minimum of stress or trauma, otherwise lipolysis will occur *in vivo*. Discussion of appropriate surgical or collection procedures is outwith the scope of this review, but it has been dealt with in relation to plasma elsewhere [73,74]. With plant and heart or brain tissues say, where tissue enzymes are especially active, rapid extraction is essential. When this is not feasible, the tissue should be frozen as rapidly as possible, for example with dry ice or liquid nitrogen, and stored in sealed glass containers at -20°C in an atmosphere of nitrogen. A storage temperature of as low as -60°C has been recommended for plasma samples of clinical origin [73], and even lower temperatures may be required in some circumstances (see below). The process of freezing tissues will damage them irreversibly, because the osmotic shock together with formation of ice crystals disrupts the cell membranes. When the original environment of the tissue lipids is altered in this way, they encounter enzymes from which they are normally protected. Especially troublesome are the lipolytic enzymes which can hydrolyse lipids on prolonged standing, even at -20°C, and contact with organic solvents can facilitate the process. For example, rapid deacylation of phospholipids was observed in a bacterium frozen solid at -16°C, with the rate at -10°C being greater than that at 39°C in some circumstances [37]. Appreciable lipolysis also occurred in adipose tissue stored at -8°C [29], and some hydrolysis of phosphatidylethanolamine was detected in brain tissue stored for two hours at 0°C [70]. Similar results have been presented for plasma [66,71,72,87] and shrimp [92]. Slow thawing can have a devastating effect on the lipids of tissues.

Lipid peroxidation can also be troublesome in tissues stored at -20°C and even at -70°C, and it has been recommended that samples for free radical assay be stored at -196°C [115].

Phenomena of the same kind have been observed in many plant and animal tissues on storage. The presence of large amounts of unesterified fatty acids, diacylglycerols, phosphatidic acid or lysophospholipids in lipid extracts must be an indication that some permanent damage to the tissues and thence to the lipids has occurred. Such lipids are powerful surfactants and have been found to have disturbing effects on enzymes and membrane functions *in vitro* and *in vivo*, so the high concentrations sometimes reported in the literature are clearly incorrect. In plant tissues in particular, the enzyme phospholipase D is released and can attack phospholipids, so that there is an appreciable accumulation of phosphatidic acid and related compounds. For example, phosphatidylmethanol was found to be produced by phospholipase D-catalysed transphosphatidylation during extraction of developing soybean seeds with chloroform-methanol [89]. All the acyl lipids in potatoes were hydrolysed in minutes upon homogenisation [28]. Other alterations to lipids can occur that are more subtle and so are discerned less easily. For example, losses of galactolipids can take place without any obvious accumulation of partially hydrolysed intermediates [28,93]. Lipoxygenases can cause artefactual formation of oxygenated fatty acids, and autoxidation can be troublesome. Problems in stabilising plant membranes for lipid compositional analyses have been reviewed [14].

Often these changes are marginal in their overall importance, since alterations to the main lipid components may be small. On the other hand, they can make a crucial difference to the concentrations of some important lipid metabolites. The precise free fatty acid and 1,2-diacylglycerol concentrations of tissues are recognised to be key metabolic parameters. In an especially thorough study, Kramer and Hulan [49] observed that the free fatty acid concentrations obtained when heart tissue was frozen rapidly and pulverised at dry ice temperatures before extraction were only about 15% of the values when similar tissues were extracted by more widely used techniques, *i.e.* by extracting directly with a homogeniser of the rotating blade type at 0°C. With the latter, autolysis was presumed to occur during extraction. The levels of diacylglycerols were also three fold higher when the latter technique was used, a finding later confirmed by others [1]. Similarly, lysophosphatidylcholine, which had earlier been reported to be a major constituent of chromaffin granules in the adrenal gland, was found to be absent when the tissues were frozen in liquid nitrogen immediately after dissection [4]. Increases in the concentration of this lipid were found in heart tissues stored at 0°C or -20°C [63], but experiments involving studies of the concentration of lysophosphatidylcholine in intact tissues by means of ^{31}P-nuclear magnetic resonance spectroscopy confirmed that it was essentially absent [62,65].

It has also been demonstrated that enzymic oxidation can cause not only losses of unsaturated fatty acids, but apparently of intact lipids [81]. Hydroperoxide groups of oxidised lipids apparently reacted to form covalent bonds with the proteins of membranes, from which they were released only on treatment with bacterial proteases. Presumably, similar effects would be seen with autoxidised lipids in tissues.

Various pre-treatments have been suggested for de-activating enzymes so that tissues can be stored for longer periods. The lipases in small samples of plant [36] or animal [29] origin have been denatured by plunging into boiling water for short periods, and the shelf-life of samples treated in this way was reportedly prolonged to a considerable extent. Boiling with dilute acetic acid solution appeared to have a similar effect [82,83]. However, there is a need for practical re-evaluation of these procedures in the light of modern knowledge before they can be recommended. Boiling plant tissues before extraction certainly de-activated lipoxygenases and increased the recovery of linoleic and linolenic acids [107]. Conventional freeze drying and perchloric acid pre-extraction of tissues were found to produce artefacts from phospholipids; acetone desiccation did not, but this would cause losses of simple lipids [62].

The use of appropriate solvents for extraction *per se* can also limit enzymic degradation of plant lipids (see Sections C and F.4 below).

While it has sometimes been recommended that tissues be stored in saline solution, it is probably better to keep them dry in an atmosphere of nitrogen in all-glass containers or in bottles with Teflon™-lined caps at the low temperatures discussed above. Endogenous tissue antioxidants generally provide sufficient protection against oxidation under these conditions, although this may not be true for serum [71,91]. For example, appreciable decreases in the content of polyunsaturated fatty acids in plasma were observed on prolonged storage at -20°C, and these were inversely correlated with vitamin A (antioxidant) levels [91]. Plastic bags, vials or other containers should be avoided scrupulously for storage purposes. As soon as possible, tissues should be homogenised and extracted with solvent at the lowest temperature practicable and certainly without being allowed to thaw. Safe storage of plasma lipoproteins, where both the protein and lipid components are potentially unstable, is a major topic in its own right and cannot be discussed here.

Similar precautions must be taken for the storage of lipids, after they have been extracted from tissues. In this circumstance, the principle danger is a loss of unsaturated fatty acid components through autoxidation. For example, rapid autoxidation of cardiolipin was found to occur on storage in chloroform [80]. Lipid extracts should therefore be stored at the lowest practical temperature, in an inert atmosphere, in an apolar solvent and in the presence of antioxidants [17].

C. EXTRACTION OF TISSUES WITH SOLVENTS

In addition to storage considerations, there are two main facets to any practical procedure for extracting lipids from tissues, *i.e.* firstly, exhaustive extraction and solubilisation of the lipids in organic solvents and secondly, removal of non-lipid contaminants from the extracts. They are discussed in this and the next section respectively.

Many different solvents will dissolve pure single lipid classes, but they are only suitable for extracting lipids from tissues if they can overcome the strong forces of association between tissue lipids and other cellular constituents, such as proteins and polysaccharides. However, even polar complex lipids, which do not normally dissolve easily in non-polar solvents, can sometimes be extracted with these when they are in the presence of large amounts of simple lipids such as triacylglycerols. Therefore, the behaviour of a given solvent as a lipid extractant for a specific tissue cannot always be predicted. In order to release all lipids from their association with cell membranes or with lipoproteins, the ideal solvent or solvent mixture must be fairly polar. Yet, it must not be so polar that it reacts chemically with the lipids nor that triacylglycerols and other non-polar simple lipids do not dissolve and are left adhering to the tissues. If chosen carefully, the extracting solvent may have a function in preventing any enzymatic hydrolysis, but *vice versa* it should not stimulate any side reactions.

There is an increasing awareness of the potential toxicity of solvents to analysts, and this is another factor that must be taken into consideration when selecting a solvent mixture, especially if the laboratory is not adequately equipped with fume hoods or other ventilation. No solvent is completely safe.

Those factors affecting the extractability of lipids by solvents have been reviewed comprehensively by Zahler and Niggli [123]. The two main structural features of lipids controlling their solubility in organic solvents are the hydrophobic hydrocarbon chains of the fatty acid or other aliphatic moieties and any polar functional groups, such as phosphate or sugar residues, which are markedly hydrophilic. Any lipids lacking polar groups, for example triacylglycerols or cholesterol esters, are very soluble in hydrocarbons such as hexane, toluene or cyclohexane and also in moderately polar solvents such as diethyl ether or chloroform. In contrast, they are rather insoluble in a polar solvent such as methanol. The solubility of such lipids in alcohols increases with the chain-length of the hydrocarbon moiety of the alcohol, so they tend to be more soluble in ethanol and completely soluble in butan-1-ol. Similarly, lipids with fatty acyl residues of shorter chain-length tend to be more soluble in more polar solvents; tripalmitin is virtually insoluble in methanol but tributyrin dissolves readily. Unless solubilised by the presence of other lipids, polar lipids, such as phospholipids and glycosphingolipids, are only slightly soluble in hydrocarbons, but they dissolve readily in more polar solvents like methanol, ethanol or chloroform. Such solvents with high dielectric constants and polarity are required to overcome ion-dipole interactions and hydrogen bonding. Tabulated data on the solubilities of a limited range of "typical" lipids are available [98].

Analysts should be aware that water is also a solvent for lipids, contrary to some definitions of the term, and water in tissues or that used to wash lipid extracts, for example, can alter the properties of organic solvents markedly. Most complex lipids are slightly soluble in water and at least form micellar solutions, and lipids such as gangliosides, polyphosphoinositides, lysophospholipids, acylcarnitines and coenzyme A esters are especially soluble (see Section G).

Lipids exist in tissues in many different physical forms. The simple lipids are often part of large aggregates in storage tissues, such as oil bodies or adipose tissue, from which they are extracted with relative ease. In contrast, complex lipids are usually constituents of membranes, where they occur in a close association with such compounds as proteins and polysaccharides, with which they interact, and they are not extracted so readily. These interactions are only very rarely through covalent bonds, and in general weak hydrophobic or Van der Waals forces, hydrogen bonds and ionic bonds are involved. For example, the hydrophobic aliphatic moieties of lipids interact with the non-polar regions of the amino acid constituents of proteins, such as valine, leucine and isoleucine, to form weak associations. Hydroxyl, carboxyl and amino groups in lipid molecules, on the other hand, can interact more strongly with biopolymers *via* hydrogen bonds. Lipids such as the polyphosphoinositides are most likely bound to other cellular biopolymers by ionic bonds, and these are not easily disrupted by simple solvation with organic solvents. It is usually necessary to adjust the pH of the extraction medium to effect quantitative extraction in this instance. In addition, purely mechanical factors can limit the extractability of lipids. The helical starch (amylose) molecules in cereals form inclusion complexes with lysophosphatidylcholine, for example, limiting its accessibility to solvents. Also, cell walls in some microorganisms are rather impermeable to solvents, especially in the absence of water, which must be added to cause swelling of cellular polysaccharides. Some fatty acid or other alkyl moieties may indeed be linked directly to proteins or polysaccharides by covalent bonds, and then the optimum isolation procedure is likely to be one more suited to the analysis of the biopolymer rather than of the lipid.

In order to extract lipids from tissues, it is necessary to use solvents which not only dissolve the lipids readily but overcome the interactions between the lipids and the tissue matrix, and it is essential to perturb both the hydrophobic and polar interactions at the same time. No single pure solvent appears to be suitable as a general-purpose lipid extractant, although evidence has been presented that ethanol (20 mL/g tissue) under reflux for five minutes will extract essentially all the lipids from liver homogenates [57]. As with much other interesting work, the method was not extended to other tissues, and there is some danger of transesterification occurring as a side reaction under these conditions (see Section E below). While there are limitations to its use and alternatives are frequently suggested, most lipid analysts have accepted that a mixture of chloroform and methanol in the ratio of 2:1 (v/v) will extract lipids more thoroughly from animal, plant or bacterial tissues than other simple solvent combinations (the endogenous water in the tissues should be treated as a ternary component of the system). Since the publication of a classic paper on the subject by Folch, Lees and Stanley [26] in 1957, this has become the standard against which other methods are judged (although there are earlier applications of these solvents [25,106]).

Schmid [96] has addressed the problem of why chloroform-methanol (2:1, v/v) is such a good lipid extractant. The capacity of chloroform to associate with water molecules, presumably by weak hydrogen bonds, is a key property. Provided that the ratio of chloroform-methanol to tissue (assumed to be mainly water) is greater than 17:1, the equivalent of 5.5% water can be solvated and remain in a single phase. In contrast, mixtures of methanol with carbon tetrachloride or tetrachlorethylene, which lack the active proton of chloroform, were only able to solubilise relatively small amounts of water. Many practical methods have been developed for chloroform-methanol extraction and these are discussed in Section F below.

There are disadvantages, however. In addition to the toxicity problem, which is controllable in a well-ordered laboratory, the mixture is a potent irritant to skin. Neither chloroform nor methanol is completely stable and both were found to generate acidic by-products, which could catalyse esterification of free fatty acids or transesterification of lipids (see also Section E below) [99].

Chloroform-propan-2-ol mixtures have been recommended for the extraction of erythrocytes [9,27].

Dichloromethane-methanol (2:1, v/v) was found to give identical results to chloroform-methanol in the extraction of plasma and liver in one study [12], and the lower toxicity of dichloromethane was regarded an advantage.

Propan-2-ol-hexane (3:2, v/v) mixtures have probably been the second most frequently used solvent for the extraction of lipids from animal tissues, partly because of the good extractive properties *per se* but largely for the much lower toxicity [34,84]. Better recoveries of prostaglandins were reported with this mixture than with the Folch procedure [94], but it does not extract gangliosides quantitatively. Others used propan-2-ol-hexane (20:78, v/v) for extraction as part of a procedure for the isolation of cerebrosides and sulphatides from brain tissues [121]. Methanol-hexane (1:1, v/v) appears to be an odd choice for the extraction of lipids, because of the lack of complete miscibility, but it was reported to give excellent results with leaf tissue, and indeed higher concentrations of sensitive lipids such as phosphatidylinositol were found than with the Folch procedure [105]. Hexane-ethanol (5:2, v/v) was recommended for the extraction of ubiquinone [41,108], and may have wider applications [11]. Similarly, heptane-ethanol with the surfactant sodium dodecyl sulphate added has been recommended for determining vitamin E/lipid ratios in animal tissues [10].

Systematic studies of the solubilities of certain lipids in toluene-ethanol mixtures indicated that this combination might have superior properties to chloroform-methanol, but it does not appear to have been tested adequately with complex lipids or with samples of real biological interest [97]. Benzene-ethanol [61], benzene-methanol [104] and propan-2-ol-benzene-water (2:2:1, v/v) [2] also gave excellent results with some samples, but such potentially toxic mixtures are best avoided.

Butan-1-ol saturated with water has been recommended for the extraction of cereals or wheat-flour [60,68], in which the lipids may be in close association with starch, some in the form of inclusion complexes. The structure of the starch granules appears to be the most important factor, however [67]. This solvent may have wider uses, for example for quantitative extraction of lipids which are relatively soluble in water, such as lysophospholipids [7,31] or acylcarnitines [56,69]; hexan-2-ol has even been recommended for the latter [69]. Butan-1-ol-diisopropyl ether (2:3, v/v) was reported to be suitable for complete extraction of lipids from plasma, without causing denaturation of the proteins [13].

It is well known that diethyl ether or chloroform alone are good solvents for lipids, yet they are poor extractants of lipids from tissues. They can, however, have some practical value for the isolation of the non-polar lipids from triacylglycerol-rich tissues, such as oil seeds or adipose tissue, as they do not extract significant amounts of non-lipid contaminants at the same time. When they are used to extract plant tissues, these solvents also enhance the action of phospholipase D [42] unfortunately, as does butan-1-ol [20]. Propan-1-ol and propan-2-ol strongly inhibit this reaction and the latter, which has the lower boiling point, has been recommended for use with plant tissues, as a preliminary extractant especially [43,77,78].

While simple lipids and glycolipids dissolve readily in acetone, it will not dissolve phospholipids readily and indeed is often used to precipitate them from solution in other solvents, in effect as a crude preparative procedure. The lipid mixture is usually dissolved in diethyl ether and then four volumes of cold anhydrous acetone is added to precipitate the phospholipids [33]. On the other hand, endogenous water and the solubilising effects of other lipid components may permit acetone to extract more phospholipids from animal or plant tissues than might be predicted from a knowledge of the solubility of lipid standards in the pure solvent. For example, ethyl acetate-acetone-water (2:1:1 by volume) has been recommended for extracting lipids from cultures of human cells [102,103]. Acetone has also been recommended as a preliminary extraction solvent, before conventional chloroform-methanol extraction [59]. A disadvantage is that it can react with certain lipids to produce artefacts (see Section E below). As glycolipids are soluble in acetone, chromatographic solvents containing this solvent are frequently utilised in the separation of glycolipids from phospholipids.

In recent years, supercritical fluids have been evaluated as extractants for lipids (reviewed elsewhere [6,50]). While these appear to hold promise for selected simple lipids, there appears to be little prospect for more general use at the moment.

As an alternative to conventional solvent extraction, the technology of column chromatography has been adapted to the purpose in specific circumstances (see Section G.3 below, for example, and reviewed elsewhere [19]).

D. REMOVAL OF NON-LIPID CONTAMINANTS FROM EXTRACTS

When polar organic solvents are used to extract lipids from tissues, they tend to co-extract appreciable amounts of natural non-lipid materials, such as amino acids, carbohydrates, urea and even salts, as contaminants. A variety of procedures have therefore been developed to eliminate these, ideally without causing losses of lipids. For example, one of the simplest methods consists in evaporating the polar solvents, followed by dissolving the residual lipids in a small volume of a relatively non-polar solvent, such as hexane-chloroform (3:1, v/v), leaving many of the extraneous non-lipid substances behind [57]. Such a clean-up is rarely complete so the procedure is little used, although it should not be overlooked when large numbers of similar samples have to be purified for routine analysis by less demanding techniques, such as thin-layer chromatography (TLC). Other procedures which have been tried but with limited success only, include dialysis, adsorption and cellulose column chromatography, electrodialysis and electrophoresis.

Much of the contaminating material can be removed from chloroform-methanol (2:1, v/v) mixtures simply by shaking with one fourth the volume of water, or better a dilute aqueous salt solution (*e.g.* 0.88% potassium chloride) as described in the classic paper by Folch, Lees and

Stanley [26]. The solvents then partition into two layers or phases, the lower consisting of chloroform-methanol-water in the ratio 86:14:1 (by volume) with an upper phase in which the proportions are respectively 3:48:47. The purified lipid is contained in the lower phase, which amounts to about 60% of the total volume, while the upper phase contains the non-lipid contaminants. Unfortunately, any gangliosides which may have been present also partition into the upper layer. These are minor compounds and their analysis is rather specialised, so a simple washing procedure of this kind yields satisfactory lipid samples for most purposes. When they are required for further analysis, gangliosides can be recovered from the Folch upper phase by dialysis followed by lyophilisation (see Section G.3 below) [40].

It is not always recognised how important it is that the proportions of chloroform, methanol and water in the combined phases should be as close to 8:4:3 by volume as is practicable. When it is necessary to wash the lower phase again to ensure the elimination of all the non-lipid contaminants, methanol-water (1:1, v/v), *i.e.* a mixture similar in composition to the upper phase, should be used in order to maintain the correct proportions, otherwise some loss of polar lipids can occur.

In a successful adaptation of the above method, that is especially suited to large samples with a high water content, Bligh and Dyer [8] took into account the water already present in the samples when adding further water in the washing step. As this procedure uses smaller volumes of chloroform and methanol, it is both economical and convenient.

A more time-consuming method of removing non-lipid contaminants, if more elegant and complete, consisted in carrying out the washing procedure by liquid/liquid partition chromatography on a column. The aqueous phase (the 'Folch' upper layer) is immobilised on a column of a hydrophilic dextran gel, such as Sephadex G-25™, while the organic lower phase is passed through the column. While this type of lipid purification procedure was first developed by Wells & Dittmer [113], a modification described by Wuthier [120] is simpler and thus more suitable for large numbers of samples. In brief, chloroform, methanol and water in the ratio 8:4:3 by volume were partitioned as in a conventional 'Folch' wash, the column of Sephadex G-25™ was packed in suspension in the upper phase and the crude lipid extract was applied to the column in a small volume of the lower phase. While lipids were eluted rapidly by further lower phase, contaminants of low molecular weight remained on the column. Gangliosides and non-lipids could be recovered from the column by washing with 'Folch' upper phase, and the column regenerated for further use. The technique can also be used to remove acid, alkali and salts from lipid samples. An alternative simple procedure was described by others [117]. In contrast, a more complicated Sephadex™ column procedure was described [101] in which larger amounts of lipids were purified and gangliosides were obtained in a discrete fraction free of non-lipid contaminants. As various bile acids were obtained in distinct fractions from the conventional lipids, the method appeared to be particularly suited to the analysis of bile lipids [90]. However, all of these methods are very time-consuming, and as a result they have never achieved widespread use. Such column procedures are rather different in principle from a solid-phase extraction method described below (Section G.3).

An alternative approach consisted in pre-extraction of tissues with 0.25% acetic acid, which deactivated the enzymes (see Section D above), and simultaneously removed all potential contaminants of the lipid extracts [82,83]. Although this procedure appears only to have been applied to brain tissue and soybeans to date, it might repay further investigation.

E. ARTEFACTS OF EXTRACTION PROCEDURES

The production of free fatty acids, diacylglycerols, phosphatidic acid or lysophospholipids, due to faulty storage of tissues prior to extraction is discussed in Section B above. Extraneous substances can be introduced into lipid extracts from innumerable sources, and these are discussed elsewhere in this volume (Christie, Chapter 2). Plasticisers are especially troublesome.

When chloroform-methanol or indeed any alcoholic extracts, that contain lipids, are heated or stored for long periods in the presence of small amounts of sodium carbonate or bicarbonate of tissue origin, base-catalysed transesterification can occur and appreciable amounts of methyl esters may be found in the extracts [55]. By adjusting the pH of the extraction medium to 4 to 5, the problem can be circumvented [55,109]. Similar findings are often described, and it is possible that both acidic and basic non-lipid contaminants may catalyse the same reaction. Tissue acyltransferases can also catalyse ethyl ester formation [64]. However, small amounts of methyl and ethyl esters do appear to occur naturally in some tissues, and when they are detected confirmation should be obtained of their natural origin. Artefacts of extraction or storage can be eliminated simply by extracting the tissues with solvents that do not contain any alcohol, such as diethyl ether [23], hexane [44], acetone [46] or acetone-chloroform [54], and repeating the analysis for methyl esters on this material.

While 6-*O*-acyl-galactosyldiacylglycerols are known to be natural components of some plant tissues, it appears that they may also be formed as artefacts by acyl transfer from other lipids when cells are disrupted; they are found in much smaller amounts when the tissues are homogenized in the presence of the extracting solvent [38]. Similar difficulties may be encountered with homogenates of bacteria [100].

Some rearrangement of plasmalogens may occur when they are stored for long periods in methanol [110].

Acetone should not be used in the analysis of tissues rich in polyphosphoinositides, such as brain lipids, as it causes rapid dephosphorylation [22,114]. An acetone derivative (imine) of phosphatidylethanolamine was reported to be formed *in vitro* during extraction of freeze-dried tissue with acetone [3,39].

F. PRACTICAL EXTRACTION PROCEDURES

1. Some General Comments

All solvents contain small amounts of lipid-like contaminants and only those of highest quality need not be distilled routinely before use. Plastic apparatus and containers (other than those made from Teflon™) should be avoided at all costs, because plasticisers (usually diesters of phthalic acid) are leached out surprisingly easily and may appear as spurious peaks on chromatograms and affect UV spectra, for example. Organic solvents will extract substantial amounts of these compounds, and wet animal tissues alone in contact with plastic can extract small amounts.

Polyunsaturated fatty acids will autoxidise very rapidly if left unprotected in air. Although natural tissue antioxidants such as tocopherols may afford some protection, it is advisable to add an additional antioxidant such as BHT ("butylated hydroxy toluene" or 2,6-di-*tert*-butyl-*p*-cresol) at a level of about 25 mg/L to the solvents used for storage purposes. This need not interfere with later chromatographic analyses, as it tends to evaporate on removal of solvents or appears in conveniently empty regions of chromatograms [18]. Whenever practicable, extraction procedures should be carried out in an atmosphere of nitrogen, and both tissues and tissue extracts should be stored at -20°C under nitrogen. It is helpful to deaerate solvents by flushing them with nitrogen or helium before use.

If a blender is used to homogenise tissues, it should be one in which the drive to the knives or grinders is from above, so that there is no contact between solvent and any greased seals or bearings. Lyophilised tissues are particularly difficult to extract and it may be necessary to rehydrate them before extraction to ensure quantitative recovery of lipids.

When the aqueous and organic phases are partitioned in an extraction procedure, it should be noted that centrifugation may be necessary or of assistance in ensuring complete separation of the layers. This also serves to compact the interfacial layer, rendering it easier to recover for further extraction if need be.

Solvents should be removed from lipid extracts under vacuum in a rotary film evaporator at a temperature only a little above ambient and no higher than 40°C. When a large amount of solvent must be evaporated, the extract should be concentrated to a small volume and then transferred to as small a flask as is convenient so that the lipids do not dry out as a thin film over a large area of glass. While there is no need to bleed nitrogen in continuously during the evaporation process, as the solvent vapours effectively displace any air, the vacuum should be broken eventually with nitrogen. Lipids should not be left in the dry state, but should be taken up in or covered by an inert non-alcoholic solvent such as hexane. The last traces of water may be removed by co-distillation with ethanol or toluene.

As cautioned earlier, it should never be forgotten that chloroform and methanol are highly toxic and should only be used in well-ventilated areas. The single operation most likely to introduce appreciable amounts of solvent vapour into the atmosphere is filtration. Mixtures of chloroform and methanol, especially, are powerful irritants when they come into contact with the skin. (A colleague of the author was taught a painful lesson when he spilt a large volume down his trousers!).

2. The "Folch" Procedure

A large number of different extraction procedures, varying in the nature of the solvents, the method of homogenising, removal of contaminants and many other aspects, have been described in the literature. Unfortunately, a high proportion of these have been tried in only one laboratory or with a limited range of tissues or organisms. Only rarely have they been tested exhaustively and compared rigorously with a widely used procedure, such as that of Folch *et al.* [26] or Bligh and Dyer [8]. Until this is done, whatever the virtues of alternatives, the latter two procedures are likely to continue in their position of pre-eminence. An extraction procedure must be selected to give the optimum yield of representative lipids in as practical a manner as possible. While a method chosen for routine analysis of the cholesterol content of large numbers of clinical samples say must be accurate and reproducible, it need not incorporate modifications to recover every trace of such complex lipids as gangliosides or inositides. If the latter compounds are required for analysis, however, there is no alternative but to use complex and exhaustive extraction procedures.

For consistent results with any method, a strict protocol must be adopted that follows the principles laid down by the originators of the method. With the procedure of Folch *et al.* [26], it is essential that the ratio of chloroform, methanol and saline solution in the final mixture be close to 8:4:3. On the other hand, some variation in the approach to attaining the optimum concentrations may be possible. The tissue can be homogenised initially in the presence of both solvents, for example, although better results are often obtained if the methanol (10 mL/g tissue) is added first, followed after brief blending by chloroform (20 volumes). More than one extraction may be needed, but with most tissues the lipids are removed almost completely after two or three treatments. Generally, there is no need to heat the solvents with the tissue homogenates, but this may sometimes be necessary, for example with wet bacterial cells [111]. Treatment with acid [79] or proteases [86] may also be of benefit in this instance. However, the extractability of tissue is variable and depends both on the nature of the tissue and of the lipids. For example, the extractability of gangliosides from brain is highly dependent on the concentration of monovalent cations in the tissue. It may be necessary in some instances to employ more stringent extraction procedures; Ways and Hanahan [112] recommended separate re-extraction with chloroform on its own followed by methanol alone, while others [90] advised that a five-stage extraction procedure, using both acidic and basic solvent systems, was required for exhaustive recovery of lipids from tissues.

3. *The Bligh and Dyer Method*

The Bligh and Dyer method [8] is probably the second most common extraction procedure for lipids, and it is probably the most misunderstood and abused. It was developed as an economical method of extracting the lipids from large volumes of wet tissue, from frozen fish specifically, with the minimum volume of solvent. In essence, the endogenous water in the tissue was considered as a ternary component of the extraction system and sufficient chloroform and methanol were added to give a single phase system for homogenisation. After filtering, the residue was re-homogenised with fresh chloroform to ensure that the simple lipids were extracted completely, and the combined organic layers were added to fresh saline solution which produced a biphasic mixture. The chloroform layer contained the lipid of course.

If applied correctly to wet tissues, this procedure gives good recoveries of the more important lipid classes. It is rapid and therefore suited to many routine applications. However, too much should not be expected of it and incomplete recovery of the metabolically important but quantitatively minor acidic lipids is inevitable. When it has apparently been found wanting, the author has sometimes noted that the strict protocol laid down by Bligh and Dyer [8] has not always been followed.

4. *Extraction of Plant Tissues*

The lipids of plant material and photosynthetic tissues especially are liable to undergo extensive enzyme-catalysed degradation when extracted with chloroform-methanol mixtures, as discussed in Section E above. The problem is best overcome by means of a preliminary extraction with propan-2-ol [43]. The favoured method [77,78] is to extract the plant tissue with 100 volumes of propan-2-ol, and then to re-extract the solid residue with chloroform-methanol (2:1, v/v). After evaporation of the solvent, the crude extract is taken up in fresh chloroform-methanol and given a "Folch" wash (see section F.2 above) to eliminate non-lipid contaminants. This appears to be a milder method than extraction with ethanol-trichloroacetic acid (3:1, v/v) as has been suggested [45].

G. SOME SPECIAL CASES

Problems are most often encountered with lipids that contain highly hydrophilic functional groups. With the common biphasic extraction systems, these lipids may partition into the aqueous layer and can be lost. Improved recoveries can then be obtained by utilising monophasic extraction media, by using salting-out procedures to force these lipids into the organic phase or by recovering the organic material from the aqueous phase.

1. Acyl-Coenzyme A Esters

A monophasic extraction medium has been recommended for quantitative recovery of acyl-coenzyme A esters by Christiansen [16]. In essence, a conventional chloroform-methanol (2:1, v/v) extract was taken to dryness without the washing step, and the residue was divided into chloroform- and methanol-soluble fractions, which were only combined at the time of analysis. A comparable approach has been suggested for phospholipids [48]. On the other hand, it has been demonstrated that butan-1-ol is capable of extracting acyl-CoA esters rather efficiently, without formation of emulsions [5]. Others have used a salting out approach, *i.e.* a preliminary extraction with propan-2-ol-phosphoric acid, partition with hexane to remove simple lipids, addition of a saturated ammonium sulphate solution and then extraction with chloroform-methanol (1:2, v/v) [58,119].

Rosendahl and Knudsen [88], in contrast, have indicated that the problem may be more difficult than has hitherto been realised. They were able to attain consistent yields of about 55% of the total coenzyme A esters only by combining two-phase extraction with salting out and the addition of acyl-CoA-binding protein.

2. Lysophospholipids and polyphosphoinositides

Poor and inconsistent yields can also be obtained in the extraction of strongly hydrophilic glycerolipids, for similar reasons to those discussed in the previous section, unless precautions are taken. When lysophospholipids are major components of tissue extracts, it has been recommended that acid (best avoided as it may be harmful to any plasmalogens present) or inorganic salts be added during extraction with chloroform-methanol, or better that water-saturated butan-1-ol be used to extract the lipid [7,31].

With polyphosphoinositides, it is especially important to ensure that tissues are stored in a manner such that enzymatic degradation is minimized. Strategies such as single phase extraction, salting out or acidification are then used to optimise the recoveries, although there does not appear to be a consensus as to which of these is best for the purpose. For example, it has been suggested that tissues should be extracted with chloroform-methanol in the presence of calcium chloride initially and subsequently after acidification [35]. Most groups appear to favour exhaustive extraction procedures involving repeated treatment of the tissue with various chloroform-methanol mixtures together with 1M hydrochloric acid [15,21,32,95]. In an alternative method, an ion-pairing reagent, tetrabutylammonium sulphate, was added to the medium and gave apparently much higher recoveries of polyphosphoinositides [30]. However, this approach does not seem to have been followed by others.

3. Gangliosides

Gangliosides are so soluble in water that they remain entirely in the aqueous phase during a 'Folch' extraction. The established procedure for isolation of these compounds has involved dialysis of the 'Folch' upper phase to remove ions and other compounds of low molecular weight followed by lyophilisation [40]. As an alternative to the 'Folch' system, partition of lipid extracts between diisopropyl ether, butan-1-ol and 50 mM aqueous sodium chloride has been suggested, with the gangliosides again being retained in the aqueous layer [51,52].

Large scale column procedures for isolating gangliosides are described in Section D above. Williams and McCluer [118] were among the first to adapt solid-phase extraction methodology to the analysis of lipids, and described the use of Sep-Pak™ ODS cartridges for the isolation of gangliosides from tissue extracts. The method involved simply passing the upper aqueous phase from a Folch extract through a column, packed with a bonded octadecylsilyl phase, which retained the gangliosides for subsequent recovery by elution with chloroform-methanol. This procedure with occasional minor variations appears now to be widely accepted for the purpose [122].

Of course, such preparations do contain other materials, and further purification by chromatographic and chemical means is essential for many purposes [24]. The best method for this is a subject of some debate among the experts [53,116,122].

ACKNOWLEDGEMENT

This review is published as part of a programme funded by the Scottish Office Agriculture and Fisheries Dept.

REFERENCES

1. Abe,K. and Kogure,K., *J. Neurochem.*, **47**, 577-582 (1986).
2. Allen,P.C., *Anal. Biochem.*, **45**, 253-259 (1972).
3. Ando,N., Ando,S. and Yamakawa,T., *J. Biochem. (Tokyo)*, **70**, 341-348 (1971).
4. Arthur,G. and Sheltawy,A., *Biochem. J.*, **191**, 523-532 (1980).
5. Banis,R.J., Roberts,C.S., Stokes,G.B and Tove,S.B., *Anal. Biochem.*, **73**, 1-8 (1976).
6. Bartle,K.D. and Clifford,T.A., in *Advances in Applied Lipid Research*, Vol. 1, pp. 217-264 (1992) (edited by F.B. Padley, JAI Press, London).
7. Bjerve,K.S., Daae,L.N.W. and Bremer,J., *Anal. Biochem.*, **58**, 238-245 (1974).
8. Bligh,E.G. and Dyer,W.J., *Can. J. Physiol.*, **37**, 911-917 (1959).
9. Broekhuyse,R.M., *Clin. Chim. Acta*, **51**, 341-343 (1974).
10. Burton,G.W., Webb,A. and Ingold,K.U., *Lipids*, **20**, 29-39 (1985).
11. Cabrini,L., Landi,L., Stefanelli,C., Barzanti,V. and Sechi,A.M., *Comp. Biochem. Physiol.*, **101B**, 383-386 (1992).
12. Carlson,L.A., *Clin. Chim. Acta*, **149**, 89-93 (1985).
13. Cham,B.E. and Knowles,B.R., *J. Lipid Res.*, **17**, 176-181 (1976).
14. Chapman,D.J. and Barber,J., *Methods Enzymol.*, **148**, 294-319 (1987).
15. Christensen,S., *Biochem. J.*, **233**, 921-924 (1986).
16. Christiansen,K., *Anal. Biochem.*, **66**, 93-99 (1975).
17. Christie,W.W., *Lipid Analysis (Second Edition)*, Pergamon Press, Oxford (1982).
18. Christie, W.W., *Gas Chromatography and Lipids*, The Oily Press, Ayr (1989).
19. Christie,W.W., in Advances in Lipid Methodology - One, pp. 1-17 (1992) (edited by W.W. Christie, Oily Press, Ayr).
20. Colborne,A.J. and Laidman,D.L., *Phytochem.*, **14**, 2639-2645 (1975).
21. Creba,J.A., Downes,C.P., Hawkins,P.T., Brewster,G., Michell,R.H. and Kirk,C.J., *Biochem., J.*, **212**, 733-747 (1983).
22. Dawson,R.M.C. and Eichberg,J., *Biochem. J.*, **96**, 634-643 (1965).
23. Dhopeshwarkar,G.A. and Mead,J.F., *Proc. Soc. Exp. Med.*, **109**, 425-429 (1962).
24. Fidelio,G.D., Ariga,T. and Maggio,B., *J. Biochem. (Tokyo)*, **110**, 12-16 (1991).
25. Folch,J., Ascoli,I., Lees,M., Meath,J.A. and LeBaron,F.N., *J. Biol. Chem.*, **199**, 833-841 (1951).
26. Folch,J., Lees,M. and Stanley,G.H.S., *J. Biol. Chem.*, **226**, 497-509 (1957).
27. Freyburger,G., Heape,A., Gin,H., Boisseau,M. and Cassagne,C., *Anal. Biochem.*, **171**, 213-216 (1988).
28. Galliard,T., *Phytochem.*, **9**, 1725-1734 (1970).
29. Geyer,K.G. and Goodman,H.M., *Proc. Soc. Exp. Biol. Med.*, **133**, 404-406 (1970).
30. Grove,R.I., Fitzpatrick,D. and Schimmel,S.D., *Lipids*, **16**, 691-693 (1981).
31. Hajra,A.K., *Lipids*, **9**, 502-505 (1974).
32. Hajra,A.K., Fisher,S.K. and Agranoff,B.W., in *Neuromethods 7. Lipids and Related Compounds*, pp. 211-225 (1988) (edited by A.A. Boulton, G.B. Baker & L.A. Horrocks, Humana Press, Clifton).
33. Hanahan,D.J., Turner,M.B. and Jayko,M.E., *J. Biol. Chem.*, **192**, 623-628 (1951).
34. Hara,A. and Radin,N.S., *Anal. Biochem.*, **90**, 420-426 (1978).
35. Hauser,G. and Eichberg,J., *Biochim. Biophys. Acta*, **326**, 201-209 (1973).
36. Haverkate,F. and van Deenen,L.L.M., *Biochim. Biophys. Acta*, **106**, 78-92 (1965).
37. Hazlewood,G.P. and Dawson,R.M.C., *Biochem. J.*, **153**, 49-53 (1976).
38. Heinz,E., *Biochim. Biophys. Acta*, **144**, 333-343 (1967)
39. Helmy,F.M. and Hack,M.H., *Lipids*, **1**, 279-281 (1966).
40. Kanfer,J.N., *Methods Enzymol.*, **14**, 660-664 (1969).
41. Katayama,K., Takada,M., Yuzuriha,T., Abe,K. and Ikenoya,S., *Biochem. Biophys. Res. Commun.*, **95**, 971-977 (1980).
42. Kates,M., *Can. J. Biochem. Physiol.*, **34**, 967-980 (1956).
43. Kates,M. and Eberhardt,F.M., *Can. J. Bot.*, **35**, 895-905 (1957).
44. Kaufmann,H.P. and Viswanathan,C.V., *Fette Seifen Anstrichm.*, **65**, 925-629 (1963).
45. Kaul,K. and Lester,R.L., *Plant Physiol.*, **55**, 120-129 (1975).
46. Kinnunen,P.M. and Lange,L.G., *Anal. Biochem.*, **140**, 567-576 (1984).
47. Klein,R.A. and Kemp,P., in *Methods in Membrane Biology*. Vol. 8, pp. 51-217 (1977) (edited by E.D. Korn, Plenum Press, New York).
48. Kolarovic,L. and Fournier,N.C., *Anal. Biochem.*, **156**, 244-250 (1986).

49. Kramer,J.K.G. and Hulan,H.W., *J. Lipid Res.*, **19**, 103-106 (1978).
50. Laakso,P., in *Advances in Lipid Methodology - One*, pp. 81-119 (1992) (edited by W.W. Christie, Oily Press, Ayr).
51. Ladisch,S. and Gillard,B., *Anal. Biochem.*, **146**, 220-231 (1985).
52. Ladisch,C. and Gillard,B., *Methods Enzymol.*, **138**, 300-306 (1987).
53. Ledeen,R.W. and Yu,R.K., *Methods Enzymol.*, **83**, 139-191 (1982).
54. Leikola,E., Nieminen,E. and Solomaa,E., *J. Lipid Res.*, **6**, 490-493 (1965).
55. Lough,A.K., Felinski,L. and Garton,G.A., *J. Lipid Res.*, **3**, 478-480 (1962).
56. Lowes,S. and Rose,M.E., *Trends Anal. Chem.*, **8**, 184-187 (1989).
57. Lucas,C.C. and Ridout,J.H., *Prog. Chem. Fats other Lipids*, **10**, 1-150 (1970).
58. Mancha,M., Stokes,G.B. and Stumpf,P.K., *Anal. Biochem.*, **68**, 600-608 (1975).
59. McGrath,L.T. and Elliot,R.J., *Anal. Biochem.*, **187**, 273-276 (1990).
60. Mecham,D.K. and Mohammad,D.K., *Cereal Chem.*, **32**, 405-415 (1955).
61. Melton,S.L., Moyers,R.E. and Playford,C.G., *J. Am. Oil Chem. Soc.*, **56**, 489-493 (1978).
62. Meneses,P., Para,P.F. and Glonek,T., *J. Lipid Res.*, **30**, 458-461 (1989).
63. Mock,T., Pelletier,M.P.J., Man,R.Y.K. and Choy,P.C., *Anal. Biochem.*, **137**, 277-281 (1984).
64. Mogelson,S. and Lange,L., *Fed. Proc.*, **42**, 2046 (1983).
65. Mogelson,S., Wilson,G.E. and Sobel,B.E., *Biochim. Biophys. Acta*, **619**, 680-688 (1980).
66. Moilanen,T. and Nikkari,T., *Clin. Chim. Acta*, **114**, 111-116 (1981).
67. Morrison,W.R. and Coventry,A.M., *Starch*, **41**, 21-23 (1989).
68. Morrison,W.R., Tan,S.L. and Hargin,K.D., *J. Sci. Food Agric.*, **31**, 329-340 (1980).
69. Morrow,R.J. and Rose,M.E., *Clin. Chim. Acta*, **211**, 73-81 (1992).
70. Moscatelli,E.A. and Duff,J.A., *Lipids*, **13**, 294-296 (1978).
71. Mueller,H.W., Loeffler,E. and Schmandt,W., *Clin. Chim. Acta*, **124**, 343-349 (1982).
72. Muhlfellner,O., Muhlfellner,G., Zofel,P. and Kaffarnik,H., *Z. Klin. Chem. Klin. Biochem.*, **10**, 37-41 (1972).
73. Naito,H.K. and David,J.A., in *Lipid Research Methodology*, pp. 1-76 (1984) (edited by J.A.Story, A.R.Liss Inc., New York).
74. Nelson,G.J., in *Blood Lipids and Lipoproteins. Quantitation, Composition and Metabolism*, pp. 3-24 (1972) (edited by G.J. Nelson, Wiley & Sons, New York).
75. Nelson,G.J., in *Analysis of Lipids and Lipoproteins*, pp. 1-22 (1975) (edited by E.G. Perkins, American Oil Chemists' Soc., Champaign).
76. Nelson,G.J., in *Analyses of Oils and Fats*, pp. 20-59 (1991) (edited by E.G. Perkins, American Oil Chemists' Soc., Champaign).
77. Nichols,B.W., *Biochim. Biophys. Acta*, **70**, 417-422 (1963).
78. Nichols,B.W., in *New Biochemical Separations*, pp. 321-337 (1964) (edited by A.T. James and L.J. Morris, Van Norstrand, New York).
79. Nishihara,M. and Koga,Y., *J. Biochem.* (*Tokyo*), **101**, 997-1005 (1987).
80. Parinandi,N.L., Weis,B.K. and Schmid,H.H.O., *Chem. Phys. Lipids*, **49**, 215-220 (1988).
81. Parinandi,N.L., Weis,B.K., Natarajan,V. and Schmid,H.H.O., *Arch. Biochem. Biophys.*, **280**, 45-52 (1990).
82. Phillips,F.C. and Privett,O.S., *Lipids*, **14**, 590-595 (1979).
83. Phillips,F.C. and Privett,O.S., *Lipids*, **14**, 949-952 (1979).
84. Radin,N.S., *Methods Enzymol.*, **72**, 5-7 (1981).
85. Radin,N.S. in *Neuromethods 7. Lipids and Related Compounds*, pp. 1-61 (1988) (edited by A.A. Boulton, G.B. Baker and L.A. Horrocks, Humana Press, Clifton).
86. Rizzo,A.F. and Korkeala,H., *Biochim. Biophys. Acta*, **792**, 367-370 (1984).
87. Rogiers,V., *Clin. Chim. Acta*, **84**, 49-54 (1978).
88. Rosendahl,J. and Knudsen,J., *Anal. Biochem.*, **207**, 63-67 (1992).
89. Roughan,P.G., Slack,C.R. and Holland,R., *Lipids*, **13**, 497-503 (1978).
90. Rouser,G., Kritchevsky,G. and Yamamoto,A., in *Lipid Chromatographic Analysis*. Vol. 1, 99-162 (1967) (edited by G.V. Marinetti, Edward Arnold, London).
91. Salo,M.K., Gey,F. and Nikkari,T., *Internat. J. Vit. Nutr. Res.*, **56**, 231-239 (1986).
92. Sasaki,G.C. and Capuzzo,J.M., *Comp. Biochem. Physiol.*, **78B**, 525-531 (1984).
93. Sastry,P.S. and Kates,M., *Canad. J. Biochem.*, **3**, 1280-1287 (1964).
94. Saunders,R.D. and Horrocks,L.A., *Anal. Biochem.*, **143**, 71-75 (1984).
95. Schacht,J., *Methods Enzymol.*, **72**, 626-631 (1981).
96. Schmid,P., *Physiol. Chem. Phys.*, **5**, 141-150 (1973).

97. Schmid,P., Calvert,J. and Steiner,R., *Physiol. Chem. Phys.*, **5**, 157-166 (1973).
98. Schmid,P. and Hunter,E., *Physiol. Chem. Phys.*, **3**, 98-102 (1971).
99. Schmid,P., Hunter,E. and Calvert,I., *Physiol. Chem. Phys.*, **5**, 151-155 (1973).
100. Shaw,N., *Bacteriol. Revs.*, **34**, 365-377 (1970).
101. Siakotos,A.N. and Rouser,G., *J. Am. Oil Chem. Soc.*, **42**, 913-919 (1965).
102. Slayback,J.R.B., Campbell,I.M. and Vaughan,M.H., *Biochim. Biophys. Acta*, **431**, 217-224 (1976).
103. Slayback,J.R.B., Cheung,L.W.Y., and Geyer,R.P., *Anal. Biochem.*, **83**, 372-384 (1978).
104. Sobus,M.T. and Holmlund,C.E., *Lipids*, **11**, 341-348 (1976).
105. Somersalo,S., Karunen, P. and Aro,E.M., *Physiol. Plant.*, **68**, 467-70 (1986).
106. Sperry,W.M. and Brand,F.C., *J. Biol. Chem.*, **213**, 69-76 (1955).
107. Spreitzer,H., Schmidt,J. and Spiteller,G., *Fat Sci. Technol.*, **91**, 108-113 (1989).
108. Takada,M., Ikenoya,S., Yuzuriha,T. and Katayama,K., *Biochim. Biophys. Acta*, **679**, 308-314 (1982).
109. Tuchman,M. and Krivit,W., *J. Chromatogr.*, **307**, 172-179 (1984).
110. Viswanathan,C.V., Hoevet,S.P., Lundberg,W.O., White,J.M. and Muccini,G.A., *J. Chromatogr.*, **40**, 225-234 (1969).
111. Vorbeck,M.L. and Marinetti,G.V., *J. Lipid Res.*, **6**, 3-6 (1965).
112. Ways,P. and Hanahan,D.J., *J. Lipid Res.*, **5**, 318-328 (1964).
113. Wells,M.A. and Dittmer,J.C., *Biochemistry*, **2**, 1259-1263 (1963).
114. Wells,M.A. and Dittmer,J.C., *Biochemistry*, **4**, 2459-2468 (1965).
115. Whiteley,G.S.W., Fuller,B.J. and Hobbs,K.E.F., *Cryo-Letters*, **13**, 83-86 (1992).
116. Wiegandt,H. (editor), *Glycolipids. New Comprehensive Biochemistry, Vol.* 10, Elsevier, New York (1985).
117. Williams,J.P. and Merrilees,P.A., *Lipids*, **5**, 367-370 (1970).
118. Williams,M.A. and McCluer,R.H., *J. Neurochem.*, **35**, 266-269 (1980).
119. Woldegiorgis,G., Spennetta,T., Corkey,B.E., Williamson,J.R. and Shrago,E., *Anal. Biochem.*, **150**, 8-12 (1985).
120. Wuthier,R.E., *J. Lipid Res.*, **7**, 558-561 (1966).
121. Yahara,S., Kawamura,N., Kishimoto,Y., Saida,T. and Tourtellotte,W.W., *J. Neurol. Sci.*, **54**, 303-315 (1982).
122. Yates,A.J., in *Neuromethods 7. Lipids and Related Compounds*, pp. 265-327 (1988) (edited by A.A. Boulton, G.B. Baker & L.A. Horrocks, Humana Press, Clifton).
123. Zahler,P. and Niggli,V., in *Methods in Membrane Biology*, Vol. 8, pp. 1-50 (1977) (edited by E.D. Korn, Plenum Press, New York).
124. Zhukov,A.V. and Vereshchagin,A.G., *Adv. Lipid Res.*, **18**, 247-282 (1981).

Chapter 7

TANDEM MASS SPECTROMETRY IN THE STRUCTURAL ANALYSIS OF LIPIDS

Jean-Luc Le Quéré

INRA, Laboratoire de Recherches sur les Arômes, 17 rue Sully, 21034 Dijon, France

A. INTRODUCTION

Tandem mass spectrometry, also named by its acronym MS/MS, has grown considerably in the past decade, to such an extent that it can be regarded as an established method, and considered as a quasi-routine technique in most mass spectrometry laboratories. This important growth is related to instrumental development (triple quadrupoles, multi-sector instruments, hybrid instruments, ion-trap mass spectrometers), specific sample introduction and ionization methods, and to achievements in fast computing.

The diversity of MS/MS applications is notable and all the fields of research, where structural analysis of molecules is a key point, have gained considerably through the use of the technique. These include environmental studies, natural and industrial products applications, forensic chemistry and petroleum products applications, biomolecules and pharmaceutical studies.

Specificity and sensitivity improvements in MS/MS *versus* MS have gained further benefits from the capabilities of coupling MS/MS to separation techniques such as capillary gas chromatography (GC) and high-performance liquid chromatography (HPLC).

The number of real problems solved with tandem mass spectrometry is now rather important and the number of examples is growing very fast. Selected and recent applications in the area of lipid chemistry are reviewed here. However, the first part of this chapter emphasises the concepts and instrumentation requirements of tandem mass spectrometry, and is intended for scientists who have some knowledge of mass spectrometry, but access only to simple instruments and desire to learn about the technique of MS/MS. For detailed specialist reviews, two dedicated monographs may be consulted [10,54].

B. TANDEM MASS SPECTROMETRY

The presence of dissociated metastable ions in a conventional mass spectrum should be considered as the first example of a MS/MS experiment. Those ions result from the dissociation of parent ions produced in the ionization source with an excess internal energy. The parent ions are extracted from the ion source, but may then fragment prior to detection in certain field-free regions of the mass spectrometer. The percentage of metastable ions relative to all ions in a mass spectrum is generally very small, but they can be informative.

Dissociation of metastable ions obeys a single basic concept, depicted in equation (1) where m_1^+ is the parent ion, m_2^+ is the fragment ion generally called the daughter ion,

$$m_1^+ \longrightarrow m_2^+ + m_n \qquad (1)$$

and m_n is the neutral fragment formed in the dissociation reaction. The basic MS/MS experiment is based on this unique concept.

1. Concepts and principles.

If mass spectrometry is defined as the study of the fragmentation of a molecular ion produced by ionization of a molecule, then tandem mass spectrometry should be considered as the study of the fragmentation of a particular fragment-ion present in a mass spectrum. Thus, a MS/MS spectrum is simply a mass spectrum of a particular ion in a mass spectrum.

Selection of the particular ion (the parent ion) is effected by a first mass

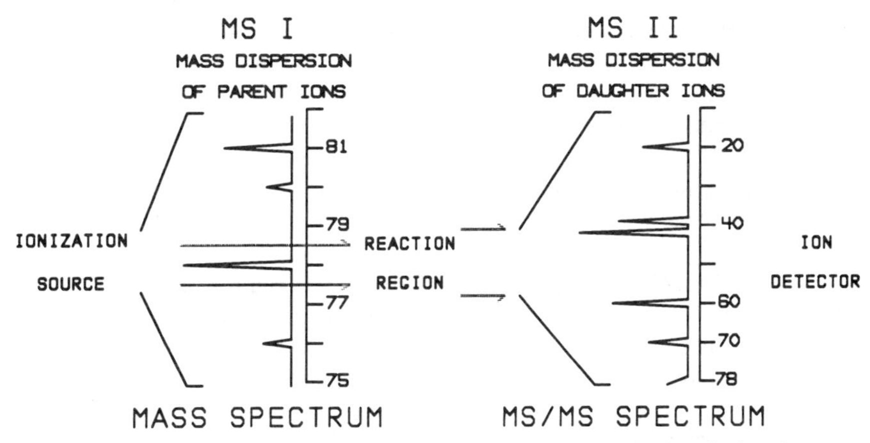

Figure 7.1. Scheme of an MS/MS experiment, where a spectrum can be described as the mass spectrum of a particular ion in a mass spectrum (daughter ion spectrum). (Reproduced by kind permission of the authors and of VCH Publishers [10]).

analyser. The mass-selected parent ion enters a field-free reaction region where it dissociates. The resulting fragments (the daughter ions) are then analysed by a second mass analyser to give a MS/MS spectrum. The daughter ions are generally characteristic of the structure of the parent ion and the MS/MS spectrum allows the determination of that structure. This MS/MS experiment is known as the daughter ion scan (Figure 7.1). It is the most common mode used for MS/MS measurements, and it is used in most of the analytical applications.

To increase the percentage of fragmenting ions in the reaction region, one must add internal energy to the parent ions. The most common method is to admit a collision gas to the reaction region to activate the parent ion (collisional activation, CA) and induce dissociations by collision of the parent ion with the neutral target gas (collision-induced dissociation, CID, or collisionally activated dissociation, CAD).

In this process, depicted by equation (2), the extra internal energy of parent ions obtained by collision with the neutral gas (G), allows fragmentations of ions that would have been stable without collisional activation.

$$m_1^+ \xrightarrow{G} m_2^+ + m_n \quad (2)$$

Any of the defined parameters in equation (2), *i.e.* m_1^+, m_2^+ and m_n, can be designated as independent variables in a tandem mass spectrometry experiment [10]. If the parent ion is the independent variable, a MS/MS experiment leads to the daughter ion scan described above. This mode produces the MS/MS spectrum of all the daughter ions of a particular parent.

A parent ion scan, where the daughter ion is now the independent variable, leads to a MS/MS spectrum of all the parent ions that produce a particular daughter ion.

A constant neutral loss scan is a MS/MS experiment where the spectrum obtained contains all the parent ions that lose a particular neutral molecule on fragmentation. This mode is actually a variation on the parent scan, where the neutral fragment is the independent variable, instead of the daughter ion.

In these MS/MS experiments, interferences are reduced to a minimum. This makes the technique highly specific. The specificity of detection permits reduction in sample purification and allows direct analysis of mixtures. Early works on MS/MS often compared daughter ion MS/MS to gas chromatography-mass spectrometry (GC/MS) [54]. In these descriptions, the first mass analyser separates the components of the mixtures, as the gas chromatograph does in GC/MS. However, the separation steps are different in nature : separation in time, based on chemical properties, in GC/MS *versus* separation in space, based on physical properties, in MS/MS. Nevertheless, both meet the requirements for quantitative analysis. A complete elimination of chromatographic separation for complex mixtures is, however, impossible and coupling MS/MS to separation techniques (GC, HPLC) adds a certain degree of selectivity.

Moreover, such MS/MS experiments as neutral loss scans and parent ion scans have no counterparts in GC/MS. These are specific in their ability to look for specified compound classes in unknown mixtures. They identify all the parent ions that fragment by loss of a specified neutral fragment or by formation of a stable ion characteristic of a particular substructure.

Another advantage of MS/MS *versus* MS is the sensitivity of the technique. The detection limit of a mass spectrometer is limited by chemical noise, *i.e.* the background contamination of the sample or of the system. The first step of MS/MS filters this background and improves the signal-to-noise ratios, despite the unavoidable loss of signal in the transmission of ions in the second stage of mass analysis.

2. *MS/MS Instrumentation*

Historically, the first MS/MS experiments were performed on two-sector, double-focusing, instruments composed of an electric sector (E) and a magnetic sector (B) arranged either in forward (EB) or reverse (BE) geometry. Daughter ion scans are advantageously performed using a reverse geometry BE instrument, in which the parent ion selected with the magnet, dissociates in a collision cell situated in the intersector region. Fragment ions are then analysed by scanning the electric sector voltage.

More complex scans are needed to obtain other parent/daughter ions relationships in these reverse geometry instruments. These complex scans, the so-called "linked scans", are also needed in other instruments of forward geometry for MS/MS measurements. Mathematical treatments of these linked scanning techniques for double-focusing mass spectrometers may be found in dedicated treatises [9,10,54], and are out of the scope of this chapter. However, the following paragraphs, presenting only the most common scanning experiments, emphasise the potential analytical applications of these methods.

a. MS/MS experiments with double-focusing mass spectrometers

Double-focusing instruments are common in mass spectrometry facilities. They were first designed for accessing high resolution mass separation and are excellent compromises between resolving power, sensitivity and mass range. Instruments of reverse geometry allowed the first parent-daughter ions relationships to be performed through the acquisition of mass-analysed ion kinetic energy (MIKE) spectra.

(i). *Mass-analysed ion kinetic energy spectra (MIKES).* In a typical MIKES experiment, the ions formed in the source are accelerated and travel through the magnetic sector. The magnetic field is fixed in order to select an ion of a particular mass to charge ratio, the parent ion. The mass-selected precursor ion dissociates in the collision cell, situated at the focal point in the second field-free region located between the magnetic and the electric sector. The electric sector filters ions according to their kinetic energy. The kinetic energy of a given daughter ion formed in the second field-free region is a fraction of the kinetic energy of the parent ion. Therefore, a scan of the electric field strength allows the transmission of different ions formed by dissociation of a mass-selected precursor ion. This ion kinetic energy spectrum can be converted to a mass-analysed daughter ion spectrum because a linear relationship exists between the electric field strength necessary to pass through the analyser and the daughter ion mass [10,54].

The mass resolution of the parent ion depends only on the resolving power of the magnet and could be reasonably high. In contrast, the resolution of the daughter ion peak decreases substantially, because of a certain amount of kinetic energy release occurring in the dissociation of the parent ion. It should be stressed that the MIKES method is occasionally subject to artifacts. Artifact peaks appear when ions of higher mass than the parent ion fragment in the first field-free region before entering the magnet and form daughter ions of the same momentum-to-charge ratio than the selected parent ion. However, those artifacts appear to be much less important than in linked-scan methods [9].

(ii). *Linked scans.* When the collision cell is situated in the first field-free region (*i.e.* after the accelerating region but before the analysers) of instruments of either conventional or reverse geometry, other MS/MS scans are possible. Covariant scans, in which two of the three values (accelerating field, V, electric field, E, and magnetic field, B) are changed, but always maintain a specified mathematical relationship [54], the so-called linked scans, can provide daughter-parent ions filiations.

- *The B/E scan.* This linked scan, in which B and E are scanned simultaneously in such a way that the ratio B/E is kept constant, provides daughter ion spectra and is probably the most widely used. The basis of this daughter ion scan is that, on fragmentation in a reaction region, if the fragment ions are not post-accelerated, the velocity of the daughter ions is the same as that of the parent ion [10]. If the ratio of the magnetic and electric fields is maintained constant, only ions of the same velocity can pass through both sectors, and daughter ions of a given parent ion (*i.e.* a given velocity) can reach the detector.

As the dissociation occurs in the first field-free region, and as the discriminating parameter is the velocity of ions filtered by the constant B/E scan, the kinetic energy release degrades the resolution of a daughter ion to a much smaller extent than in the MIKES method. A comparison of the two kinds of spectra is depicted in Figure 7.2 for 9-octadecenoic acid. However, the release of internal energy during fragmentation limits the resolution of the parent ion and artifact peaks originating from fragmentations within the electric field sector may arise [54].

- *The* B^2/E *scan.* The linked scan, in which the ratio of the electric field to the square of the magnetic field is maintained constant, leads to a parent ion MS/MS spectrum. All ions that dissociate to a chosen daughter ion in the first field-free region of a two-sector instrument of either geometry are thus detected. As the discriminating parameter is now the daughter ion mass, there is no velocity discrimination and kinetic energy release gives rise to broad peaks as in MIKES [9,10,54], but degradation of the resolution then concerns parent ions.

- *The* $B^2(1\text{-}E)/E^2$ *scan.* This rather complex scan, the last useful linked scan to be mentioned in this section, allows the detection of all the parent ions that lose a constant neutral fragment in the first field-free region. This constant neutral loss scan is particularly useful in mixture analysis as it allows the analyst to choose a neutral fragment that is lost in fragmentation of different parent ions, which incidentally could be different molecular species (screening of homologous compound classes in a mixture for example).

b. MS/MS instruments

As outlined above, MS/MS experiments can be performed on any modern double-focusing mass spectrometer. However, these two-sector instruments could not be considered as true tandem mass spectrometers, because the two sectors are compulsory parts of a single mass spectrometer. Instruments where one or two analysers are added to double-focusing

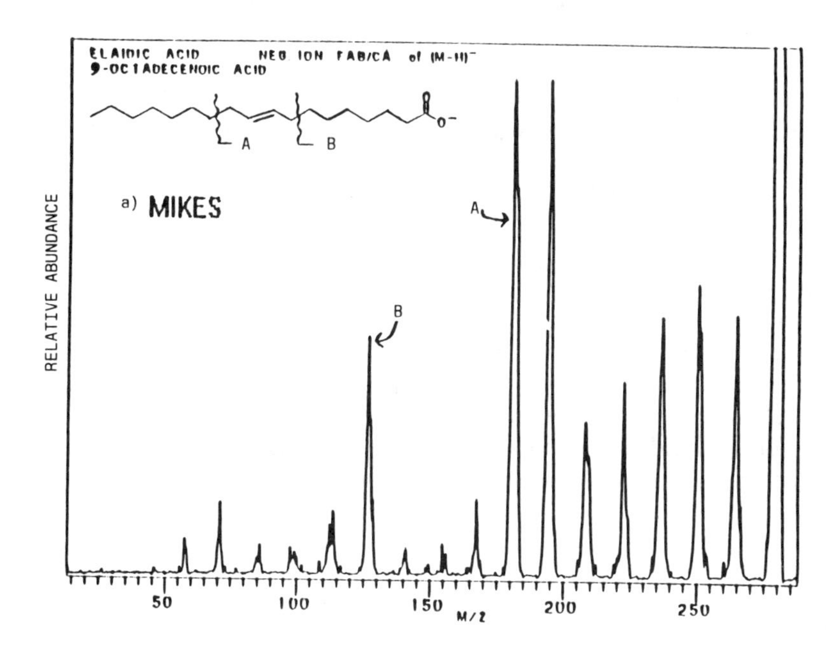

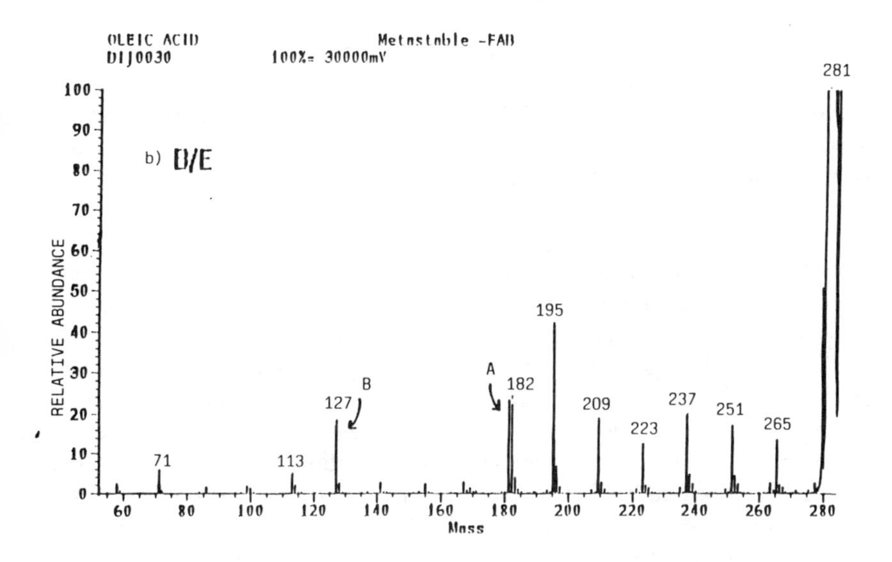

Figure 7.2. Daughter ion MS/MS spectra of 9-octadecenoic acid after CAD of the carboxylate anion obtained by fast atom bombardment. The upper spectrum, a, is a MIKE spectrum of elaidic acid, while the lower spectrum, b, is the B/E linked scan spectrum of oleic acid. (Spectrum a is reproduced by kind permission of the authors and of the *Journal of the American Chemical Society* [65]).

mass spectrometers have been designed in order to overcome some unresolved MS/MS problems.

(i). *Three- and four-sector instruments.* The main reason for adding one sector to a double-focusing instrument is to provide high mass resolution for the first stage of MS/MS experiments. Isobaric parent ions can thus be resolved. Another advantage is the introduction of a third field-free reaction region, allowing examination of consecutive reactions in MS/MS/MS experiments where isomeric parent ions can be differentiated. The most common triple sector instrument consists in a conventional two-sector instrument (of either forward or reverse geometry) as the first stage, and an electric sector as the second stage. Those EBE or BEE instruments, commercially available, are versatile for providing high-resolution parent ion selection and MS/MS/MS experiment capabilities.

Another arrangement, a BEB geometry, presents some additional advantages. It allows high-resolution analysis of daughter ions, when the first magnetic sector is used to select a parent ion, and observation of dissociations of either low or high energy in the second and third reaction regions [10].

Acquisition of high-resolution separation for both the parent ion and the daughter ion, requires two double-focusing mass spectrometers on both sides of a field-free reaction region, *i.e.* a four-sector instrument. However, the principal advantage of tandem high-resolution instruments, now commercially available, is the very good transmission of high masses achieved at high resolution by the first mass spectrometer, and concomitant versatility of MS/MS/MS applications for large molecules of biological interest. Detailed mechanistic studies on ion dissociations are also accessible with these instruments.

(ii). *Triple quadrupole instruments.* Quadrupole mass spectrometers are probably the most widely used type. Most of the bench-top instruments profit by the quadrupole technology, and they are used as universal and selective detectors for separation techniques (GC, HPLC).

The main advantages of quadrupoles are : rapid scanning capabilities, unit mass resolution, small size, simplicity of operation under facile computer control and a relative low cost. Their main drawbacks are the limited mass range and the lack of high resolution possibilities. Another characteristic of quadrupole mass analysers is that mass analysis requires low kinetic energies (eV range), in contrast to sector instruments that operate in the kV range.

A triple quadrupole (QQQ) consists simply in three sequential quadrupoles. The first and third of these are operated in the normal way for mass analysis (combination of radio-frequency (rf) and direct-current (dc) voltages). The second quadrupole is operated in the rf-only mode and is used as a high-pressure collision cell. In this mode, the second quadrupole acts to focus the fragment ions scattered by the collision process from the central axis of the ion beam. Conversely to tandem sector instruments, in triple quadrupole instruments, CAD is a low energy process, but the dissociation is quite efficient at these low energies [54].

The major advantage of the QQQ instrument is its operational simplicity for MS/MS measurements. A daughter ion scan is obtained simply by setting the first quadrupole to pass the parent ion and scanning the last quadrupole. A parent ion scan is obtained by operating the tandem instrument in the opposite way; the last quadrupole is set for the chosen daughter ion while the first quadrupole is scanning. A constant neutral loss scan is obtained by scanning both the mass analysers in concert at the same rate, but with a mass offset equal to the selected neutral fragment loss. All the spectra are obtained with unit mass resolution under facile computer control. This makes the triple quadrupole the instrument of choice for routine GC/MS/MS. However, the triple quadrupole system does not allow high-energy CAD nor MS/MS/MS experiments to be performed.

(iii). *Hybrid instruments.* Hybrid instruments, in which sectors and quadrupoles are combined in single MS/MS instruments, have been developed in order to take advantage of the best performances of both analysers. Moreover, the cost of such instruments is much lower than that of four-sector instruments.

The main difference in the CAD process between tandem quadrupole and tandem sector instruments resides in the collision-energy range. The sector instruments operate with keV ion kinetic energies and use keV CAD. Quadrupoles operate with ion energies in the eV range and use low-energy CA (typically 1 to 100 eV). Generally, the MS/MS spectra obtained in the two energy ranges are similar. But differences, due to the different internal energy distributions deposited to the ions in the CA process, may arise [10]. Moreover, some MS/MS data are only obtainable with one of the collision-energy range (*vide infra*). This difference in operational ion kinetic energy involves some accelerating, or decelerating, device to be positioned between the two types of analysers in hybrid instruments. Nevertheless, the possibility to access both CA energy ranges with a single instrument is another advantage of hybrids.

Many hybrid arrangements have been achieved [10]. Among them, the most common are those in which a double-focusing sector instrument precedes a double quadrupole system where the first quadrupole is an rf-only quadrupole gas cell (EBQQ or BEQQ geometry). These instruments, commercially available, allow high resolution parent ion selection and unit mass resolution of daughters. The last configuration to be mentioned is the QEB arrangement. Its most important feature resides in its ability to achieve high-resolution mass measurements on both daughter and parent ions.

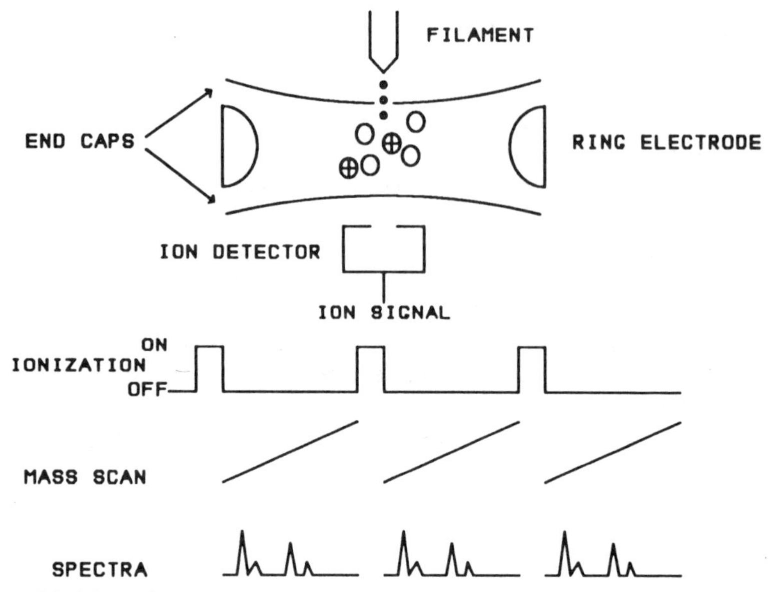

Figure 7.3. Schematic diagram of an ion-trap mass spectrometer and fundamental operation mode. (Reproduced by permission of Finnigan MAT Sàrl).

Hybrid instruments need rather complicated scan laws, in order to take advantage of all the reaction regions operating at low or high collision-energy. These scan modes are, however, well documented [10,30].

Like triple quadrupole instruments, hybrid tandem mass spectrometers are largely free of the artifacts encountered with double-focusing mass spectrometers.

(iv). *Ion-trap instruments.* The ion trap is conceived as a three-dimensional quadrupole consisting of hyperbolic cross-section electrodes, a ring electrode and two end-caps. An rf voltage of variable amplitude is applied to the ring electrode. In the ion trap, ionization and mass analysis occur in the same place. Ionization occurs with the injection of pulsed electrons. The ring electrode is held at a low rf amplitude to trap all the ions that acquire stable trajectories within the electrodes. The rf amplitude is then scanned upward, trajectories of ions of increasing mass become unstable and the ions are sequentially ejected from the trap through an end cap into the electron multiplier (Figure 7.3).

The sequence for an MS/MS experiment using an ion trap includes ionization, ejection of all ions except the parent ion, activation of the parent ion and trapping followed by ejection of the daughter ions. While conventional MS/MS spectrometers are tandem-in-space, the ion trap is a tandem-in-time instrument. Tandem mass spectrometry is achieved by applying a supplementary rf voltage (resonant excitation) across the end caps. The parent ion is thus accelerated, increasing its kinetic energy and trajectory, and undergoing more energetic collisions with the buffer gas normally present in the ion trap (the ion trap operates with a background gas pressure of about 10^{-3} Torr). Multistage tandem mass spectrometry (MS^n) is achievable by insertion of additional ejection and excitation steps.

Considered in the early days as a very sensitive, low cost, mass detector for gas chromatography, the ion-trap is now able to perform high-mass analysis (mass limits of 70 000 Daltons have been demonstrated) and high resolution (resolution up to 10^6 has been achieved) [17,51]. MS/MS experiments (and MS^n) using unit mass resolution for both parent and daughter ions for high-molecular weight compounds are thus possible [17]. The instrumentation and operation of the quadrupole ion trap have been described in the recent articles cited above [17,51] and a monograph details its fundamentals and applications [52]. Despite this impressive development, the instrument technology has not been exported yet from the specialist to the analytical laboratory.

The main feature of the ion trap is the fact that the ions are trapped, and the MS/MS experiments are separated in time, rather than in space as in conventional MS/MS instruments where the ions move as a beam. This feature belongs also to another type of instrument, the Fourier transform-ion cyclotron resonance (FT-ICR) mass spectrometer. Even though the MS/MS performances of an FT-ICR are conceptually similar to those of a simpler ion-trap (MS^n achievement capabilities), and the number of new FT-MS has continued to grow in recent years, FT-ICR MS instruments represent less than 1 % of all mass spectrometers world-wide [53]. The instrumentation cost and complexity certainly explain that FT-MS is only available in specialists laboratories. Detailed descriptions of the technique may be found elsewhere [10,53].

c. Ionization methods

The different principles described above considered apart, MS/MS has gained considerably through the use of novel ionization methods. Particularly, the desorption-ionization techniques such as fast atom bombardment (FAB) and secondary ion mass spectrometry (SIMS) complement conventional ionization methods, electron ionization (EI) and chemical ionization (CI), for thermally labile or non-volatile compounds. Desorption-ionization (FAB or SIMS) is clearly the method of choice for generating mass spectra of large polar organic molecules. It should be stressed that this ionization method may be coupled to HPLC via a continuous-flow FAB probe [11,12]. Recently, another method capable of producing ions from liquids into the gas phase, electrospray ionization, has developed rapidly [63]. This ion evaporation technique employs a strong electric field to vaporize highly charged liquid droplets from a small flow of sprayed solutions. Electrospray forms multiply charged ions and very high masses are thus measurable with good precision on conventional mass spectrometers. Electrospray is also compatible with separation techniques, such as HPLC, and with MS/MS experiments.

C. APPLICATIONS OF MS/MS IN THE ANALYSIS OF LIPIDS

1. Fatty acids

Many lipids from natural sources contain polyunsaturated fatty acids or structurally modified fatty acids. The structural modifications may include branching, hydrocarbon rings, epoxy substituents and hydroxy and alkoxy groups. In general, determining the location of double bonds in carbon chains and the position of particular structural features has been a challenging problem. Mass spectrometry of intact or derivatized fatty acids or esters has been extensively used for this purpose. Comprehensive reviews on the subject appeared recently [34,40]. However, there is no universal method for determining structural modifications of polyunsaturated fatty acids and this constitutes a stimulating challenge.

One of the first example of use of MS/MS for locating double bonds was the MIKES/CID study of monounsaturated fatty acids from the bacterium *Mycobacterium phlei* [13]. The double bonds were derivatized into amino alcohols *via* epoxy methyl esters. MIKES/CID spectra of the pseudomolecular ions formed by chemical ionization, using ammonia as reagent gas, showed intense signals allowing the location of the amino groups, and thence the position of the original double bond. However, this method used a chemical modification of the double bonds and presents no additional advantage over more conventional derivatization methods used in GC/MS analysis of unsaturated fatty acid esters [40].

The discovery of a peculiar fragmentation pattern in the high-energy CAD spectra of the negative ions $[M-H]^-$ of fatty acids by Bambagiotti and co-workers [6,7] and by Gross and co-workers [65] led to an interesting MS/MS approach for locating double bonds and structural modifications in fatty acids. Thus, the MIKE spectra obtained after collisional activation of the carboxylate anion of monounsaturated fatty acids desorbed by fast atom bombardment show interesting features [65]. On the MIKE/CAD spectrum of elaïdic acid (Figure 7.2a), the pattern observed shows three low intensity peaks between two enhanced peaks. The two enhanced peaks correspond to allylic cleavages of the carbon chain, and hence they allow the determination of the position of the double bond. This fragmentation pattern has been studied and rationalized in terms of charge-remote fragmentation induced by high-energy collisional activation [44].

In this fragmentation, the cleavage reactions are not charge-mediated. The bond cleavages occur at sites in the ion that are remote from the charge site and the mechanism of fragmentation does not involve any significant intervention of the charge. Charge-remote fragmentations of ions are mechanistic analogies to gas-phase thermal decompositions of

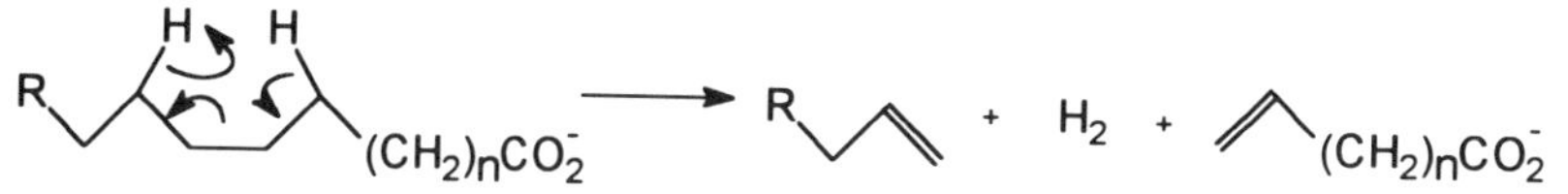

$$R\text{-}CH_2\text{-}CH(H)\text{-}CH_2\text{-}CH_2\text{-}CH(H)\text{-}(CH_2)_nCO_2^- \longrightarrow R\text{-}CH_2\text{-}CH{=}CH_2 + H_2 + CH_2{=}CH\text{-}(CH_2)_nCO_2^-$$

Scheme 1

neutral molecules. Bonds distant from the charge-site are cleaved *via* simple rearrangements and homolytic bond dissociations governed by classical thermochemistry [1]. The mechanism involves a hydrogen rearrangement and the elimination of H_2 and a neutral alkene (Scheme 1), the charge remaining stable at the carboxylate end [3,41,44]. Comprehensive reviews on this subject are available [1,31-33,40].

Constant B/E linked scanning, offering a greater resolution than MIKES, has been applied to the characterization of fatty acid methyl esters [5]. The spectra obtained via high energy CAD of the carboxylate anions $[RCOO]^-$, produced by negative ion chemical ionization, contained the charge-remote fragmentation information. Thus, charge-remote fragmentations may be obtained with any double-focusing mass spectrometer. The daughter ion spectrum of oleic acid, obtained *via* B/E linked scanning after CAD of its carboxylate anion desorbed by FAB, reveals some features allowing a better understanding of the fragmentation process (Figure 7.2b). The enhanced resolution shows that the first allylic cleavage peak (A) is a doublet. A simple β-cleavage gives a stabilized radical anion at m/z 182 (first peak of doublet A) and 1,4-elimination of H_2 gives the second peak in the doublet A at m/z 181 and the second allylic cleavage peak (B) at m/z 127 (Scheme 2).

The only structural requirement for inducing charge-remote fragmentations is that the charge is stable and localized at a specific functional group. In this respect, it may be advantageous to substitute positive ions for carboxylate anions. Positive ions could be metal cationized species [23], or ions formed by protonations at sites of high-proton affinity, such as the nitrogen-containing group in picolinyl esters [24,25], or a triphenyl-phosphonium group [14]. Cationization with lithium, for instance, is a prerequisite for the analysis of polyunsaturated fatty acids containing four or more double bonds. Upon collisional activation, the carboxylate anion of docosahexaenoic acid decomposes to give no charge-remote fragmentation (Figure 7.4). Activating the di-lithiated species, on the other hand, gives charge-remote fragmentations allowing the determination of each site of unsaturation (Figure 7.4) [2].

Analytical applications of charge-remote fragmentation are numerous in the fatty acids field. These include structural characterization of branched fatty acids [38], fatty acids with cyclopropane or cyclopropene rings and hydroxyl groups [66], and identification of homoconjugated octadecadienoic acids [21]. Recent reviews exist on the subject [1,33].

When complex mixtures of isomeric forms of fatty acids are being

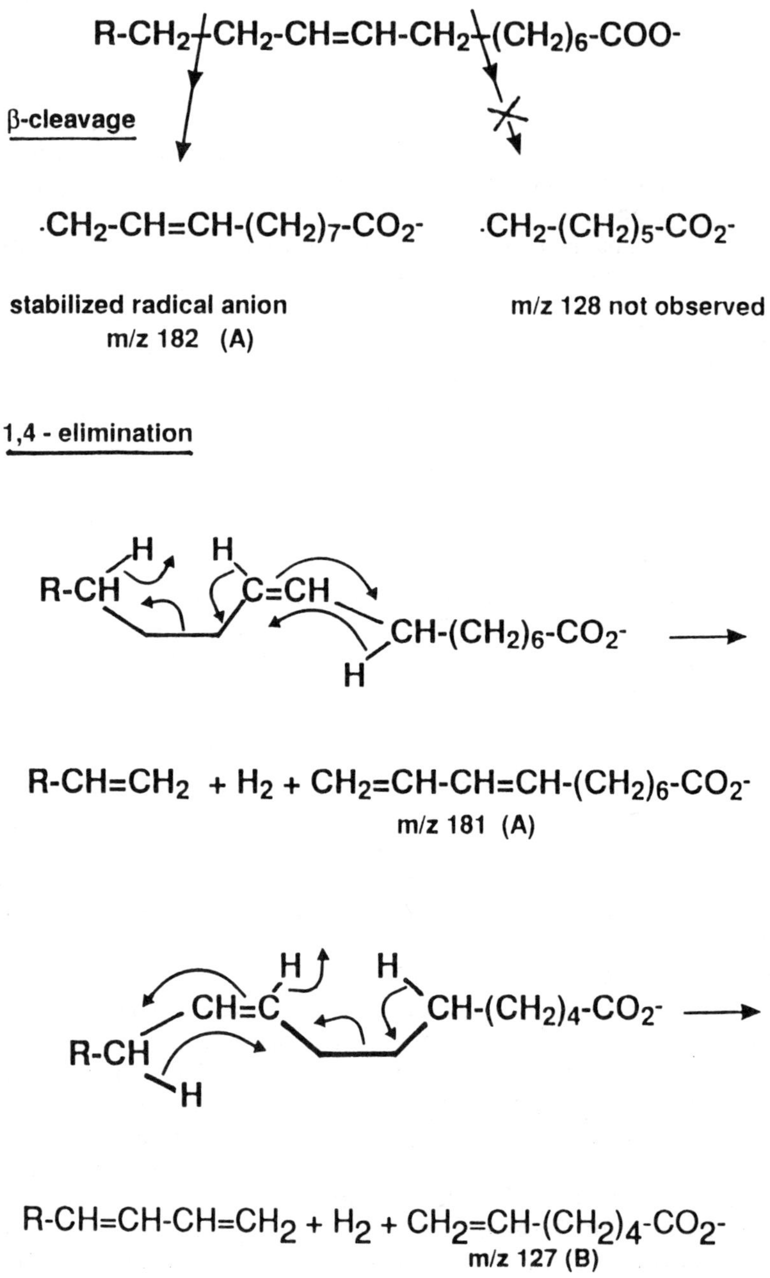

$R\text{-}CH_2\text{-}CH_2\text{-}CH{=}CH\text{-}CH_2\text{-}(CH_2)_6\text{-}COO^-$

β-cleavage

$\cdot CH_2\text{-}CH{=}CH\text{-}(CH_2)_7\text{-}CO_2^-$ $\cdot CH_2\text{-}(CH_2)_5\text{-}CO_2^-$

stabilized radical anion
m/z 182 (A)

m/z 128 not observed

1,4 - elimination

$R\text{-}CH(H)\text{-}CH_2\text{-}CH_2\text{-}C(H){=}CH\text{-}CH(H)\text{-}(CH_2)_6\text{-}CO_2^- \longrightarrow$

$R\text{-}CH{=}CH_2 + H_2 + CH_2{=}CH\text{-}CH{=}CH\text{-}(CH_2)_6\text{-}CO_2^-$
m/z 181 (A)

$R\text{-}CH(H)\text{-}CH{=}C(H)\text{-}CH_2\text{-}CH_2\text{-}CH(H)\text{-}(CH_2)_4\text{-}CO_2^- \longrightarrow$

$R\text{-}CH{=}CH\text{-}CH{=}CH_2 + H_2 + CH_2{=}CH\text{-}(CH_2)_4\text{-}CO_2^-$
m/z 127 (B)

Scheme 2

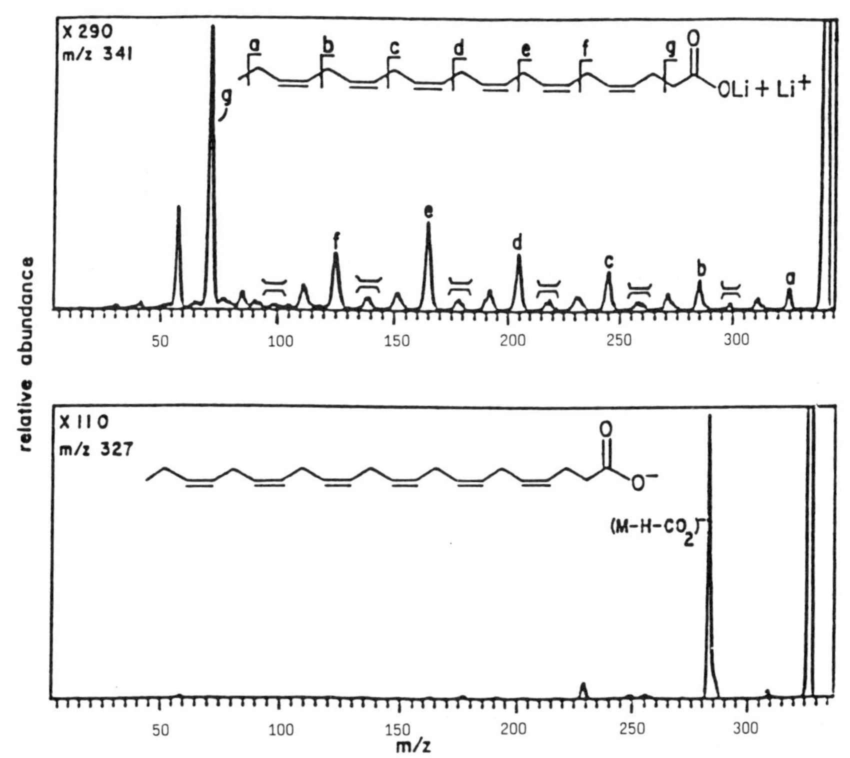

Figure 7.4. CAD-MIKE spectra of *cis*-4,7,10,13,16,19-docosahexaenoic acid $[M + 2Li - H]^+$ ion and $[M - H]^-$ anion. (Reproduced by kind permission of the authors and of *Analytical Chemistry* [2]).

analysed, a chromatographic separation is necessary. Gas phase carboxylate anions of fatty acids may be produced in very high yield from electron capture ionization of pentafluorobenzyl esters (equation 3).

$$RCOOCH_2C_6F_5 + e^- \longrightarrow RCOO^- + {}^{\cdot}CH_2C_6F_5 \qquad (3)$$

Collisional activation of these carboxylates is the basis of a GC/MS/MS method developed by Promé and co-workers [56]. The chromatographic separation of pentafluorobenzyl esters is possible for most positional isomers of mono-unsaturated fatty acids. It was applied successfully to the characterization of polyunsaturated fatty acids originated from the bacteria *Mycobacterium phlei* [4]. These "phleic acids" possess up to 40 carbon atoms and six double bonds that are ethylene interrupted. Characteristic charge-remote fragmentation patterns were observed that allowed location of all the unsaturations (Figure 7.5). Methyl branching may also be localized with this technique [19].

Examples of this approach taken from our own work are the charge-remote fragmentations of cyclopentane- and cyclohexane- containing fatty

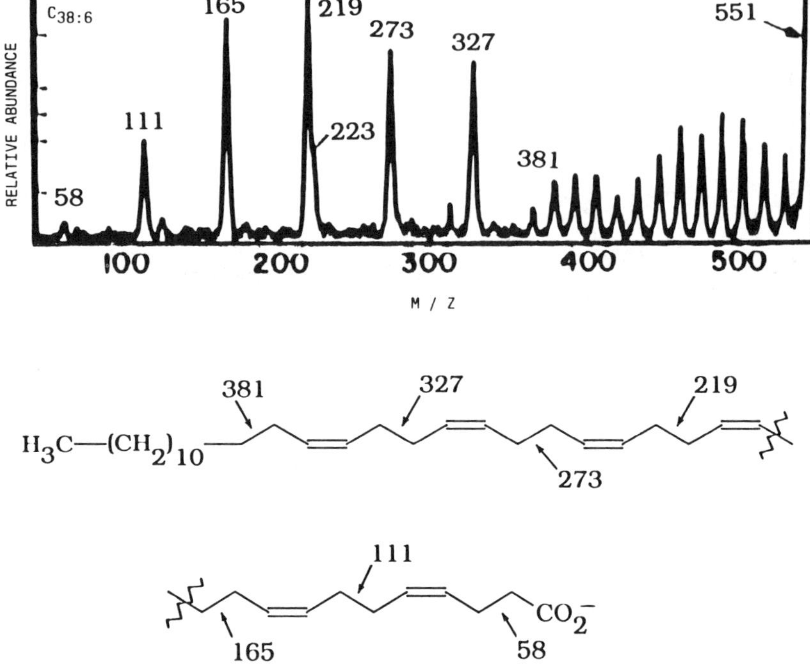

Figure 7.5. CAD-MIKE spectrum of the molecular carboxylate anion of a "phleic" acid having 38 carbon atoms and 6 double bonds, obtained from dissociative electron capture of its corresponding pentafluorobenzyl ester. (Reproduced by kind permission of the authors and of *Rapid Communications in Mass Spectrometry* [4]).

acids isolated from heated vegetable oils [49]. For these compounds, the rings may be located and the ring size established through observation of enhanced peaks corresponding to favoured cleavages at the exocyclic bonds of the rings (see Figure 7.6 for an example). It should be emphasized here that picolinyl esters should also be amenable to GC/MS/MS and that methyl esters themselves give an abundant carboxylate anion in negative chemical ionization conditions [5].

High-performance liquid chromatography has been successfully coupled to MS/MS experiments through the use of a coaxial continuous-flow fast-atom-bombardment interface [26]. HPLC/MS/MS spectra of a variety of compounds, including fatty acids, phospholipids and steroids were obtained.

Low-energy tandem mass spectrometry using triple quadrupole mass spectrometers has been recently used in the analysis of branched and unsaturated fatty acids. CAD of the molecular ions of several methyl-branched saturated fatty acid methyl esters revealed enhanced cleavage at the branching position [71]. A unique regular series of carbomethoxy ions was observed for these saturated fatty acid methyl esters, and the spectra obtained lacked any other CID ions. On the contrary, the CID of

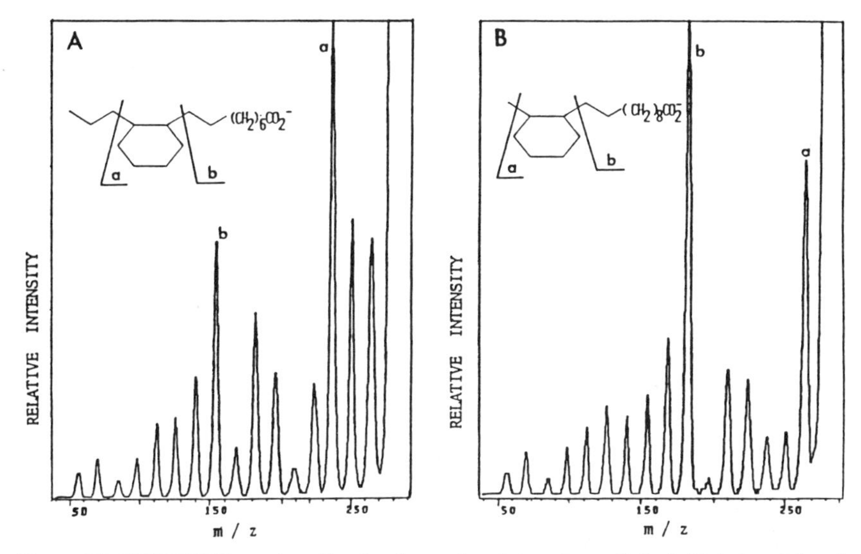

Figure 7.6. CAD-MIKE spectra of molecular carboxylate anions of disubstituted cyclohexyl C_{18} acids, isolated from heated sunflower oil, obtained by GC/MS/MS of their pentafluorobenzyl esters.

unsaturated, keto- and hydroxy- fatty acid methyl esters produced multiple series of ions, from which it was not possible to locate the structural modifications. The authors do not believe that the mechanism is similar to the charge-remote one and claim that radical cation sites are driving the fragmentation process.

Parent ion scanning has been used to locate the double bond in monounsaturated fatty acid picolinyl esters using a triple quadrupole instrument [57]. The method adds selectivity and specificity to the established GC/MS investigations of fatty acid picolinyl esters [15]. The presence of a basic site in the picolinyl derivatives allowed formation of MH^+ species under chemical ionization conditions. CID-MS/MS of the protonated molecules at low collision energy did not produce charge-remote fragments, but only charge-mediated decompositions from which the position of the double bonds could not be determined [57].

2. *Prostaglandins and other eicosanoids*

Prostaglandins and thromboxanes are cyclo-oxygenated metabolites of arachidonic acid. They possess important biological activities, and are thought to participate actively in cell-cell communication. Prostaglandin E_2 (PGE_2) for example plays an important role as a local mediator and modulator of renal blood flow and excretory function, and thromboxane A_2 (TXA_2) is a potent vasoconstrictor and platelet activator. Moreover, prostaglandins are pharmacologically active substances and many investigations have been devoted to their potential use as active agents. It

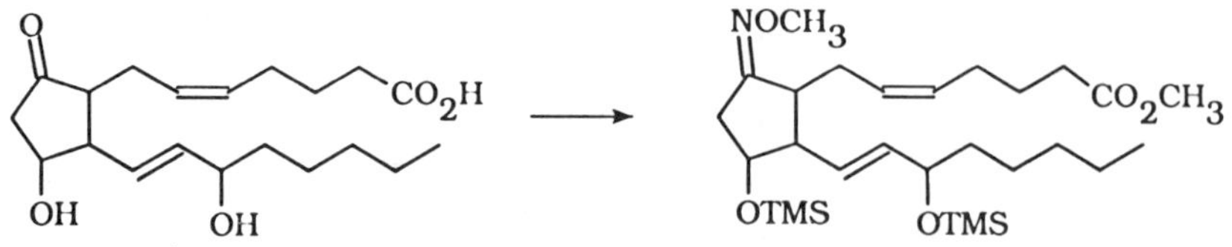

Scheme 3

is thus very important to determine the levels of these prostanoids in biological fluids, and mass spectrometry has been used for both their structural characterization and quantification.

Schweer and co-workers addressed the problem of precise quantification of naturally occurring prostanoids in biological fluids [61]. GC/MS, using stable isotope derivatives as internal standards for isotope dilution analysis, is an established procedure for quantification of these prostanoids in biological fluids. This requires a triple derivatization procedure to methyl ester/methyloxime/bis-trimethylsilyl ether derivatives (see scheme 3 for PGE_2). The tedious purification steps that follow do not preclude some interfering and co-eluting peaks, which prevent reliable quantifications. GC/MS/MS studies in CAD of specific ions of the prostaglandins were performed after normal electron ionization of the methyl ester/methyloxime/TMS ether derivatives with a triple-quadrupole instrument. GC/MS/MS adds selectivity and specificity, as a prostaglandin-specific decomposition was chosen for the CAD experiment. The sensitivity was found to be ten times lower than in GC/MS, however, but was claimed to be high enough for the concentrations found in urine [61]. Schweer and co-workers investigated all available deuterated standards of prostaglandins and demonstrated unambiguous assignments of the major parent ion fragments [59].

Pentafluorobenzyl (PFB) esters give intense $[M\text{-}PFB]^-$ ions in negative ion chemical ionization which may be used as parent ions for daughter ion experiments. 11-Dehydrothromboxane B_2 (11-dehydro TXB_2), the index metabolite of TXA_2 in urine, was studied as PFB ester/TMS derivative [58] as were other prostanoids [60]. For the 11-dehydro TXB_2, interfering peaks with the same *m/z* values were observed in GC/MS, even in Selected Ion Monitoring (SIM) after extensive purification of the sample. Because of the selectivity of the tandem mass spectrometric measurements, the purification procedure has been reduced compared to the conventional GC/MS method [58]. CAD mass spectra of the $[M\text{-}PFB]^-$ ions of major prostanoids, in a triple quadrupole instrument, showed loss of the derivatizing groups primarily and only a few ions of low intensity arising from the fragmentation of the carbon skeletons [60]. Nevertheless, in spite of this lack of structural information, the MS/MS method, combining sensitivity and higher specificity, and facilitating sample clean-up, was found more reliable than the direct SIM-GC/MS approach [60]. The differences between the values obtained by GC/MS and GC/MS/MS were found to be rather large (from 4 to 199% for 11-dehydrothromboxane B_2 and from 137 to 291% for 2,3-dinorthromboxane B_2) [67]. Therefore, quantitative measurements using single stage GC/MS were not considered feasible.

Combining immunoaffinity chromatography in the sample preparation to GC/MS/MS with negative ion CI of PFB-prostaglandin derivatives allowed the clean-up procedures to be shortened while simplifying the preparation of the sample, a basic requirement for routine analysis [50]. Chromatograms that were almost free of interferences could thus be obtained for PGE_2, $PGF_{2\alpha}$ and 6-oxo$PGF_{1\alpha}$.

Attempts to obtain GC/MS/MS assays of prostaglandins using a BE sector instrument were hindered by the low efficiency of the high energy CAD process [64]. MS/MS of PGE_2 using an ion-trap mass spectrometer showed much higher CAD efficiency (50 % versus 1 %). Remote-charge cleavages of the carbon skeleton were observed in both cases, but with a considerably improved sensitivity for the ion trap instrument [64]. Moreover, selected reaction monitoring (SRM) provided no benefit over collecting complete daughter ion spectra, because of the rapid scanning rate of the ion trap mass spectrometer, which is moreover more compatible with capillary GC than the scanning rate of the older BE instruments [64].

Structural information for selected prostanoids has been obtained recently by high-energy CID of their pseudo-molecular ions generated by dynamic (continuous-flow) and static FAB [70]. Bariated carboxylates [M - H + Ba^+] and carboxylate anions undergo charge-remote fragmentations. The patterns obtained are characteristic of the substitutions present along the carbon chain of the eicosanoids. However, the structural information obtained from the barium salts and from the carboxylate anions are substantially different. The $[M - H + Ba]^+$ from PGE_2, for instance, decomposes upon CA to give an abundant ion at *m/z* 263 resulting from a cleavage in the α-position of the cyclopentyl ring (Figure 7.7). Loss of water is characteristic for all eicosanoids containing hydroxyl substituents. The B/E linked spectrum of the carboxylate anion is characterized by an ion that is formed by the loss of water and hexanal (Figure 7.7). Thus, the structural information obtained from each process (positive and negative ions) is complementary [70]. The bariated $[M - H + Ba]^+$ ion of thromboxane B_2 (TXB_2) produced upon CA an abundant daughter ion at *m/z* 307, probably involving a rearrangement of the hemiacetal ring (Figure 7.8). The major daughter ions formed in the CID of the carboxylate anion of TXB_2 (Figure 7.8) may involve extensive rearrangements of the hemiacetal ring. Those postulated rearrangements were supported by ^{18}O labelling [70].

Positions of substituents in prostanoids are best located by CID of the bariated species, whereas activations of the carboxylate anion are more

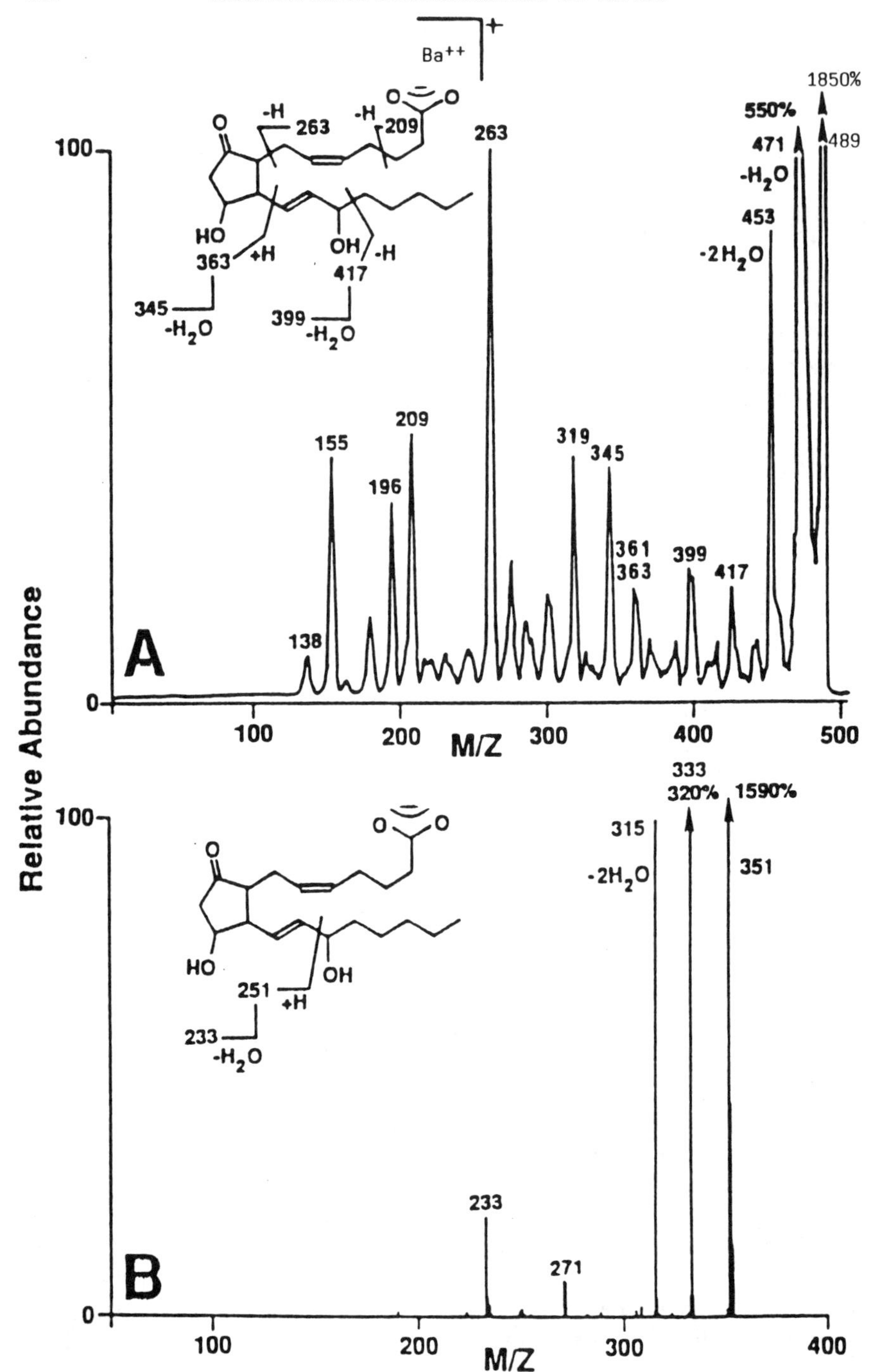

Figure 7.7. (A), CAD-MIKE spectrum of bariated PGE_2 (*m/z* 489). (B), CAD-B/E spectrum of the carboxylate anion of PGE_2 (*m/z* 351). (Reproduced by kind permission of the authors and of *Journal of the American Society for Mass Spectrometry* [70]).

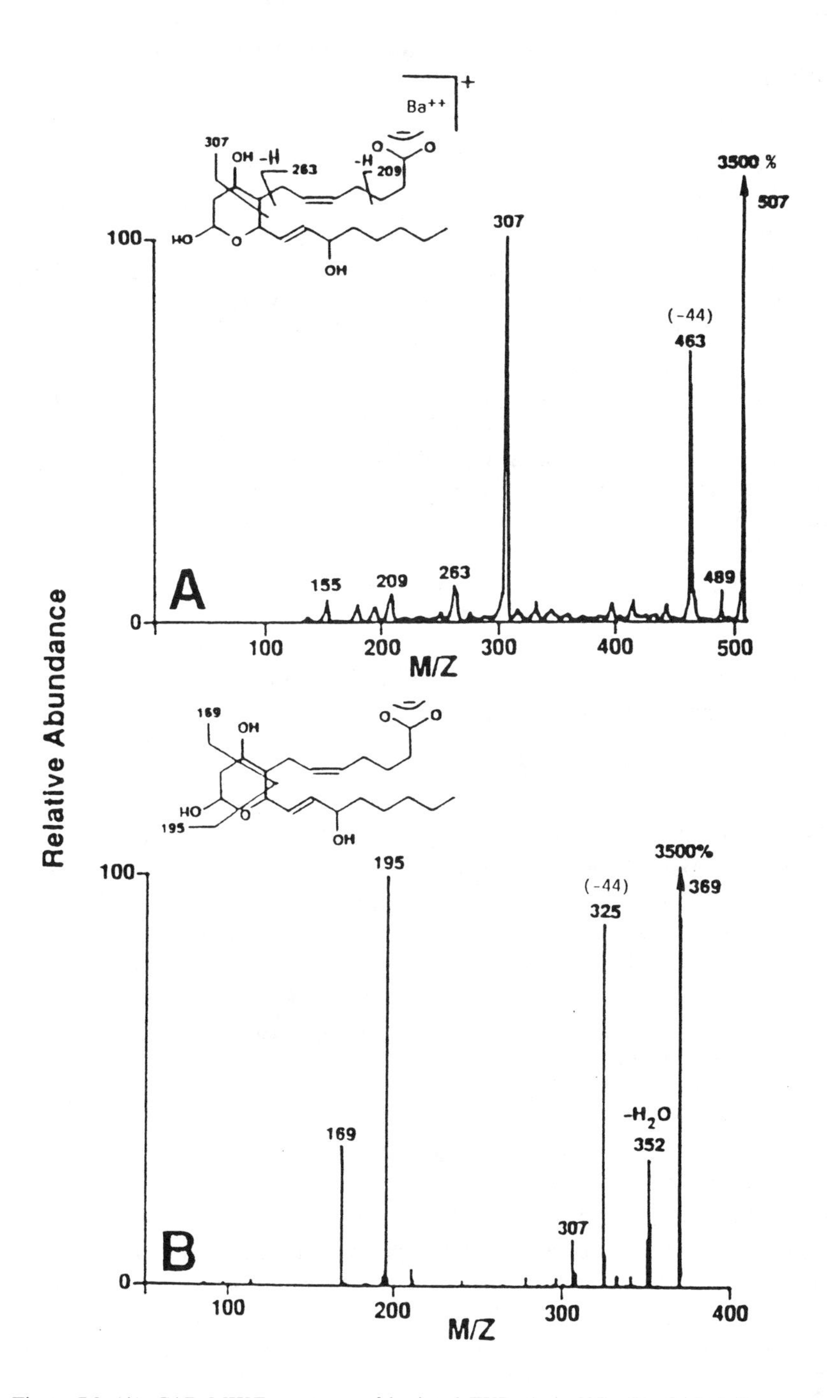

Figure 7.8. (A), CAD-MIKE spectrum of bariated TXB_2 (*m/z* 507). (B), CAD-B/E spectrum of the carboxylate anion of TXB_2 (*m/z* 369). (Reproduced by kind permission of the authors and of *Journal of the American Society for Mass Spectrometry* [70]).

useful for determining structural features in the terminal eicosanoid carbon chain.

3. *Triacylglycerols*

An MS/MS study of intact triacylglycerols (TAG) of castor bean have been performed on a EBQQ hybrid instrument through CID of the $[M-CH_3]^+$ ions produced by electron impact of the trimethylsilyl (TMS) derivatives of the hydroxy fatty acids in the neutral lipids [36]. This method surpasses the transmethylation experiments generally used to elucidate the acyl composition of storage lipids through fatty acid GC-profiles. Much more specific structural information on the intact neutral lipids could be obtained, and chain length and degree of unsaturation for each acyloxy group were indicated by the specific $[RCO + 74]^+$ ion on the CID spectra. Structural information on the TAG of a crude oil, ranging from palmitoyl-diricinoleoyl to arachidoyl-diricinoleoyl glycerols, could be obtained without purification, even though prior HPLC fractionation is advantageous to generate daughter spectra free of isotopic contamination and to concentrate minor components [36].

Kallio and co-workers have studied triacylglycerols of various origins by ammonia negative-ion chemical-ionization tandem mass spectrometry using a triple quadrupole instrument [22,45,46]. CAD of the intense $[M-H]^-$ parent ions, selected one at a time by the first MS stage, gave essentially the RCO_2^- and $[M-H-RCO_2H-100]^-$ daughter ions, allowing identification of the individual fatty acid constituents in the TAG. The relative intensities of the $[M-H-RCO_2H-100]^-$ ions could be used for determining the regiospecific position of the corresponding fatty acyl group in the TAG [45,46]. The contribution of the fatty acid at the *sn*-2 position to this diagnostic daughter ion is either very low or non existent. In the TAG of Baltic herring flesh oil, 17 molecular weight species containing up to 3 double bonds, corresponding to 67 fatty acid combinations, were studied. Palmitic acid appeared to have a slight preference for location at the *sn*-2 position [45].

In low erucic acid turnip rapeseed oil, the five major fatty acid combinations represented two thirds of the total TAG, and contained oleic, linoleic and α-linolenic acids with linoleic acid preferentially at the *sn*-2 position [46]. In human milk, palmitic acid occupied preferentially this *sn*-2 position, and the 18:1-16:0-18:1 TAG, comprising 9% of the total, was shown to be the most abundant species [22]. However, the authors pointed out that the method presents some limitations : the positional isomers of fatty acids cannot be distinguished, discrimination of the fatty acids at *sn*-1 and *sn*-3 positions is not possible and in complex TAG mixtures, information on the *sn*-2 position is difficult to assess [46]. Interfacing a chromatographic technique to the MS/MS system is therefore advisable.

A triple quadrupole MS/MS method based on electrospray and CAD of the TAG pseudomolecular ions $[M+NH_4]^+$ has been introduced [27]. The most abundant daughter ions resulted from loss of fatty acids, and the acylium ions of the fatty acids were also detected. Structural information for the acylglycerols could be deduced, but positions of structural modifications of the fatty acid chains could not be located, apparently because the site of unsaturation migrated during CAD [27].

More recently, liquid chromatography coupled to plasmaspray tandem mass spectrometry has been used to separate and identify common glycerolipids including important phospholipids [68].

4. Phospholipids

Phospholipids are an important class of lipids found in abundance in the cell walls and membranes of living organisms. They are recognized as important mediators of the cell biology, and are also the source of specific fatty acids. Glycerophospholipids consist of four primary functional groups: a glycerol-3-phosphate backbone on which two fatty acids (R and R') are esterified on the free hydroxyl groups in the *sn*-1 and *sn*-2 positions, and a second alcohol esterified to the phosphate group. This polar head group determines the phospholipid class.

Because of their importance in lipid biochemistry, the determination of their structures is important and includes characterization of molecular species, of phospholipid class and of fatty acyl composition. Classical methods to achieve this goal are time- and material-consuming. Tandem mass spectrometry allows both separation and identification of molecular species from a mixture. Applications of MS/MS to phospholipid determination have been the subject of two recent non-exhaustive reviews [33,39], and the technique, particularly in the FAB ionization mode, has developed rapidly in the years.

Crawford and Plattner [20] determined phospholipid molecular species using isobutane chemical ionization and claimed that the molecular species with the same molecular weight could be quantified by tandem mass spectrometry from the RCO^+ daughter ions of the parent diglyceride ion [20]. Sherman and co-workers [62] demonstrated that the fatty acid composition of isomeric phosphatidylinositols (PI) could be determined by analysing the daughter ion spectra of the deprotonated molecular anion desorbed by FAB. To resolve fatty acyl daughter ions when the parent ion contains isobaric species, it was necessary to use a triple sector BEB instrument, as MIKE spectra do not give the necessary unit resolution. For example, two isobaric PI with fatty ester compositions 18:1/18:1 and 18:0/18:2 could be differentiated [62].

The unit resolution available in the QQ part of a hybrid BEQQ instrument also allowed the determination of the combinations of 16:0/18:3 and 16:1/18:2 acids in isobaric phosphatidylcholine species [55]. The quantification of the fatty acid composition of individual molecular PI species was examined by comparison with GC of the fatty acid methyl esters obtained by methanolysis of soybean and bovine PI [62]. The data obtained agreed to a limited extent. Some of the discrepancy observed could be the result of differences in fragmentation yield from the variable positional and fatty ester chain composition [62].

In the low-energy CAD spectra of the $[M-CH_3]^-$ ion of phosphatidylcholine (PC) species desorbed by FAB with a BEQQ instrument, Münster and Budzikiewicz found that the intensity of the carboxylate signal originating from the *sn*-1 acid was always smaller than that associated with the *sn*-2 position [55]. This was also found in a previous study on high-energy CAD of the $[M-CH_3]^-$ ion of a limited number of PC [42] and in a more recent study on high-energy CAD of the $[M-15]^-$ ion of 1-hexadecanoyl-2-arachidonoyl-*sn*-glycerophosphocholine [69]. However, in a recent study involving a much larger number of phosphatidylcholines, the rule was found to have been violated when the fatty acid esterified at *sn*-2 was small or polyunsaturated [37]. Moreover, in some marginal cases, the ratio of the intensities of the *sn*-2/*sn*-1 carboxylate daughters reversed over an extended scanning time [37]. When the $[M-86]^-$ ion (*i.e.* [M-choline] ion, equivalent to the analogous phosphatidic acid) is selected as the parent ion, daughter ions arising from loss of the two free fatty acids (*i.e.* $[M-86-RCOOH]^-$) are obtained. In every case these authors examined, the ion formed by loss of the fatty acid from the *sn*-2 position was more abundant [37]. Because the selected parent ion is equivalent to phosphatidic acid, the technique should be applicable to other phospholipid classes. Preliminary results on diacylglycerylphosphoserine and diacylglycerylphosphoinositol seem to confirm the hypothesis [37].

If the nature of the hydrophilic head group requires confirmation, a positive ion FAB-MS/MS method can be used. Thus, the FAB spectra of phospholipids containing a secondary, tertiary and quaternary nitrogenous base attached to the phosphate group of 1,2-dipalmitoylglycerol-3-phosphate contain an ion derived from the intact nitrogenous base [28]. The CA-MIKE spectrum produced from each of these ions was found to be characteristic of each particular base [28]. The nitrogenous base of sphingomyelin isolated from a patient with Niemann-Pick disease (sphingomyelin abnormality) was thus confirmed to be phosphocholine [28].

Phospholipid class (nature of the head group), as well as positional substitution data, are thus available from CAD spectra. The structures of the two fatty acids remain to be established. Each phospholipid class submitted to FAB produces high mass ions characteristic of the molecular species ($[M-1]^-$ for PI and $[M-15]^-$ for PC for instance) as well as ions that correspond to the carboxylate moieties. Collisional activation of these carboxylate anions allows the determination of structural features. The carboxylate anions undergo specific fragmentations, of the charge-remote type, that are entirely consistent with the charge-remote fragmentations of carboxylate anions desorbed from free acids [43]. The positions of structural modifications within the acyl chains (unsaturation, branching) can thus be determined.

Alternatively, as CAD of any of the high mass ions produces the carboxylate anions of the constituent fatty acids, a MS-MS-MS strategy can be used. Activation of one high mass ion produces the two carboxylates in the first stage, and in a second step, the carboxylate arising unambiguously from one lipid species can be activated. This MS-MS-MS method adds specificity when analysing complex mixtures of phospholipids [33] but requires a four-sector tandem instrument for high energy CAD.

Interestingly, low energy collisional activation of pseudo-molecular ions from phospholipids containing epoxidized arachidonic acid allowed the identification of the epoxidized arachidonates [8]. For example, the low-energy CAD decomposition of the $[M-15]^-$ ion desorbed by FAB from a 11,12-epoxyeicosatrienoyl-glycerophosphocholine (EET-PC) yielded abundant carboxylate anions (Figure 7.9) as expected. However, numerous fragment ions were also observed at masses lower than the expected acyl ions. These fragment ions at *m/z* 167, 179 and 208 are identical to the ions observed in the CAD of the molecular species of pure 11,12-epoxyeicosatrienoic acid, and characteristic of the position of the epoxy substituent [8]. It is most likely that the acyl daughter ion produced in the second quadrupole of the triple quadrupole instrument underwent further CAD to produce the diagnostic ions. The pressure of the collision gas in the rf-only quadrupole was high enough to produce secondary collisional activation [8].

Fredrickson and co-workers [29] analysed intact polar ether lipids extractable from archaebacteria. The molecular structure of these polar lipids differs fundamentally from all other biota and are important for classification of the archaebacteria [29]. The analyses *via* FAB-MS/MS experiments were performed without further purification. The presence of structural features such as *O*-methylation (Figure 7.10) and cyclic isoprenoid chains, which are difficult to ascertain with classical methods, were determined [29]. The technique was used for screening mixtures of ether-linked polar lipids from halophilic archaebacteria for the presence of novel chemical structures [48].

Fast atom bombardment of lysed cells and crude lipid extracts of bacteria produces selective desorption of phospholipids characteristic of bacterial species [35]. Constant neutral loss (CNL) linked scanning has been used to increase the selectivity and the specificity for particular phospholipid classes [35]. Fragmentation of phospholipids occurs with charge retention on the glycerol moiety with positive ion FAB and a phosphatidic acid fragment is observed in the negative ion spectra. This

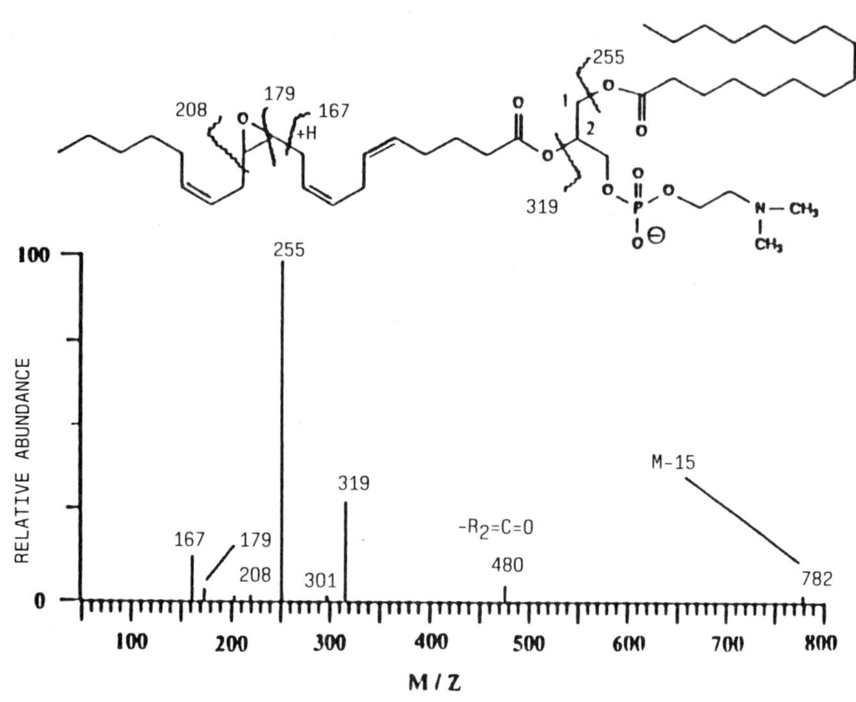

Figure 7.9. Analysis of a major glycerophosphocholine molecular species containing 11,12-epoxyeicosatrienoic acid (EET). CAD of the major negative ion at high mass (*m/z* 782, M - 15) resulted in the formation of abundant carboxylate anions (*m/z* 255 and 319) and the decomposition ions characteristic of 11,12-EET. This molecular species was identified as 16:0/11,12-EET/GPC. (Reproduced by kind permission of the authors and of *Analytical Biochemistry* [8]).

Figure 7.10. Characteristic fragmentations obtained for a major polar lipid analogue of 2,3-di-*O*-phytanylglycerol, featuring unusual *O*-methylation. [29].

represents a mass-invariant loss from a neutral species for a given phospholipid class. A loss of 141 amu from the protonated MH^+ ion, for instance, is characteristic of phosphorylethanolamine [35]. CNL provides increased selectivity over normal scans where ions with differing head and fatty acyl groups overlap. The same strategy has been used recently to identify the glycerophosphoethanolamine phospholipids in human polymorphonuclear leukocytes [47]. Moreover, identification of all arachidonate-containing phosphatidylethanolamine (PE) species was carried out through parent ion scans of the arachidonoyl carboxylate anion [47]. This parent ion scanning experiment was used to estimate the relative concentrations of arachidonic acid-containing PE molecular species in human neutrophils. The results obtained were not statistically different from the published results obtained with more conventional methods [47].

Cole and Enke [16] recently introduced another strategy to differentiate the phospholipid classes. FAB-MS, employing ion-molecule reactions with ethyl vinyl ether in a triple quadrupole mass spectrometer was used [16]. The protonated molecular ion of a particular phospholipid was mass-selected by the first quadrupole, reacted with ethyl vinyl ether in the second quadrupole and the resulting product ions were analysed by the third quadrupole. Each phospholipid class gave a unique pattern of reaction products [16], as the addition reactions occur on the polar head group. Both ion-molecule reaction and CID data could be obtained in a single full daughter scan of m/z values above and below the parent ion. Specific neutral gain scans, selective for each particular phospholipid class, can be performed (Figure 7.11).

Neutral gain scans greatly increase specificity and signal-to-noise ratio, improving the detection limit. These ion-molecule reaction neutral gain scans provide the same data as those obtained in neutral loss scan experiments, but are advantageous when neutral loss data are ambiguous or unavailable [16].

5. *Other complex lipids*

The difficulties encountered in the structural determination of glycolipids have been addressed through tandem mass spectrometry methods. A recent review article on the subject intended to establish a systematic nomenclature for fragments observed in the mass spectra of glycolipids and presented suitable derivatization strategies for glycolipid structural determinations [18]. Fragmentations of ceramides, the ceramide part and the carbohydrate portion of glycosphingolipids have been rationalized [18]. When underivatized compounds were studied in the FAB mode, positive ion MS/MS spectra of the $[M + H]^+$ ions revealed information on the structure of the aglycon moiety and charge-remote fragmentations of the hydrocarbon chains were observed. On the other hand, negative ion CAD spectra displayed characteristic fragments of the carbohydrate moiety of glycolipids [18]. Negative ion spectra are particularly useful for the analysis of underivatized gangliosides and of sulphate- and phosphate-containing glycolipids. This method have been used successfully to characterize sulphated triglycosylated lipids from halophilic archaebacteria [48].

To improve ion formation and fragmentation, it is very often desirable

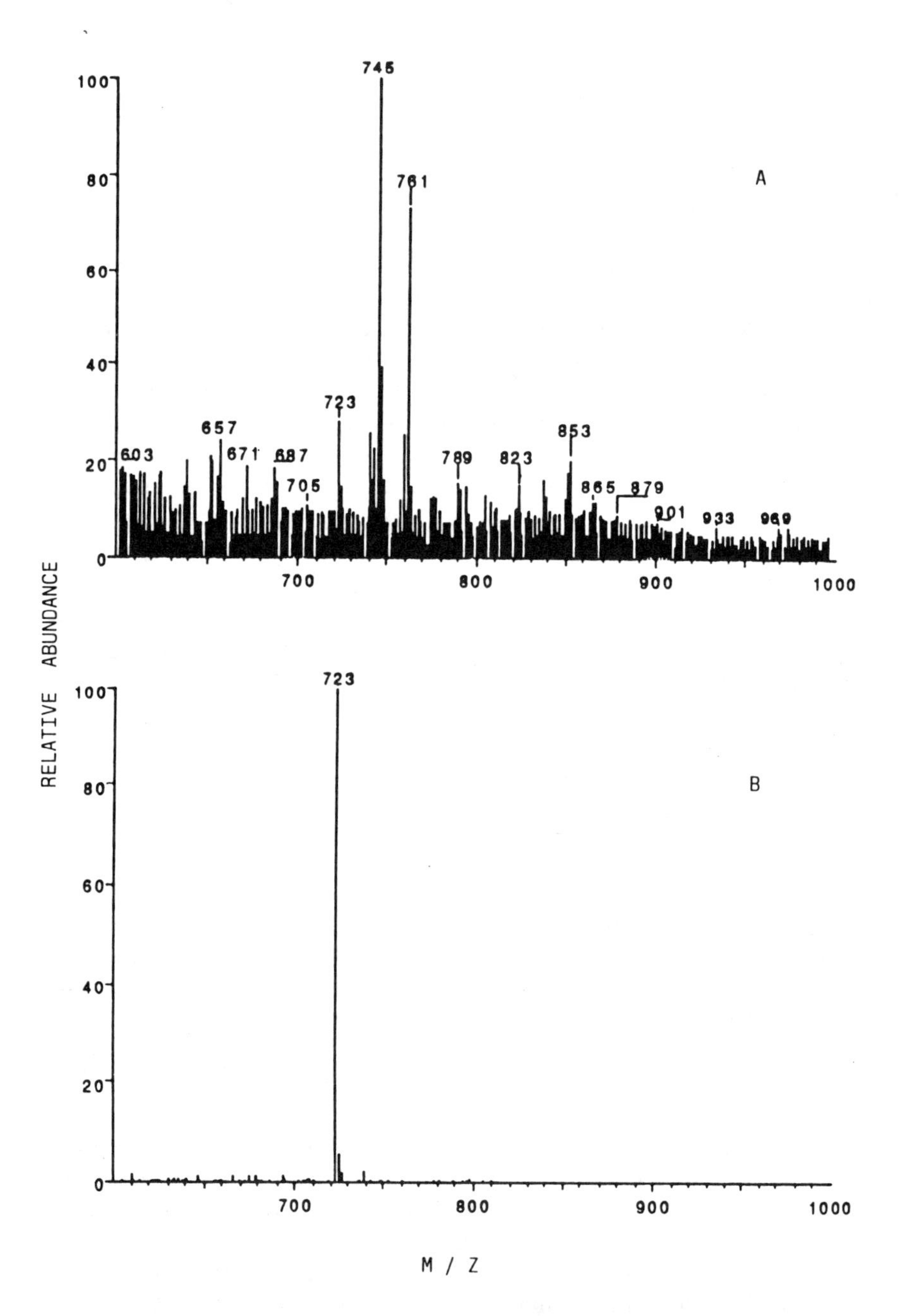

Figure 7.11. (A), Conventional mass spectrum of a sample containing phosphatidylglycerol. (B), Neutral gain scan of 26 amu of the same sample. This addition is specific for phosphatidylglycerol. (Reproduced by kind permission of the authors and of *Journal of the American Society for Mass Spectrometry* [16]).

to use a derivatization procedure prior to MS/MS experiments [18]. Permethylation and borane reduction of the amide group of the ceramide to an amine allow an important increase in sensitivity, while hydroboration permits the location of double bonds in the ceramide chains [18]. Other few examples may be found in a recent review on charge-remote fragmentation [33].

D. CONCLUSION

Tandem mass spectrometry can be considered as a powerful new approach for the determination of lipid structures. Structural details can be determined precisely, especially when only minute amounts of compounds are available, or when key components occur in complex mixtures, including biological fluids. All the lipid classes ranging from common fatty acids to more complex lipids are amenable to MS/MS and the number of real problems solved with this technique is becoming important, as this review has attempted to demonstrate.

Abbreviations

B, magnetic sector; CA, collisional activation; CAD, collisionally activated dissociation; CID, collision induced dissociation; E, electric sector; FAB, fast atom bombardment; GC, gas chromatography; HPLC, high-performance liquid chromatography; MIKES, mass-analysed ion kinetic energy spectra; MS, mass spectrometry; MS/MS, tandem mass spectrometry; PC, phosphatidylcholine; PE, phosphatidylethanolamine; PG, phosphatidylglycerol; PI, phosphatidylinositol; Q, quadrupole analyser; rf, radio-frequency; SIM, selected ion monitoring; SIMS, secondary ion mass spectrometry; SRM, selected reaction monitoring; TAG, triacylglycerol.

REFERENCES

1. Adams,J., *Mass Spectrom. Rev.*, **9**, 141-186 (1990).
2. Adams,J. and Gross,M.L., *Anal. Chem.*, **59**, 1576-1582 (1987).
3. Adams,J. and Gross,M.L., *J. Am. Chem. Soc.*, **108**, 6915-6921 (1986).
4. Aurelle,H., Treilhou,M., Promé,D., Savagnac,A. and Promé,J.C., *Rapid Commun. Mass Spectrom.*, **1**, 65-66 (1987).
5. Bambagiotti,M.A., Coran,S.A., Vincieri,F.F., Petrucciani,T. and Traldi,P., *Org. Mass Spectrom.*, **21**, 485-488 (1986).
6. Bambagiotti,M.A., Coran,S.A., Giannellini,V., Vincieri,F.F., Daolio,S. and Traldi,P., *Org. Mass Spectrom.*, **19**, 577-580 (1984).
7. Bambagiotti,M.A., Coran,S.A., Giannellini,V., Vincieri,F.F., Daolio,S. and Traldi,P., *Org. Mass Spectrom.*, **18**, 133-134 (1983).
8. Bernstrom,K., Kayganich,K. and Murphy,R.C., *Anal. Biochem.*, **198**, 203-211 (1991).
9. Boyd,R.K., *Spectroscopy (Ottawa)* **1**, 169-200 (1982).
10. Busch,K.L., Glish,G.L. and McLuckey,S.A., *Mass Spectrometry/Mass Spectrometry : Techniques and Applications of Tandem Mass Spectrometry* (VCH Publishers, Inc., New York, 1988).
11. Caprioli,R.M., *Anal. Chem.*, **62**, 477 A-485 A (1990).

12. Caprioli,R.M. and Suter,M.J.F., *Int. J. Mass Spectrom. Ion Proc.*, **118/119**, 449-476 (1992).
13. Cervilla,M. and Puzo,G., *Anal. Chem.*, **55**, 2100-2103 (1983).
14. Chang,Y.S. and Watson,J.T., *J. Am. Soc. Mass Spectrom.*, **3**, 769-775 (1992).
15. Christie,W.W., Brechany,E.Y. and Holman,R.T., *Lipids*, **22**, 224-228 (1987).
16. Cole,M.J. and Enke,C.G., *J. Am. Soc. Mass Spectrom.*, **2**, 470-475 (1991).
17. Cooks,R.G., Hoke,S.H., Morand,K.L. and Lammert,S.A., *Int. J. Mass Spectrom. Ion Proc.*, **118/119**, 1-36 (1992).
18. Costello,C.E. and Vath,J.E., *Methods Enzymol.*, **193**, 738-768 (1990).
19. Couderc,F., Aurelle,H., Promé,D., Savagnac,A. and Promé,J.C., *Biomed. Environ. Mass Spectrom.*, **16**, 317-321 (1988).
20. Crawford,C.G. and Plattner,R.D., *J. Lipid Res.*, **25**, 518-522 (1984).
21. Crockett,J.S., Gross,M.L., Christie,W.W. and Holman,R.T., *J. Am. Soc. Mass Spectrom.*, **1**, 183-191 (1990).
22. Currie,G.J. and Kallio,H., *Lipids*, **28**, 217-222 (1993).
23. Davoli,E. and Gross,M.L., *J. Am. Soc. Mass Spectrom.*, **1**, 320-324 (1990).
24. Deterding,L.J. and Gross,M.L., *Anal. Chim. Acta*, **200**, 431-445 (1987).
25. Deterding,L.J. and Gross,M.L., *Org. Mass Spectrom.*, **23**, 169-177 (1988).
26. Deterding,L.J., Moseley,M.A., Tomer,K.B. and Jorgenson,J.W., *Anal. Chem.*, **61**, 2504-2511 (1989).
27. Duffin,K.L., Henion,J.D. and Shieh,J.J., *Anal. Chem.*, **63**, 1781-1788 (1991).
28. Easton,C., Johnson,D.W. and Poulos,A., *J. Lipid Res.*, **29**, 109-112 (1988).
29. Fredrickson,H.L., de Leeuw,J.W., Tas,A.C., van der Greef,J., LaVos,G.F. and Boon,J.J., *Biomed. Environ. Mass Spectrom.*, **18**, 96-105 (1989).
30. Glish,G.L. and McLuckey,S.A., in *Advances in Mass Spectrometry*, Vol. 11A, pp. 274-275 (1989) (edited by P. Longevialle, Heyden & Son, London).
31. Gross,M.L., in *Advances in Mass Spectrometry* , Vol. 11A, pp. 792-811 (1989) (edited by P. Longevialle, Heyden and Son Ltd, London).
32. Gross,M.L., *Mass Spectrom. Rev.*, **8**, 165-197 (1989).
33. Gross,M.L., *Int. J. Mass Spectrom. Ion Proc.*, **118/119**, 137-165 (1992).
34. Harvey,D.J., in *Advances in Lipid Methodology - One*, pp. 19-80 (1992) (edited by W.W. Christie, Oily Press, Ayr).
35. Heller,D.N., Murphy,C.M., Cotter,R.J., Fenselau,C. and Uy,O.M., *Anal. Chem.*, **60**, 2787-2791 (1988).
36. Hogge,L.R., Taylor,D.C., Reed,D.W. and Underhill,E.W., *J. Am. Oil Chem. Soc.*, **68**, 863-868 (1991).
37. Huang,Z.-H., Gage,D.A. and Sweeley,C.C., *J. Am. Soc. Mass Spectrom.*, **3**, 71-78 (1992).
38. Jensen,N.J. and Gross,M.L., *Lipids*, **21**, 362-365 (1986).
39. Jensen,N.J. and Gross,M.L., *Mass Spectrom. Rev.*, **7**, 41-69 (1988).
40. Jensen,N.J. and Gross,M.L., *Mass Spectrom. Rev.*, **6**, 497-536 (1987).
41. Jensen,N.J., Tomer,K.B. and Gross,M.L., *Anal. Chem.*, **57**, 2018-2021 (1985).
42. Jensen,N.J., Tomer,K.B. and Gross,M.L., *Lipids*, **21**, 580-588 (1986).
43. Jensen,N.J., Tomer,K.B. and Gross,M.L., *Lipids*, **22**, 480-489 (1987).
44. Jensen,N.J., Tomer,K.B. and Gross,M.L., *J. Amer. Chem. Soc.*, **107**, 1863-1868 (1985).
45. Kallio,H., in *Contemporary Lipid Analysis, 2nd Symposium Proceedings*, pp. 48-62 (1992) (edited by N.U. Olsson & B.G. Herslof, LipidTeknik, Stockholm).
46. Kallio,H. and Currie,G., *Lipids* **28**, 207-215 (1993).
47. Kayganich,K.A. and Murphy,R.C., *Anal. Chem.*, **64**, 2965-2971 (1992).
48. Klöppel,K.D. and Fredrickson,H.L., *J. Chromatogr.*, **562**, 369-376 (1991).
49. Le Quéré,J.L., Sébédio,J.L., Henry,R., Couderc,F., Demont,N. and Promé,J.C., *J. Chromatogr.*, **562**, 659-672 (1991).
50. Mackert,G., Reinke,M., Schweer,H. and Seyberth,H.W., *J. Chromatogr.*, **494**, 13-22 (1989).
51. March,R.E., *Int. J. Mass Spectrom. Ion Proc.*, **118/119**, 71-135 (1992).
52. March,R.E. and Hughes,R.J., *Quadrupole Storage Mass Spectrometry* (John Wiley & Sons, New York, 1989).
53. Marshall,A.G. and Schweikhard,L., *Int. J. Mass Spectrom. Ion Proc.*, **118/119**, 37-70 (1992).
54. McLafferty,F.W. (Editor), *Tandem Mass Spectrometry* (John Wiley & Sons, New York, 1983).

55. Münster,H. and Budzikiewicz,H., *Rapid Commun. Mass Spectrom.*, **1**, 126-128 (1987).
56. Promé,J.-C., Aurelle,H., Couderc,F. and Savagnac,A., *Rapid Commun. Mass Spectrom.*, **1**, 50-52 (1987).
57. Rubino,F.M. and Zecca,L., *J. Chromatogr.*, **579**, 1-12 (1992).
58. Schweer,H., Meese,C.O., Fürst,O., Kühl,P.G. and Seyberth,H.W., *Anal. Biochem.*, **164**, 156-163 (1987).
59. Schweer,H., Seyberth,H.W. and Meese,C.O., *Biomed. Environ. Mass Spectrom.*, **15**, 129-138 (1988).
60. Schweer,H., Seyberth,H.W., Meese,C.O. and Fürst,O., *Biomed. Environ. Mass Spectrom.*, **15**, 143-151 (1988).
61. Schweer,H., Seyberth,H.W. and Schubert,R., *Biomed. Environ. Mass Spectrom.*, **13**, 611-619 (1986).
62. Sherman,W.R., Ackermann,K.E., Bateman,R.H., Green,B.N. and Lewis,I., *Biomed. Mass Spectrom.*, **12**, 409-413 (1985).
63. Smith,R.D., Loo,J.A., Edmonds,C.G., Barinaga,C.J. and Udseth,H.R., *Anal. Chem.*, **62**, 882-899 (1990).
64. Strife,R.J., Kelley,P.E. and Weber-Grabau,M., *Rapid Commun. Mass Spectrom.* **2**, 105-109 (1988).
65. Tomer,K.B., Crow,F.W. and Gross,M.L., *J. Am. Chem. Soc.*, **105**, 5487-5488 (1983).
66. Tomer,K.B., Jensen,N.J. and Gross,M.L., *Anal. Chem.*, **58**, 2429-2433 (1986).
67. Uedelhoven,W.M., Meese,C.O. and Weber,P.C., *J. Chromatogr.*, **497**, 1-16 (1989).
68. Valeur,A., Michelsen,P. and Odham,G., *Lipids*, **28**, 255-259 (1993).
69. Zirrolli,J.A., Clay,K.L. and Murphy,R.C., *Lipids*, **26**, 1112-1116 (1991).
70. Zirrolli,J.A., Davoli,E., Bettazzoli,L., Gross,M. and Murphy,R.C., *J. Am. Soc. Mass Spectrom.*, **1**, 325-335 (1990).
71. Zirrolli,J.A. and Murphy,R.C., *J. Am. Soc. Mass Spectrom.*, **4**, 223-229 (1993).

Chapter 8

THE ANALYSIS OF ACYLCARNITINES

Barbara M. Kelly, Malcolm E. Rose and David S. Millington*

Department of Chemistry, The Open University, Milton Keynes, MK7 6AA, UK and **Duke University Medical Centre, Durham, North Carolina 27710, USA*

A. INTRODUCTION

1. Carnitine and acylcarnitines: structure and function

Carnitine, **1** (L-3-hydroxy-4-aminobutyrobetaine or L-3-hydroxy-4-*N*-trimethylaminobutanoic acid) and its *O*-acyl esters, **2**, are key substances in the metabolism of fatty acids. Their detection in biological fluids can be used for the diagnosis of a number of metabolic disorders. It is this latter feature which has in recent years led to an increase in interest in detection

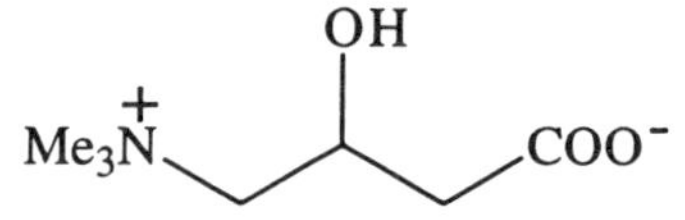

$$\underset{\mathbf{1}}{Me_3\overset{+}{N}CH_2CH(OH)CH_2COO^-} \qquad \underset{\mathbf{2}}{Me_3\overset{+}{N}CH_2CH(OCOR)CH_2COO^-}$$

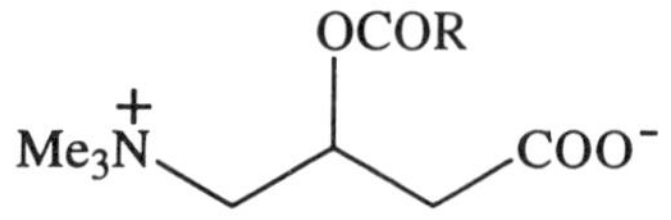

Scheme 1

and characterization of carnitine and its *O*-acyl esters, using a wide variety of methods and instrumentation. The aim of many of these investigations is to determine the presence of abnormal metabolites in the biological fluids which may be indicative of specific enzyme defects.

The role of carnitine is to act in the transport and metabolism of fatty acids, to maintain a balance between free and esterified CoA, and to remove any excess of acyl groups (RCO) from mitochondria. An accumulation of acyl CoA intermediates is potentially toxic, causing inhibition of enzymes, so carnitine is important because it is involved in their removal from the mitochondria *via* acylcarnitines **2** [50]. Carnitine is found in an omnivorous diet but a biosynthetic pathway is available through methylation of certain protein-bound lysine residues by *S*-adenosyl methionine [26].

Transport of acyl groups across mitochondrial membranes takes place in several stages and, in the process, carnitine is assisted by a number of enzymes. Initially, activation of fatty acids occurs in the cytosol, in an acylation reaction that is dependent on adenosine 5'-triphosphate. The acyl groups are bound to coenzyme A forming highly polar thiol esters, acyl-CoA. The activation step is carried out by acyl-CoA synthetases which are associated with the outer mitochondrial membrane and which are chain-length specific. However, a long-chain fatty acyl-CoA cannot cross the inner mitochondrial membrane directly. Its acyl group is first transferred to carnitine. Conjugation of long-chain fatty acyl groups with carnitine is brought about by the enzyme carnitine palmitoyl transferase (CPT I) and the acylcarnitines generated enter the mitochondria where the CPT II enzymes facilitate the regeneration of carnitine and acyl-CoA. The carrier system into and out of the mitochondrion is thought to be identical. The enzymes involved in this carrier system have overlapping chain-length specificities [26]. Carrier enzymes exist for the transport of specific chain-length acylcarnitines, with equimolar amounts of each transferase being found in the mitochondrial membrane [15]. Acylcarnitine translocase, a transmembrane protein, is responsible for the passage of the carnitine *O*-acyl esters through the inner mitochondrial membrane. Under normal circumstances, once inside the cell transesterification occurs yielding the starting compounds: carnitine and acyl-CoA. The latter is dismembered, two carbons at a time, by undergoing β-oxidation, producing acetyl-CoA [4,164]. The acetyl-CoA can then be transported out of the mitochondrion as a carnitine conjugate formed in a reaction mediated by carnitine acetyl transferase (CAT). This process is shown in simplified form in Figure 8.1.

The breakdown of the fatty acyl chains is governed by a number of enzymes and co-factors and takes place through the closely coupled enzyme system of β-oxidation, including a group of enzymes known as acyl-coenzyme A dehydrogenases. These enzymes have overlapping chain-length specificities and act on particular chain-length substrates. These are referred to as short-chain (SCAD), medium-chain (MCAD) and long-chain acyl-CoA dehydrogenases (LCAD). Disorders which manifest themselves through deficiencies in these enzymes cause an accumulation of a specific chain-length acyl-CoA which can have toxic effects. Carnitine acts through conjugation with these acyl moieties to form acylcarnitines which can be identified in biological fluids at abnormally high levels and the chain lengths of the acylcarnitines detected will be indicative of a particular enzymic disorder.

2. History and biosynthesis of carnitine

In 1904 Franz Knoops first proposed that fatty acids were oxidised at the β-carbon position. This early work concentrated on the use of chemical labelling to trace metabolic pathways. It was not until the 1950s with the discovery of coenzyme A, the isolation of fatty acids and the elucidation of the mechanisms involved that Knoops' proposal was confirmed [166]. Much of the mechanism of carnitine's interactions within the β-oxidation pathway is now understood. In 1905 [58], the empirical formula ($C_7H_{15}NO_3$) was assigned to a compound discovered in meat extract. It took a further twenty-two years for the structural formula to be proven as L-3-hydroxy-4-*N*-trimethylaminobutanoic acid [159]. The later discovery of carnitine in insects was of great interest as previously carnitine had only been identified from vertebrate muscle. In 1951 Carter *et al.* [30] established carnitine as Vitamin B_T with the first assay being carried out on the mealworm, *Tenebrio molitor*, as Vitamin B_T was considered essential for its growth. Their assay for the presence of carnitine, The Tenebrio Test, was then applied to a range of biological materials.

In 1953 Fraenkel [48] was the instigator of the first carnitine assay applied to human urine and blood. This assay was then widely used for the analysis of biological samples and carnitine was found to be distributed, with a few exceptions, throughout nature. Carnitine levels in mammalian tissue were found to vary between 0.1 and a few millimoles per litre [94], with the highest levels recorded in the heart and skeletal muscle. In 1957 Fraenkel and Friedman [49] proposed that, if a compound was so ubiquitous and appeared so important functionally to the organism, it should have been identified earlier if it was not

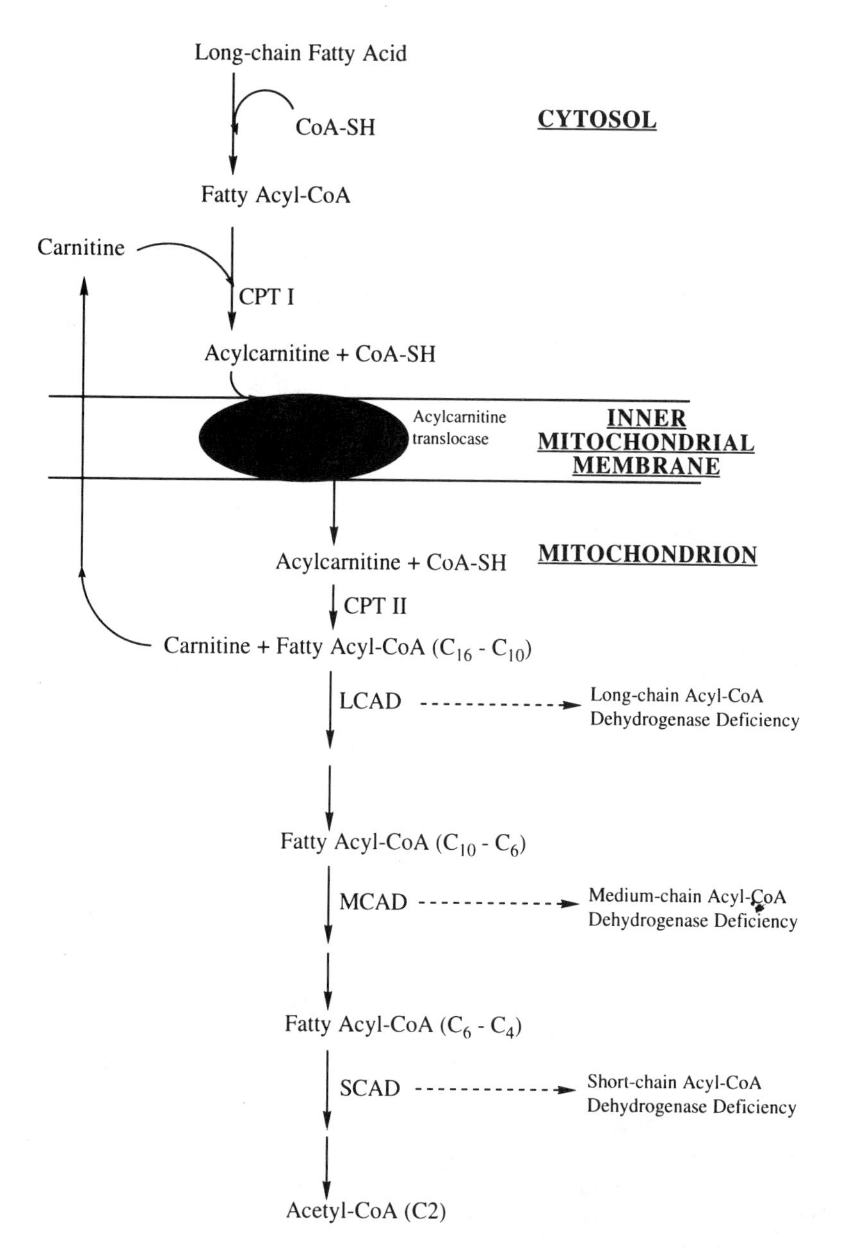

Fig. 8.1. Simplified view of the activation, membrane crossing and β-oxidation of long-chain fatty acids.

synthesised by the organism. Thereafter work began on discovering the endogenous biosynthetic pathway in mammals.

Elucidation of the biosynthetic pathway of carnitine in mammals began in 1961 when studies revealed that the methyl groups of the quaternary ammonium functionality were derived from methionine [172]. The precursors for the rest of the carnitine molecule, however, remained unknown with work continuing in this area for the next decade. In 1962 the conversion of γ-butyrobetaine to carnitine was described [24], as was the later discovery of lysine as a precursor of butyrobetaine [64].

In 1973 a biosynthetic pathway of carnitine was elucidated. In animals protein-bound lysine (stored mainly in muscle tissues) becomes available as peptide residues, and is methylated by *S*-adenosylmethionine and a protein methylase before proteolysis liberates ε-*N*-trimethyllysine (TML). Oxidation through a further three enzyme-dependent steps converts TML to γ-butyrobutaine aldehyde [39]. Cytosolic hydroxylase then mediates in the final hydroxylation step to carnitine. This takes place in the liver, brain and human kidney tissue [136], though the liver is the primary site for carnitine synthesis in humans. Tissues lacking the cytosolic hydroxylase enzyme can use the blood circulation to export the γ-butyrobutaine precursor to the hydroxylating tissue, but rely on newly synthesized product or dietary intake for their supply of carnitine. Four other micronutrients are required as co-factors by the enzymes involved in biosynthesis, these are vitamin C, niacin, vitamin B_6 and iron [18]. Deficiencies of these micronutrients as well as of methionine have been shown to reduce carnitine levels in plasma and/or tissue [135].

Other studies on rat liver mitochondria [25], led to the hypothesis that the inner mitochondrial membrane was impermeable to CoA and acetyl CoA, so that carnitine was required to transport the acetyl groups in the form of acetylcarnitine across the mitochondrial membrane. Carnitine was also shown to stimulate the oxidation of long-chain fatty acids (palmitate) which led to the theory that carnitine played a role in the transport of other acyl groups [51]. This transport was shown to take place through carnitine translocating activated long-chain fatty acids into the mitochondrial matrix for β-oxidation (Figure 8.1).

Though the passage of long-chain fatty acids into the β-oxidation system occurs in this way, carnitine's main role for acetyl and other shorter chain fatty acids is thought to be the formation of carnitine *O*-acyl esters for the removal of toxic acyl groups from the mitochondrial matrix. Fatty acid oxidation is responsible for providing energy for the cell, particularly at times when the level of cellular glucose is low. Mitochondrial β-oxidation is therefore the main pathway for the metabolism of fatty acids and is the only carnitine-dependent route for the oxidation of long-chain fatty acids.

A number of inherited metabolic diseases can be characterized by the presence of acylcarnitines in the blood and urine of neonates. Although low levels of acylcarnitines occur in healthy individuals, in patients with metabolic disorders the amounts of certain acyl-CoA compounds being transported are increased and the elevated levels of the corresponding acylcarnitines are significant. If this change in the profile of acylcarnitines can be recognized it is then possible to identify at which stage of the β-oxidation pathway the breakdown has occurred and thus the disease involved.

Overviews of the analysis of biological samples for acylcarnitines have been published before [13,14,83,95,113]. This review concerns itself largely with analytical methods that are capable of distinguishing different acylcarnitines. However, it is appropriate first to consider briefly methods for the determination of free and total carnitine.

B. FREE AND TOTAL CARNITINE

Carnitine is frequently measured in biochemical, clinical or nutritional studies. Such measurements are relevant here because an estimation of total acylcarnitines involves hydrolysis of all acylcarnitines to carnitine followed by its determination. The detection and measurement of carnitine have been reviewed elsewhere [1,3,92,106].

In humans, plasma concentrations of carnitine are relatively stable at 46 $\pm$ 10 μmol/L of which about 15% is in the form of acylcarnitines, mostly acetylcarnitine [26]. Reports of normal concentrations of total carnitine and free carnitine in the serum of healthy men were 62.2 $\pm$ 4.4 and 55.9 $\pm$ 4.0 μmol/L, respectively, and those of healthy women were 55.9 $\pm$ 6.3 and 46.6 $\pm$ 7.1 μmol/L respectively [109]. Whole blood is recorded as containing 50% more carnitine than is plasma [20]. The levels of free carnitine in plasma are thought to reflect the carnitine tissue levels and are considered abnormally low if they fall to less than 20 μmol/L. Deficiency of carnitine can cause a number of clinical symptoms including myopathy, hypotonia and hypoglycemia [71]. Most methods for determining total carnitine involve the hydrolysis of the acylcarnitines to carnitine which can be monitored *via* a number of methods including enzymic assay, spectrophotometry, chromatography and mass spectrometry.

1. Centrifugal and spectrophotometric analysis

Centrifugal analysis provides an enzymic and spectrophotometric assay of free and total carnitine in plasma ultrafiltrates, which may be suitable for routine application in many hospital laboratories [33,109,143]. Typically, the method initially involves the hydrolysis of acylcarnitines to free carnitine. In the presence of acetyl-CoA and carnitine acetyltransferase (CAT), the carnitine is converted to acetylcarnitine, producing CoA. The CoA reduces added 5,5'-dithiobis-2-nitrobenzoic acid (DTNB) to the yellow 5-thio-2-nitrobenzoate anion in proportion to the amount of L-carnitine (Figure 8.2). The anion is measured at 412 nm, providing good linearity up to 500 μmol/L of L-carnitine [109]. Batch analysis using a centrifugal analyser allows samples to be examined for both free and total carnitine within 90 min and enables analysis of approximately 100 samples per day [143]. Centrifugal analysis has also been utilized for the determination of serum L-carnitine [32]. The recovery of carnitine, spiked into a serum sample, has been reported to be 93% [33]. The method compared favourably with a radioenzymic method because spectrophotometric assays involving DTNB have low reagent costs, and are fast, simple, and reproducible. The results correlated well although the centrifugal analyser method gave values proportionally greater by 10 to 25% for samples of plasma, dialysis fluid, urine, and muscle tissue. However, the centrifugation methods were not as sensitive and probably not as specific.

The analytical approach employing DTNB has been utilized for a spectrophotometric method of measuring free and total carnitine in human tissues [173]. Methods involving the further reaction of CoA, produced from the CAT enzyme reaction above, with 2-oxoglutarate to produce succinyl-CoA have also been described. Catalysed by 2-oxoglutarate dehydrogenase this product reduces NAD, which is then monitored spectrophotometrically [147]. Colorimetric methods have also been described for determining carnitine and its *O*-acyl esters in human serum [101] and DL-carnitine (as its reineckate) in pharmaceutical products [125].

2. *Enzymic methods*

Enzymes have been used widely to detect compounds of the L-carnitine family in biological fluids, pharmaceuticals and a variety of foods. The most commonly used enzyme for assaying these compounds is carnitine acetyl transferase (CAT). This enzyme is responsible for the catalysis of the reversible conversion of L-carnitine to its short-chain acyl esters.

$$\text{Carnitine} + \text{Acetyl-CoA} \overset{\text{CAT}}{\rightleftharpoons} \text{Acetylcarnitine} + \text{CoA}$$

It is common to couple reactions with this CAT-mediated process and there are several means of detecting the extent of reaction including uv spectroscopy and radioisotopic exchange. Detection by these means can provide enantioselectivity, structural specificity, and, with the use of radiolabelling, highly sensitive assays [105].

There are two basic methods for the assay of carnitine. The first is a radioisotopic exchange assay (REA) which uses radiolabelled acetyl-CoA ([1-^{14}C]acetyl-CoA) as the reaction substrate for the CAT enzyme and

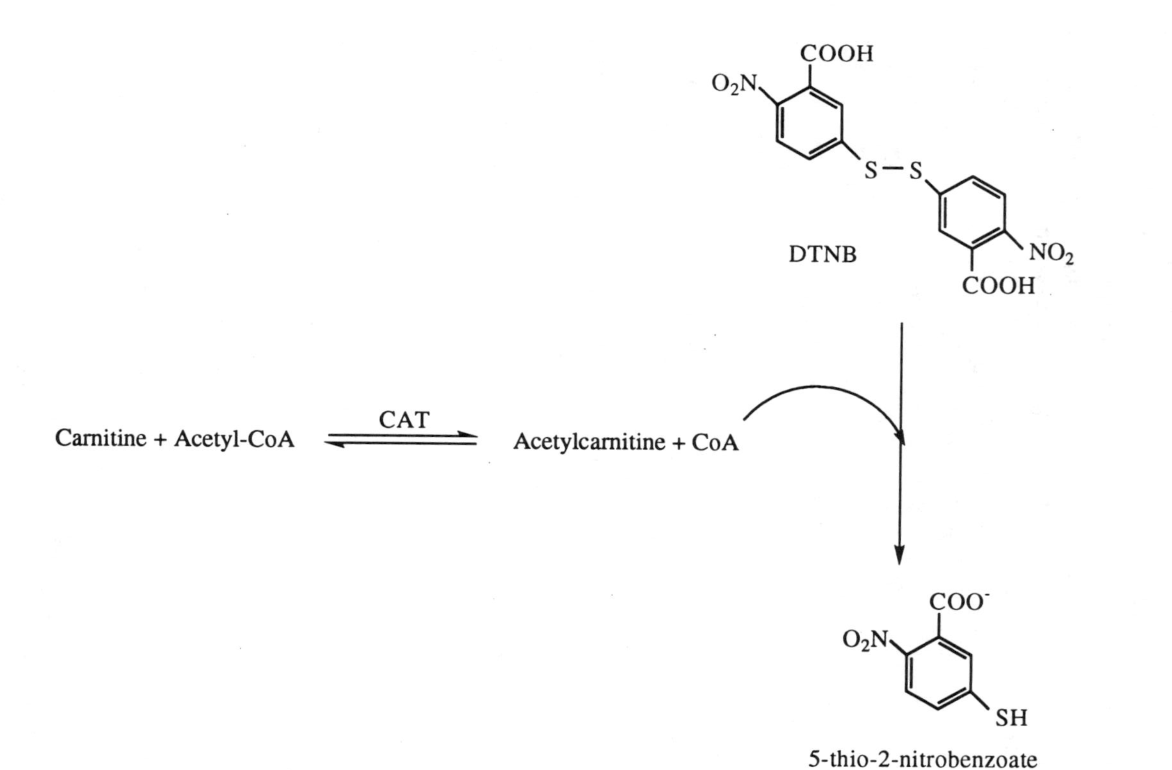

Fig. 8.2. Enzymic conversion of carnitine and acetyl-CoA into acetylcarnitine and CoA, and the further reaction of CoA with DTNB.

measures the [1-^{14}C]acetylcarnitine produced [34,99]. This method has been used and modified by a number of groups to assay the carnitine concentration in such fluids as rat bile [153], and urine, plasma and/or tissue, and human skeletal muscle from needle biopsies [31].

In a radioenzymic assay for free and esterified carnitine in plasma of healthy males and females, samples were incubated in a reaction mixture containing acetyl-CoA, carnitine acetyltransferase, and [1-^{14}C]acetyl-CoA. The radioactive acetylcarnitine was then separated from the unreacted radioactive acetyl-CoA using a column of anion-exchange resin (the principles of which are given in Section C) and centrifugation. The radioactive acetylcarnitine fraction was then quantified by scintillation counting of the supernatant liquid. The results obtained indicated that the method is reliable, precise, and reproducible for the detection of all carnitine fractions in plasma [19,144]. Alternatively, the incubation mixture following radioisotopic exchange can be mixed with Dowex 50W X8 (H^+) resin which retains the [1-^{14}C]acetylcarnitine. The unreacted radioactive acetyl-CoA can then be washed away. The radioactivity of the bound acetylcarnitine is measured by subsequent elution from the resin with ammonium hydroxide. The method is said to be convenient and to require less resin than previous radioisotopic methods [145]. The inclusion of *N*-ethylmaleimide in such radioisotope exchange reactions has been reported to ensure completion of the reaction in the presence of an excess of [1-^{14}C]acetyl-CoA, and a linear response [41,100].

This radioisotopic method has been used for the measurement of carnitine, short-chain acylcarnitine, and long-chain acylcarnitine in plasma and tissue with between- and within-batch precisions of 10.4% and 7%, respectively [10]. Radioenzymic studies have also indicated that, although carnitine is found in all particulate blood components, higher concentrations are detected in white blood cells. Red blood cells and plasma have the same concentrations of free carnitine but in the former short-chain acylcarnitine is enriched with its ratio to free carnitine being up to 1.0 [42].

A second general method is a spectrometric assay based on the reaction of liberated CoA with DTNB, as described in the previous section and shown schematically in Figure 8.2 [104,107,170]. Modifications of this method include the use of dialysis to remove interfering substances in serum and its application to the measurement of free carnitine in serum, cerebrospinal fluid, seminal fluid and human tissues [133]. The method compared well with a radioenzymic assay [154].

Plasma carnitine levels have also been determined by column chromatography [102] including DTNB detection [40].

Conversion of carnitine to resorufin, involving the hydrolysis of acylcarnitine from serum samples and the use of immobilized dehydrogenase and diaphorase enzymes, has been used for its fluorometric monitoring, at an excitation wavelength of 560 nm and an emission wavelength of 580 nm [108]. The coupling of the reaction catalysed by CAT to *N*-[*p*-(2-benzimidazolyl)phenyl]maleimide (BIPM) allows the detection of the fluorescent CoA-BIPM which is present proportionally to serum carnitine [91,103].

Variations in the buffers used for enzymic assays and the effect of deproteinization for serum samples have also been investigated. Both TRIS and HEPES buffers can be used in reactions involving the CAT enzyme. With these buffers one aim is to minimize interference with the buffers themselves and this is judged by the levels of acetylated buffer formed. In the case of HEPES, no acetylated HEPES was formed. Combined with deproteinization the TRIS buffer system provides a cheaper and faster alternative, as the deproteinization can decrease the incubation time and less reagent is required, although with a matched deproteinization method the HEPES buffer is preferred for the detection of total and free carnitine [158]. Deproteinization methods include the heating and then freezing of serum samples, with up to 60% sample volume recovery, [138] and the use of membrane filter cones [152].

Tissue labelled with [1-^{14}C]palmitate has been treated with a phospholipase allowing 2-dimensional thin-layer chromatography (TLC; acidic then basic solvent) of ^{14}C-labelled long-chain acylcarnitines. These are then hydrolysed, freeing carnitine for enzymic assay [129,130].

Enzymic methods have also been applied to the assay of carnitine and acylcarnitines in foodstuffs including milk and milk products. The method involves centrifugation of the dairy produce with 6% $HClO_4$ which facilitates the separation of the free, acid-soluble and acid-insoluble carnitines. The fractions are then incubated with the CAT enzyme and the supernatant liquid arising from centrifugation with Dowex 1-X8 resin is subjected to a radioenzymic assay [59]. Raw, pasteurized and skimmed milk, yogurt, butter and cheese are among the dairy products analysed in this manner [60].

The radiochemical method has also been used in the analysis of plasma from rats, humans [23,132] and sheep [47]. A reversed-phase high-performance liquid chromatographic technique to separate carnitine and acylcarnitines from a biological matrix utilizes a step gradient to provide baseline resolution of acylcarnitines (individually or by class) and allows quantification by using a sensitive radioenzymic assay. Application to liver tissues of rats demonstrate the validity and utility of the chromatographic method while confirming the applicability of the $HClO_4$ fractionation of acylcarnitines by functional class [63].

3. *Chromatography*

Both gas chromatography (GC) and liquid chromatography (particularly high-performance liquid chromatography, HPLC) have been used for the determination of carnitine in a variety of matrices. For GC, the involatility of carnitine has to be overcome by derivatization prior to separation. For HPLC, the lack of volatility is inconsequential but detection of low levels of carnitine by spectrometry is of concern because it is not a strong chromophore. To address this difficulty, derivatization may be employed. For example, the concentrations of carnitine and its *O*-acyl esters in biological materials were measured at low levels once the compounds of interest were subjected to solvent extraction, drying, and converted to 4-bromophenacyl ester derivatives which were separated by HPLC [84]. The method is discussed more fully in Section 5 of Part C. A similar approach, using the same derivatives and reversed-phase HPLC, has been reported for analysing urine for carnitine [121]. Results agreed with those determined by radioenzymic assay. Further details of the method have been described elsewhere [120,123].

An alternative derivatization strategy involves a pre-column reaction with 9-anthryldiazomethane to give a fluorescent ester [177]. Subsequent analysis by HPLC provided a detection limit for carnitine of about 1 pg (at a signal-to-noise ratio of 4) by fluorometric detection.

Free carnitine in tissue samples has been determined following stoichiometric generation of CoA by carnitine acetyltransferase (see above) and analysis of the CoA formed by HPLC on C_8 reversed-phase columns [2,157]. Reversed-phase HPLC has also been utilized for measuring carnitine in various pharmaceutical products [55,156,167]. Anion-exchange chromatography can be used to purify carnitine from urine or liver extracts prior to analysis by HPLC [146] and a rapid high-pressure ion-exchange chromatographic method has been developed for separating and determining quaternary ammonium compounds involved in carnitine metabolism [62].

The problem of the involatility of carnitine for GC analysis was overcome by converting it to 4-butyrolactone by reduction with sodium borohydride under basic conditions [17,21,86]. Using packed-column GC and flame ionization detection, carnitine was measured in mature rat epididymis samples [86] and in milk [21]. Following solvent extraction with chloroform/methanol (2:1), work-up and derivatization, GC was also used to determine carnitine in rabbit tissues [17].

4. Mass spectrometry

An isotope dilution assay of total and free carnitine in urine utilizes fast atom bombardment (FAB) ionization in the positive-ion mode with tandem mass spectrometry (MS/MS). Minimal sample preparation is required and it compared well with radioenzymic assay in terms of specificity, precision, and accuracy, but was much more convenient in terms of analysis time and sample throughput. The new method is also applicable to plasma analysis [81]. In physical chemistry studies [56,61,90], bombardment with atoms or ions followed by negative-ion or positive-ion mass spectrometry provides a means of monitoring the behaviour of carnitine in the solid and solution phases. In addition, the gas-phase chemistry of carnitine was studied by MS/MS.

Mass spectrometric methods which require the sample to be heated to attain the gas phase cause carnitine and its *O*-acyl derivatives to pyrolyse [67,92]. Carnitine and acylcarnitines undergo two major competing pyrolytic reactions under electron ionization conditions. Elimination of water from carnitine, or of the carboxylic acid from an acylcarnitine, precedes intramolecular displacement of trimethylamine and formation of 2(5*H*)-furanone. Secondly, the same intramolecular displacement can occur with formation of an acyloxy-substituted γ-lactone and trimethylamine. For carnitine and the higher acylcarnitines both processes were found to be prevalent in the ion source [67]. The cyclization to acyloxylactones has been optimized and adapted as a derivatization procedure prior to GC/MS analysis of acylcarnitines and hence will described in more detail in Section C.

5. *Microbiological methods*

Carnitine in human and rat fluids has been measured using a carnitine-specific mutant of the enteric yeast *Torulopsis bovina* which has a response threshold to carnitine of 100 pg/ml. A turbidimeter is used to determine growth and the method can be applied to the measurement of acid-soluble and total (acid- and alkali-soluble) carnitine [5]. Lipid-bound carnitine was assayed after precipitation with trichloroacetic acid. The average recovery of free carnitine was 95% and that of lipid-bound carnitine was in the range of 76 to 95% [161].

C. METHODS OF ANALYSIS

Some methods of analysing for acylcarnitines require minimal prior sample preparation, or even none. However, the majority of procedures include isolation and partial purification steps. These will be mentioned in the following individual sections because the type and extent of purification required before analysis are highly dependent on the technique to be applied. Even so, it is appropriate here to include brief details of the commonest method of isolating acylcarnitines: ion-exchange chromatography.

Several methods for purifying carnitine and its *O*-acyl esters have been published. For example, they can be isolated by solvent extraction from wheat with 80% methanol [53], from urine with butan-1-ol [11,126] or with hexan-2-ol [126], from human serum with chloroform/methanol [101], and from tissues also with chloroform/methanol [17]. In addition, column chromatography, molecular sieving and preparative HPLC have been applied to aid isolation, concentration and purification. However, the ion-exchange technique has dominated other chromatographic methods. Given that carnitine and its *O*-acyl esters are zwitterions (see structure **1**) and are cations in acidic solution, it is not surprising that a method based on the chemistry of ions has been widely applied.

For the aptly named ion-exchange procedure, a liquid sample is carried through a column containing a permeable solid polymer that contains fixed charged groups. These groups have varying degrees of affinity for counter-ions in the mobile phase. If the fixed charges are positive, then the resin is called an anion-exchanger. Anions in solution may have sufficient affinity to exchange with the anions originally bound electrostatically to such a resin and hence remain on the resin, while neutral compounds, cations and anions with weak affinity are washed out. In acidic solution, with the carboxylic acid group protonated, acylcarnitines are cations and elute directly from an anion-exchanger. Conversely, a cation-exchanger would contain immobile anionic groups and in acidic conditions acylcarnitines would displace the cations bound to the resin while neutral species and anions are washed through. Later, the acylcarnitines can be released by adding a solution containing cations of even greater affinity for the cation-exchange resin. In this way a purified fraction of carnitine *O*-acyl esters would be eluted. Both types of ion-exchanger have been applied to the extraction of acylcarnitines, and the use of both types in series is common, especially prior to further chromatographic analysis. It should be noted that other compounds with zwitterionic character (*e.g.* choline-containing phospholipids and some small peptides) may not be separated from acylcarnitines by this method. References to the use of ion-exchange approaches can be found in many of the publications cited in the following sections.

1. Nuclear magnetic resonance spectroscopy

Despite its inherently poor sensitivity, NMR spectroscopy is a rapid and powerful technique for monitoring metabolic perturbations in early life, provided that a high-field instrument with the best available specifications is used [27,82]. Urinary accumulations of substances that are characteristic of organic acidurias can result in millimolar levels which may be monitored by ^{1}H NMR spectroscopy. For example, under L-carnitine therapy, patients with propionic acidemia or methylmalonic aciduria were shown to excrete elevated concentrations of propanoylcarnitine and acetylcarnitine [68-70]. Whilst the method is promising for major urinary components, it is not appropriate for traces of acylcarnitines and complex matrices.

The major resonance in water-suppressed and slice-selective ^{1}H NMR imaging of healthy human liver could be assigned tentatively to a fatty acyl methylene group and minor resonances to protons in carnitine and other metabolites [6]. In a study of the relation between proton relaxation times and other resonances in the ^{1}H NMR spectra of lipids and carnitine in canine myocardial infarctions, it was concluded that analysis of the lipid and creatine peaks may serve to characterize the type of tissue [137]. ^{14}N NMR spectroscopy has also been used in health sciences. The ^{14}N resonances of carnitine, betaine, choline, phosphorylcholine and glycerophosphorylcholine could be detected and resolved in the perchloric acid extract of murine radiation-induced fibrosarcomas [52]. Such studies may give more insight into the biochemical pathways within tumours.

In a more traditional chemical study with high-resolution ^{1}H NMR spectroscopy, conformational analysis of carnitine and acetylcarnitine in water has been effected by applying a new Karplus equation with empirically derived substituent constants to the observed vicinal coupling constants. The relative energetics of extended and folded conformers suggest that binding of either carnitine or acetylcarnitine to the CAT enzyme occurs with the folded form [38]. Finally, ^{1}H NMR spectroscopy has been used, in conjunction with homochiral shift reagents, to measure the enantiomeric excess of carnitine either directly [22] or as acetylcarnitine hydrochloride [165].

2. *Thin-layer and paper chromatography*

Thin-layer chromatography has been used as a means of fractionating acylcarnitines prior to application of an analytical measurement or as a method of analysing carnitine and its *O*-acyl esters. As an example of the former type of study, purification of acylcarnitines by TLC preceded desorption chemical ionization mass spectrometry [44,45]. The latter application of TLC is illustrated by a qualitative screening method for the diagnosis of medium-chain acyl-CoA dehydrogenase deficiency [11]. In this case, reversed-phase high-performance TLC of the 4-bromophenacyl derivatives was used with CH_3OH/0.5M triethylamine phosphate (80:20) as the mobile phase for the detection of octanoylcarnitine in urine. The reversed-phase TLC system was compared with a normal adsorptive (silica gel) procedure and a combination of the two was recommended as an inexpensive means of detecting octanoylcarnitine. The same 4-bromophenacyl derivative has been used to detect [^{14}C]2-methylbutanoylcarnitine as an intermediate in ketoisoleucine degradation in rat liver mitochondria by a combination of radio-HPLC and reversed-phase high-performance TLC. After incubating with [^{14}C]ketoisoleucine, isolation of the labelled acylcarnitine by radio-HPLC, and derivatization, [^{14}C]2-methylbutanoyl-L-carnitine was the only metabolite detected by its radioactivity [9]. Picomole amounts of short-chain (acid-soluble) acylcarnitines have been reported to be quantified by a radioisotopic-exchange method using either HPLC or TLC. After separation, the radioactivity in each acylcarnitine is measured and the amount of individual acylcarnitines calculated [76]. Carnitine *O*-acyl esters have been isolated by using 2-dimensional TLC [171]. Determination of long-chain acylcarnitines labelled with ^{14}C has also been based on 2-dimensional TLC, using first an acidic solvent then a basic one. The area corresponding to acylcarnitine was scraped off, eluted, and an aliquot counted. Another aliquot was hydrolysed and the carnitine measured enzymically. Unfortunately, the method does not resolve individual acylcarnitines [129].

Overpressured-layer chromatography, a planar-layer version of HPLC that combines the advantages of high-performance TLC with those of HPLC, has been used to analyse for several quaternary ammonium compounds, including carnitine at the 3 μg level [53,54,162]. The method has been utilized particularly for studying plant metabolism and breeding, including crops like wheat [53] and maize [54]. Short-chain acylcarnitines (C_2 - C_5 acyl groups) have been separated by paper chromatography using a descending solvent system of butan-1-ol:acetic acid:water (8:1:1). The compounds were then hydrolysed and the resulting carnitine was eluted from each chromatography zone and assayed enzymically. The method allowed nanomole amounts of the acylcarnitines from biological materials to be detected [57]. Paper chromatography and bioautography have been employed for detecting L-carnitine and its acyl esters in biological materials [85].

The lack of structural information from these planar forms of chromatography, and the poor resolving power of conventional paper chromatography and TLC compared to HPLC and capillary-column gas chromatography, severely restrict their application for acylcarnitine analysis. Combined thin-layer chromatography/mass spectrometry has been described for quantifying acetylcarnitine and propanoylcarnitine in urine using a stable isotope dilution method [174] but the TLC/MS technique is not widespread and is unlikely to become routine for clinical and biochemical analysis .

3. Gas chromatography

Being involatile zwitterions, carnitine and its acyl esters will not pass unchanged through a gas chromatographic column. However, GC can be employed for their analysis because there are now several ways in which they can be converted into relatively volatile compounds that will elute through a GC column. In 1989, it was commented that derivatization to volatile analogues had been poorly investigated [95]. In the intervening years, this aspect of acylcarnitine chemistry has been explored much more thoroughly. In Section B above, it was reported that carnitine can be converted into a γ-butyrolactone prior to GC analysis [17,21,86]. Using a packed column of ethylene glycol adipate on Chromosorb™ W-AW, the method has been shown to allow the determination of carnitine and short-chain acylcarnitines, such as acetylcarnitine, in milk [21]. The same

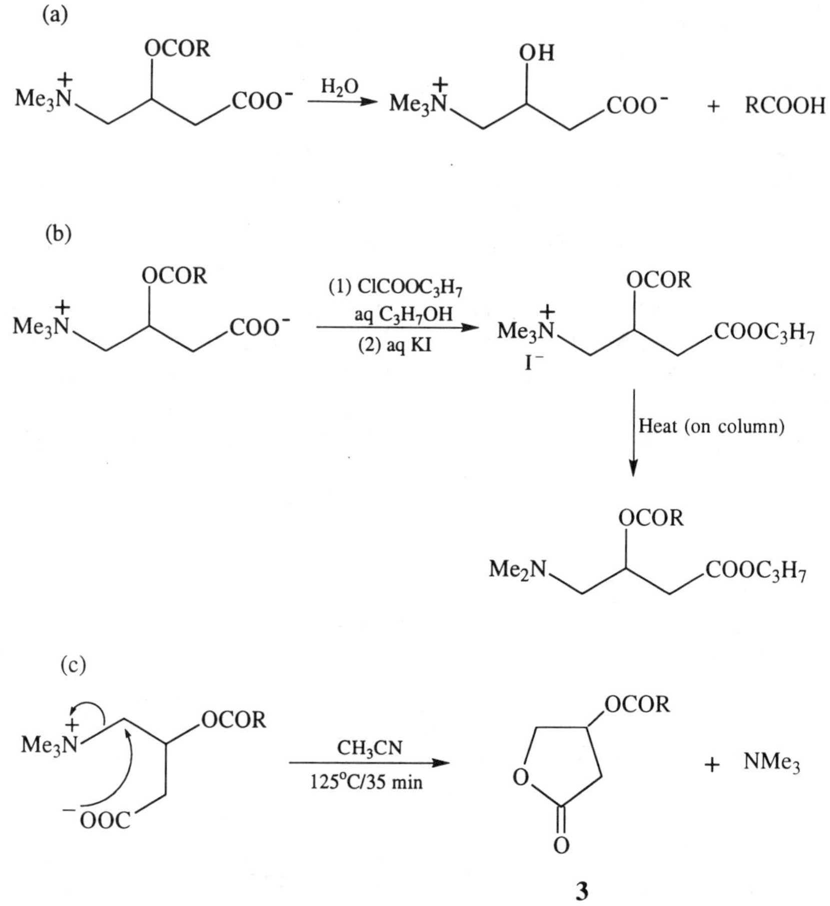

(a)

$Me_3\overset{+}{N}CH_2CH(OCOR)CH_2COO^- \xrightarrow{H_2O} Me_3\overset{+}{N}CH_2CH(OH)CH_2COO^- + RCOOH$

(b)

$Me_3\overset{+}{N}CH_2CH(OCOR)CH_2COO^- \xrightarrow[\text{(2) aq KI}]{\text{(1) ClCOOC}_3\text{H}_7\text{, aq C}_3\text{H}_7\text{OH}} Me_3\overset{+}{N}CH_2CH(OCOR)CH_2COOC_3H_7\ I^-$

Heat (on column)

$Me_2NCH_2CH(OCOR)CH_2COOC_3H_7$

(c)

$Me_3\overset{+}{N}CH_2CH(OCOR)CH_2COO^- \xrightarrow[\text{125°C/35 min}]{CH_3CN}$ **3** (OCOR-substituted γ-lactone) $+ NMe_3$

Fig. 8.3. Different schemes for converting acylcarnitines into volatile derivatives: (a) hydrolysis to carboxylic acids, (b) a dequaternization (*N*-demethylation) route, and (c) cyclization to lactones.

intramolecular displacement of trimethylamine and formation of an acyloxy-substituted γ-lactone is known to occur when carnitine and the higher acylcarnitines are subjected to electron ionization mass spectrometry [67]. Such a cyclization, brought about simply by heating, has been developed as one derivatization strategy that enables gas chromatographic examination of a wide range of acylcarnitines [96]. The derivatization is shown Figure 8.3, along with other methods of making acylcarnitines amenable to GC (or to gas chromatography/mass spectrometry, GC/MS, as described in the next section).

The first procedure shown in Figure 8.3 depicts hydrolysis of carnitine esters, following which the liberated carboxylic acids can be subjected to GC analysis [13,36]. Identification of nanomolar amounts of short-chain acyl residues in this way has several drawbacks. With the necessary extensive work-up, it is time-consuming. There may also be some ambiguity in the result because the memory of the carnitine origin of the acyl groups is not retained once hydrolysis has taken place. For instance, contaminating carboxylic acids would give misleading data. Modifications to this method are required before it can be applied to longer chain acylcarnitines [110]. Even so, the method has been applied successfully to the quantification of water-soluble acylcarnitines in rat tissues [37] and to the identification of aliphatic short-chain acylcarnitines in beef heart [12]. In both mammalian tissues, acetyl-, propanoyl-, 2-methylpropanoyl-, butanoyl-, 2-methylbutanoyl-, 3-methylbutanoyl- (isovaleryl-), tiglyl- and hexanoylcarnitine were found. Other acylcarnitines were also identified or tentatively identified in these studies.

More effective approaches are shown in Figure 8.3(b)-(c). In both of these derivatization schemes, the end product retains a memory of its origin inasmuch as a diagnostic portion of the carnitine structure occurs in the derivative. Hence, the carnitine origin of the acyl residue is unambiguous in the subsequent analysis. The application of the *N*-demethylation [66] and lactonization approaches [96,98] will be described in the next section on GC/MS.

A problem for any gas chromatographic method of analysing for acylcarnitines is the lack of structural information in the resulting chromatogram. There cannot be total confidence in structural assignments based solely on retention times, particularly if the chromatogram is a complex one. More certain assignments are made if the gas chromatograph is coupled to a mass spectrometer. For example, it has been said [7] that "there should be mass spectrometric facilities available to all laboratories offering a service for the detection of organic acidurias and that in the absence of such facilities laboratories should probably not attempt to offer a diagnostic service." Even so, it has been claimed that one gas chromatographic assay for urinary medium-chain acylcarnitines produces such readily interpretable and uncomplicated chromatograms that the requirement for mass spectrometry is circumvented [46]. In this study, GC was used directly for assay of urinary medium-chain fatty acylcarnitines, that is, the method does not require a separate derivatization step. Rather, the acylcarnitines are allowed to decompose thermally in the hot injection zone of the GC system. At 280°C, each acylcarnitine appears to undergo an ester pyrolysis reaction, giving the carboxylic acid corresponding to each acyl group. It is these acids that elute through a GC column coated with the polar stationary phase, PEGA. Given that fatty acids are the actual substances detected, it is important that free acids do not contaminate the urine extracts containing acylcarnitines. In the method described [46], carboxylic acids are extracted from the urine with chloroform prior to extraction of acylcarnitines into butan-1-ol. However, it is reported that only about 60% of the acids are so removed. Despite the inherent weaknesses, the method enabled detection of octanoylcarnitine in a symptomatic individual with medium-chain CoA dehydrogenase deficiency and in two asymptomatic siblings following administration of carnitine.

4. Gas chromatography/mass spectrometry

Once a procedure has been developed that allows GC of some suitable derivative of acylcarnitines, there are no major barriers (other than access to the required instrumentation) to employing GC/MS for the same task. Application of the coupled technique usually provides lower limits of detection and greater structural specificity compared with GC alone. As GC/MS can now be effected routinely on any one of a number of inexpensive and simple-to-use bench-top systems, this methodology falls within the realms of many biochemical and neonatal screening laboratories that may wish to obtain acylcarnitine profiles in biological matrices.

One approach to the GC/MS analysis of acylcarnitines involves direct esterification using propyl chloroformate in aqueous propan-1-ol in the presence of pyridine. This reaction requires only 5 minutes at room temperature. After addition of potassium iodide, the resulting acylcarnitine propyl ester iodides are extracted into chloroform and their subsequent *N*-demethylation can be brought about conveniently in the hot injector port of the GC/MS system (260°C), causing the formation of the volatile derivatives shown in Figure 8.3(b). These *N*-demethylated acylcarnitine propyl esters are well separated on a GC methylsilicone stationary phase (DB-1^{TM}) and are readily detected and identified by their methane chemical ionization (CI) mass spectra which are characterized by abundant $[M + H]^+$ ions and several diagnostic fragment ions. The detection limits of medium-chain acylcarnitine standards (C_4- C_{12} side-chains) were demonstrated to be below 1 ng of starting material when using selected ion monitoring of $[M + H]^+$ ions and a common fragment ion. By this method, seven acylcarnitines (with $C_{5:0}$ to $C_{10:1}$ side-chains) have been characterized in the urine of a patient suffering from medium-chain acyl-CoA dehydrogenase deficiency [66]. Only the four major acylcarnitines in the same urine sample could be detected by these authors using non-chromatographic fast atom bombardment mass spectrometry (described in Section 7). The same GC/MS method also revealed that octanoylcarnitine, not valproylcarnitine, was the most abundant medium-chain carnitine *O*-acyl ester excreted by a patient treated with valproic acid [151].

Alternatively, acylcarnitines can be extracted from urine either by ion-exchange chromatography [96,98] or by solvent extraction [126] and heated in acetonitrile for about 30 min at 125°C in the presence of *N*,*N*-diisopropylethylamine to effect cyclization to a lactone **3** (Figure

8.3(c)). Such acyloxylactones elute readily on a GC column with a stationary phase of DB5™ and can be identified by their chemical ionization and/or electron ionization mass spectra. Using this method it was found that monocarboxylic acylcarnitines from acetylcarnitine (C_2 acyl chain) to octadecanoylcarnitine (C_{18} acyl chain) can be isolated from urine with recoveries of over 80%. To obtain such recoveries, different methods of extraction had to be used for different ranges of acylcarnitines. For shorter chain acylcarnitines (C_2 to C_8 side-chains) an ion-exchange procedure was recommended. Acylcarnitines with acyl chain length C_8 to C_{12} were reported to be isolated most effectively from urine by solvent extraction with butan-1-ol as long as the urine had been acidified to about pH 2. For long-chain acylcarnitines (C_{10} to C_{18} acyl chains) solvent extraction of unacidified urine with hexan-2-ol was particularly simple and effective. This latter procedure allowed the diagnosis of long-chain acyl-CoA dehydrogenase deficiency by observation of urinary carnitine esters with acyl groups containing up to 12 carbon atoms [126]. These findings on extraction may have implications for any method of urinary acylcarnitine analysis that requires prior purification.

The lactonization and GC/MS approach has been applied to several disorders of organic acid metabolism associated with abnormalities in the levels of urinary acylcarnitines, such as medium-chain acyl-CoA dehydrogenase deficiency [96,98], propionic acidemia, isovaleric acidemia, multiple acyl-CoA dehydrogenation deficiency [98] and long-chain acyl-CoA dehydrogenase deficiency [126]. It has also been used to detect metabolites of exogenous compounds, as with 3-phenylpropanoylcarnitine in babies who had received a 3-phenylpropanoic acid load [97]. Also, in the urine of an infant undergoing valproic acid (2-propylpentanoic acid) therapy, a widely used anticonvulsant associated with carnitine deficiency [131], valproylcarnitine was detected along with a major carnitine *O*-acyl ester interpreted to be 2-propyl-3-ketopentanoylcarnitine [73,97].

The same cyclization and GC/MS approach can be applied to the detection of acylcarnitines in dried blood spots on Guthrie cards following a very simple solvent extraction with chloroform/methanol (2:1). In initial work the metabolic diseases of MCAD deficiency, propionic acidemia and methylmalonic aciduria have been identified by observation of characteristic acylcarnitines in each case, using a sensitive bench-top ion trap mass spectrometer [74].

It has been found in the course of this work that, at the derivatization stage, more prolonged heating or higher temperatures cause some decomposition to fatty acids, and that the use of heated GC injectors at over 230°C facilitates ester pyrolysis as observed by others [46]. Cold on-column injectors or split/splitless injectors at 230°C were preferred [96]. However, another worker has found that, at high injection temperatures (280°C), a degree of ester pyrolysis still occurs, resulting in peaks for carboxylic acids, but that the major reaction is lactonization as

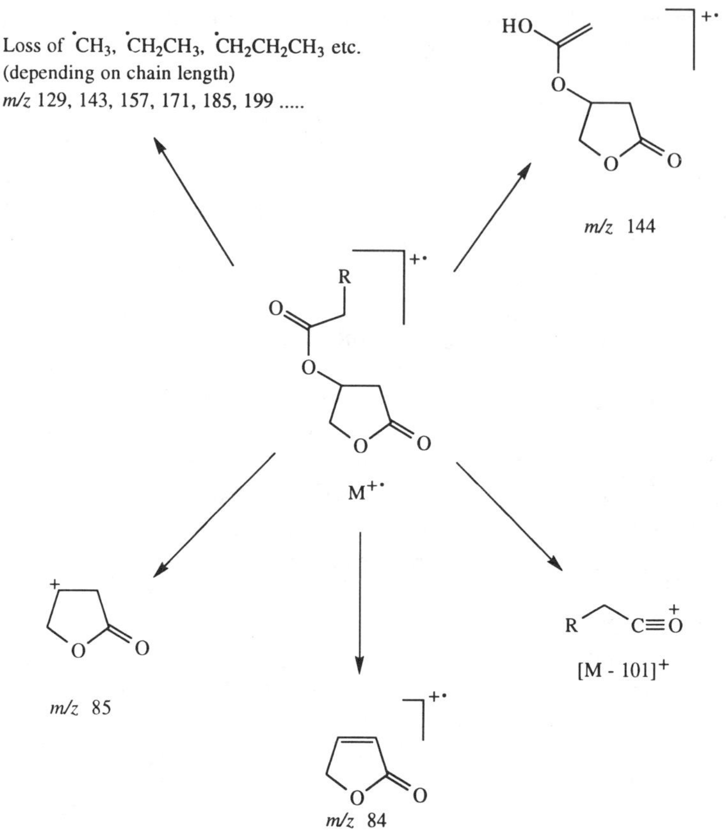

Loss of $\dot{C}H_3$, $\dot{C}H_2CH_3$, $\dot{C}H_2CH_2CH_3$ etc.
(depending on chain length)
m/z 129, 143, 157, 171, 185, 199 .....

m/z 144

$M^{+\cdot}$

m/z 85

$[M - 101]^+$

m/z 84

Fig. 8.4. Mass spectrometric fragmentation of the lactones derived from acylcarnitines.

in Figure 8.3(c) [65]. This new on-column version of the cyclization approach promises to be a fast and convenient method for GC/MS of acylcarnitines.

The electron ionization (EI) mass spectra of the acyloxylactones are not ideal for identification purposes because the molecular ions are of low abundance and often absent, particularly when dealing with trace amounts from biological samples [96,98]. However, the fragment ions are highly characteristic of structure (Figure 8.4). All of the lactones with butanoyl or larger acyl groups exhibit ions at *m/z* 85 and 144, and lower homologues yield ions at *m/z* 84. Mass chromatograms for these ions are particularly useful for locating peaks in a GC/MS analysis that are due to the acylcarnitine derivatives. Identification of a known acylcarnitine derivative is based on its GC retention time and computerized matching of its EI mass spectrum against a library of standard spectra generated from synthesised acyloxylactones. Any novel acylcarnitine is identified by measuring its relative molecular mass *via* its CI mass spectrum and by

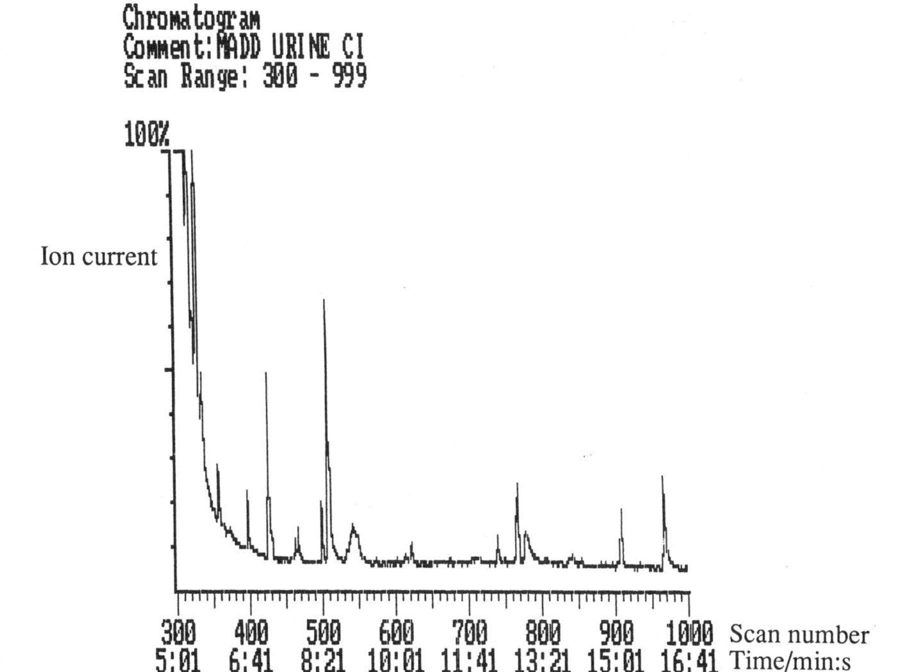

Fig. 8.5. Total ion current chromatogram for the analysis of a urine sample with a simple bench-top ion trap mass spectrometer. Prior to GC/MS analysis, the acylcarnitines are converted by heat into acyloxylactones. Chemical ionization mass spectra taken at scan numbers 507 and 767 are shown in the next figure.

interpretation of the fragmentation observed in the EI mode.

As an illustration of this GC/MS approach, Figure 8.5 shows the chromatogram obtained from 0.5 ml of urine from an infant who, on the basis of these results, was diagnosed as suffering from multiple acyl-CoA dehydrogenation deficiency (MADD). Most, but not all, of the peaks observed are due to lactones derived from acylcarnitines and these can be easily located by examination of mass chromatograms of m/z 85 and 144 in the EI mode. Examples of CI mass spectra from the chromatogram are given in Figure 8.6, along with the equivalent EI results for this sample. Note the clear $[M + H]^+$ ions for the lactones of isovalerylcarnitine (m/z 187, Figure 8.6(b)) and octanonylcarnitine (m/z 229, Figure 8.6(d)) in the CI mode, and the fragmentation available for interpretation or library searching in the EI mode (Figure 8.6(a) and (c)). Among the acylcarnitines separated and identified were: acetyl-, propanoyl-, 2-methylpropanoyl-(isobutyryl-), butanoyl-, 2-methylbutanoyl-, isovaleryl-, hexanoyl- and octanoylcarnitine, and at least two different octenoylcarnitines. It is a clear advantage of this and other methods based on chromatography that excellent specificity is obtained for identification of different acylcarnitines, including isomers.

Background Subtract
Comment: MADD URINE EI
Average of: 514 to 518 Minus: 499 to 503
100% = 4628

(a) 100% 85

Relative ion abundance

91 97 104 111 121 129 144 153 163 186

80 100 120 140 160 180 200 220

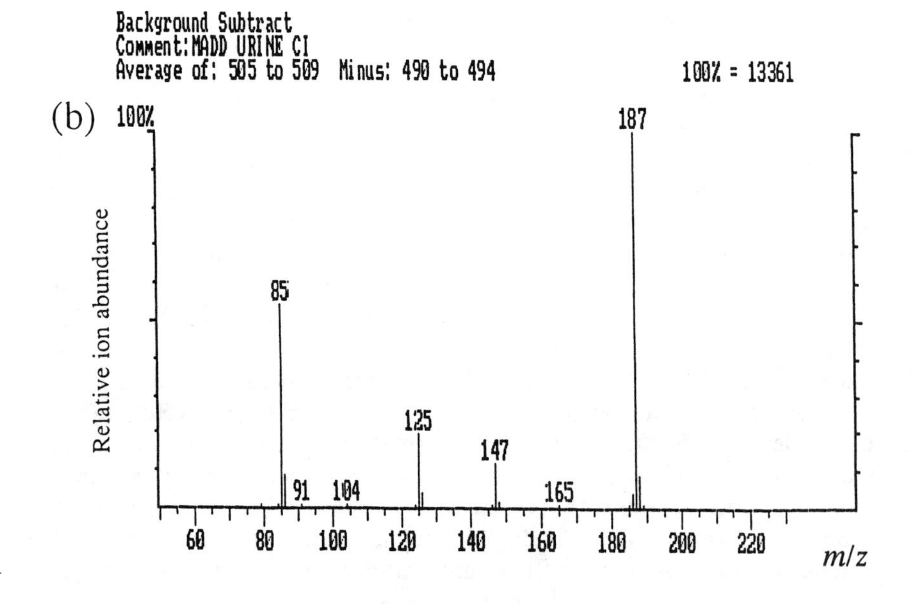

Fig. 8.6(a, b)

Background Subtract
Comment: MADD URINE EI
Average of: 775 to 779 Minus: 760 to 764 100% = 2027

(c) 100% 85

97 109 127 144 157 171 185 195 209

80 100 120 140 160 180 200 220 240

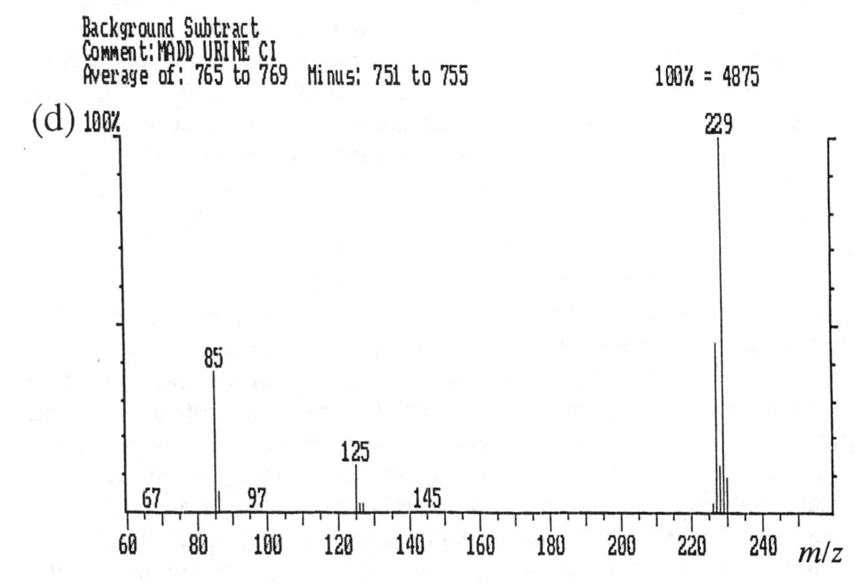

Fig. 8.6(c, d). The mass spectra of the lactones from isovalerylcarnitine, (a) and (b), and octanoylcarnitine, (c) and (d), in a urine sample from a MADD patient. The EI mass spectra are shown at the top, (a) and (c), and the CI spectra from the data in Figure 8.5 are given underneath, (b) and (d).

5. High-performance liquid chromatography

In the previous two sections, the involatility of acylcarnitines was a considerable obstacle to the intended analysis. To apply a high-resolution chromatographic technique and yet avoid such an obstacle, HPLC is a logical choice of method. A key problem for this technique is the limit of detection of acylcarnitines. The naturally occurring acylcarnitines are only weakly chromophoric (λ_{max} is about 210 nm) and are neither electrophoric nor fluorophoric to any useful degree. To enhance the detectability of eluting acylcarnitines, derivatization with a strongly chromophoric group is one option. Others are to employ radiochemical detection or to couple the HPLC column to a mass spectrometer, increasing sensitivity and, at the same time, structural specificity. All of these approaches are represented in this and the next section.

Reversed-phase step-gradient HPLC provides a simple and effective method for separating carnitine and acylcarnitines from a biological matrix (rat liver tissue) prior to quantification by radioenzymic assay. The chromatographic method, with spectrometric detection at 210 nm, also permits resolution of long-chain acylcarnitines in the presence of large excesses of carnitine and short-chain acylcarnitines [63].

For direct analysis of carnitine and acylcarnitines by HPLC, a common derivatization strategy involves the formation of chromophoric 4-bromophenacyl esters (Figure 8.7). A method for the determination of carnitine and other ω-trimethylammonium carboxylates by derivatization with 4-bromophenacyl trifluoromethanesulphonate (4-bromophenacyl triflate) in acetonitrile containing *N*,*N*-diisopropylethylamine, followed by separation by reversed-phase ion-pair HPLC [120,123], and the determination of total carnitine in human urine by base hydrolysis, ion-exchange purification of carnitine, derivatization, chromatography on Radial-Pak C_{18} of 10 μm particle diameter, and spectrophotometric detection at 254 nm have been reported [121]. Applying the same approach to acylcarnitines, their 4-bromophenacyl esters were analysed on a radially compressed column of Resolve-Pak™ C_{18}, 5 μm particle diameter, using a gradient containing varying proportions of water, acetonitrile, tetrahydrofuran, triethylamine, potassium phosphate, and phosphoric acid. Baseline separation was obtained for a standard mixture (5 nmol of each injected) containing the carnitine esters of acetic, propanoic, butanoic, valeric, hexanoic, heptanoic, octanoic, nonanoic, decanoic, lauric, myristic, palmitic and stearic acids. Nearly complete separation was obtained for a mixture of butanoyl-, isobutanoyl- (2-methylpropanoyl-), isovaleryl- (3-methylbutanoyl-) and 2-methylbutanoyl-carnitine. The method was used to generate urinary acylcarnitine profiles from patients having propionic acidemia, isovaleric acidemia, and medium-chain acyl-CoA dehydrogenase deficiency [122].

The same derivatives can be produced by reaction with 4-

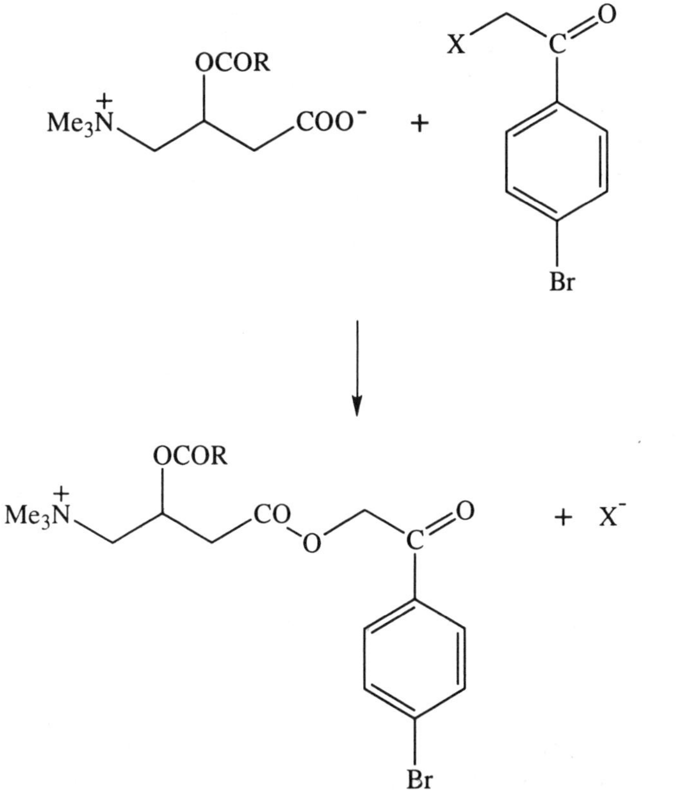

Fig. 8.7. Formation of chromophoric 4-bromophenacyl esters of acylcarnitines (X = Br or OSO_2CF_3).

bromophenacyl triflate in the presence of magnesium oxide as base. The resulting ester derivatives have been separated by HPLC on a silica column, in a mixed partition and ion-exchange mode. Using this method, carnitine and its *O*-acyl esters in biological media can be measured in 100 μl samples, with a detection limit below 1 μmol/L. The procedure allows batch analysis of more than 30 specimens and a single result is available within 2 h [84]. Urinary carnitine and acylcarnitines have been isolated on an ion-exchange column, then derivatized to 4-bromophenacyl derivatives by reaction with 4-bromophenacyl bromide in the presence of a crown ether and potassium ions under carefully controlled conditions, and subjected to reversed-phase HPLC (25 x 0.5 cm i.d. packed with 5 μm Spherisorb™ Octyl) with a mobile phase of 7% 0.25 M, pH 5.8 trimethylamine phosphate buffer in a water:acetonitrile gradient. Detection was by uv absorption at 254 nm. Urinary propanoylcarnitine was observed, as expected, in a neonate with propionic acidemia [169]. Much the same methodology has been employed to study the effects of long-chain fat loads on two asymptomatic patients with medium-chain acyl-

CoA dehydrogenase deficiency. After an overnight fast, they were given fat and urinary excretion of total carnitine was shown to be elevated in both patients. Octanoylcarnitine, hexanoylcarnitine and acetylcarnitine were the main components detected at 260 nm [160].

The 4-bromophenacyl derivatives of acetyl- to hexadecanoyl-carnitine have been analysed by reversed-phase HPLC again using a gradient of acetonitrile:water:triethylamine phosphate. Several patients suspected of suffering inherited disorders of mitochondrial fatty acid oxidation were shown to excrete octanoylcarnitine, hexanoylcarnitine, and, in some cases, a small amount of decanoylcarnitine [8,10]. Reversed-phase HPLC has also been employed for the characterization of the 4-bromophenacyl derivatives of carnitine esters of dicarboxylic acids in a study of the metabolism of [U-^{14}C]hexadecanedionoyl-mono-CoA by rat liver peroxisomes and rat skeletal muscle mitochondria [134]. A different esterification, anthrylmethyl ester derivatization, has been utilized for HPLC analysis of acylcarnitines, particularly with cation-exchange columns [155].

Reversed-phase HPLC with post-column derivatization, in an apparatus that has been termed a carboxylic acid analyser, has been investigated for studying inherited metabolic disorders by measuring urinary acylcarnitines [77-79]. The on-line derivatization involves the reaction between the carboxylic acid group in the analyte and 2-nitrophenylhydrazine in the presence of 1-ethyl-3-(3-dimethylaminopropyl)-carbodiimide hydrochloride prior to colorimetric detection. Hence, before the HPLC analysis can proceed, the interfering carboxylic acids from urine must be removed. Anion-exchange is used for the required purification step. The acylcarnitines were separated in about an hour on a CLC-ODS column (15 cm x 6 mm i.d. with a particle diameter of 5 μm) in an isocratic solvent of 5% acetonitrile:phosphate buffer. Calibration curves for carnitine, acetylcarnitine, glutarylcarnitine and propanoylcarnitine were linear over a concentration range of 30 - 1000 nmol/ml, as appropriate for urinary levels [78,79]. The technique has been used to identify glutarylcarnitine in the urine of a patient with glutaric aciduria type I [78,79], urinary propanoylcarnitine from patients with propionic acidemia and those with methylmalonic aciduria [79], and many urinary acylcarnitines, including acetyl-, butanoyl-, 2-methylpropanoyl-, isovaleryl-, glutaryl- and octanoylcarnitine, in a neonate with multiple acyl-CoA dehydrogenation deficiency [77]. Interestingly in the last case, the method was used to monitor over several days the effect of DL-carnitine therapy on the excretion of acylcarnitines by the MADD patient. In particular, the patient's dominant characteristic urinary acylcarnitine was changed from isovalerylcarnitine to 2-methylpropanoylcarnitine (isobutyrylcarnitine) during the early period of DL-carnitine therapy.

A method for detecting and quantifying acylcarnitines by HPLC using a strategy based on radioisotopic exchange has been widely described

[13,76,148-151]. Briefly, high specific activity L-[^{3}H]- or L-[^{14}C]carnitine is incorporated into the acylcarnitine pool in a sample by enzymic exchange. The acylcarnitines must be substrates for the carnitine acyltransferase(s) used and the enzyme(s) must be totally free of acyl-CoA and acylcarnitine hydrolytic activity. If such conditions are met, picomolar levels of individual acylcarnitines can be detected reliably and rapidly after isotopic equilibrium is established by subjecting the radioactive acylcarnitines to either HPLC or thin-layer chromatography. After separation, the amounts of radioactivity in the acylcarnitines are measured and the quantity of individual acylcarnitines can be calculated, for example, from the specific activity of the initial total carnitine pool.

The sensitivity and specificity of the radioisotopic exchange/HPLC method for detecting urinary medium-chain acylcarnitines have been found to be sufficient for the diagnosis of medium-chain acyl-CoA dehydrogenase deficiency. For example, over one hundred urine specimens from 75 controls and children with metabolic diseases (in the asymptomatic state without carnitine loading) were analysed in a blind experiment and all 47 patients with MCAD deficiency were correctly diagnosed using the criterion that the peak areas of octanoylcarnitine or hexanoylcarnitine are larger than those of other medium-chain acylcarnitines. However, patients receiving valproic acid or a diet enriched in medium-chain triglycerides can also test positive for MCAD deficiency by this criterion, so successful application of the method requires a knowledge of medium-chain triglyceride or valproic acid administration [151].

On-line radiochemical detection for reversed-phase HPLC also provides a means of detecting acylcarnitines that are metabolic products of radio-labelled precursors in tracer studies. For example, oxidation of [^{14}C]hexadecanoate by normal human fibroblast mitochondria [80] and by rat skeletal muscle mitochondria [168], and metabolism of [^{14}C]ketoisoleucine by rat liver mitochondria [9] have been studied in this way. Illustrative of a typical application, in normal human fibroblast mitochondria, only saturated acylcarnitine esters were detected, supporting the concept that the acyl-CoA dehydrogenase step is rate-limiting in mitochondrial β-oxidation. Incubations of fibroblast mitochondria from patients with defects of β-oxidation show different profiles of intermediates, while mitochondria from patients with defects in electron transfer flavoprotein and electron transfer flavoprotein:ubiquinone oxido-reductase are associated with slow flux through β-oxidation and accumulations of long-chain acyl-CoA esters and acylcarnitines. As expected, elevated levels of saturated medium-chain acylcarnitines were found in the incubations of mitochondria with medium-chain acyl-CoA dehydrogenase deficiency. The authors rightly state that radio-HPLC of intermediates of mitochondrial fatty acid oxidation is an important new technique to study the control, organisation and defects of the enzymes of β-oxidation [80].

In an entirely different approach, urinary carnitine *O*-acyl esters can be converted enzymically into CoA esters with carnitine acetyltransferase prior to separation of the resulting CoA esters on a radially compressed cartridge of Radial-Pak C_8 with a mobile phase containing 0.025 M tetraethylammonium phosphate in a linear gradient of 1% - 50% methanol [43]. Spectrophotometric detection at 254 nm was utilized for quantitative investigations of propionic and isovaleric acidemias and methylmalonic aciduria. The enzymic conversion approaches quantitative yields for acetyl and propanoyl esters as long as large amounts of carnitine are not present. This potential problem is not usually serious because acidemia patients produce little free carnitine in their urine [43]. A mixed chromatographic matrix of calcium phosphate supported on macroporous silica microparticles, which has similar selectivity and chemical inertness to hydroxyapatite and mechanical resistance to the pressures generally used in HPLC, has been used to separate biomolecules such as carnitine derivatives and sugars [28]. Finally in this section, a new method for the determination of acetyl-D-carnitine in the L-enantiomer by enzymic reaction and HPLC has been reported [175]. The D-isomer was converted stereoselectively by electric eel acetylcholinesterase into D-carnitine and then separated and determined by ion-pair reversed-phase HPLC.

The techniques covered in this section share with gas chromatography two key disadvantages. The chromatographic process is inherently sluggish and hence does not lend itself to population screening, and neither GC nor HPLC provides structural information on the eluting analytes. The latter deficiency can be addressed, as also in the case of gas chromatography, by coupling with mass spectrometry. In addition, the application of LC/MS obviates the need for derivatization with a chromophoric group.

6. *Liquid chromatography/mass spectrometry*

The difficulties of chromatographic analysis of carnitine and its *O*-acyl esters result from their highly polar nature and lack of direct amenability to conventional HPLC detectors (uv, fluorescence, electrochemical). In 1983, the only reported successful method of ionization for carnitine esters was fast atom bombardment (FAB) [139], and there was no device for coupling FAB with LC available at that time. The lack of a chromatographic method was severely restricting progress in the understanding of carnitine metabolism.

The first combined chromatographic/mass spectrometric method applied successfully to the analysis of intact acylcarnitines was LC/MS with thermospray ionization [176], a technique that had already proven applicable to a wide range of polar, thermally sensitive molecules of low volatility [16,87]. Other major advantages of this method were, and are still, the use of normal flow rates with standard reversed-phase columns and the lack of necessity to derivatize the molecules prior to analysis. The LC/MS interface is a heated capillary tube through which solvent is pumped, forming a supersonic jet of fine droplets with net positive or negative charges resulting from the use of a buffer. The droplets shrink as they enter the vacuum region and solute ions are produced without the use of an external ionizing technique.

In the first published example [176], free carnitine, acetylcarnitine and propanoylcarnitine were resolved on a 15 cm x 4.6 mm C_{18} reversed-phase column by isocratic elution with 5% methanol in 0.05 M ammonium acetate. Detection limits using selected-ion monitoring (SIM) were better than 0.1 nmol injected, and linear calibration curves were demonstrated using standard solutions and isotope dilution. The mass spectra showed extensive fragmentation, much of it thermally induced, but abundant protonated molecules were also observed. Subsequently, the method was employed to characterize valproylcarnitine, a novel drug conjugate [112], in the urine of patients receiving valproic acid. This was the first reported application of LC/MS in the analysis of acylcarnitines in urine, and required the physical separation of the target compound from a biological isomer, octanoylcarnitine. These isomers were resolved isocratically on the same column previously described [176] using 55% methanol: 45% 0.1M ammonium acetate solution. The urine was purified by ion-exchange chromatography and detection was facilitated by use of SIM.

Further examples of thermospray LC/MS analysis of acylcarnitines in urine were described in a review paper [88]. The excretion of disease-specific metabolites in patients with mitochondrial enzyme defects, such as propionic acidemia, methylmalonic aciduria, isovaleric acidemia, 3-hydroxy-3-methylglutaryl-coenzyme A lyase (HMG) deficiency and medium-chain acyl-coenzyme A dehydrogenase (MCAD) deficiency was confirmed by this technique. Thermospray LC/MS was again required to characterize tiglylcarnitine, a diagnostic metabolite of β-ketothiolase deficiency. The new metabolite was chromatographically distinguished from 3-methylcrotonylcarnitine, a potential biological isomer, again using isocratic LC conditions [119].

Thermospray LC/MS has made a valuable contribution to the characterization of acylcarnitines in urine in both normal and disease states. Its chief limitations are sensitivity, which is insufficient to tackle the more difficult problem of the analysis of human blood, and instability to changing conditions at the interface, such as occur with solvent gradients and minor changes in flow rate.

The successful coupling of LC with FAB was first described in 1986 [29,72]. The interface in this system was a fine capillary connected to a steel frit, allowing exposure of the LC eluate to the external FAB ion gun. The solvent is evaporated efficiently in the ion-source vacuum as the surface of the frit is continuously replenished with fresh eluate. The solutes are ionized as they elute from the column and the capillary connected to it. A specially constructed micro-LC packed column (15 cm x 0.32 mm) was employed to effect chromatography at the very low flow rates of about 5-10 μl/min required by the method.

A modification of this interface, in which the frit was replaced by a smooth hemispherical surface, was used for the first applications to acylcarnitine analysis [116,127]. Acylcarnitines were eluted from the reversed-phase micro-LC column using a steep non-linear gradient from 100% solvent A (5% methanol in 0.05M ammonium acetate plus 5% glycerol) to 30% solvent A plus 70% solvent B (95% methanol, 5% glycerol) in 5 min. Glycerol was necessary to provide stability and sensitivity at the FAB ionization stage. The rapidly changing solvent composition posed no problems, and enabled the separation and analysis of acylcarnitines of widely varying chain-length. The analysis of acylcarnitines in patients with propionic acidemia, isovaleric acidemia and MCAD deficiency was accomplished. The overall sensitivity was better with this device than with thermospray, perhaps because the mass spectra consisted almost entirely of protonated molecules with little fragmentation.

The practical difficulties of gradient micro-LC operation were effectively overcome by use of conventional LC columns (15 cm x 3.9 mm) with post-column stream-splitting [128]. Even with split-ratios of 1:50 or 1:100, the sensitivity was adequate for the analysis of biological samples. This system was employed, for example, to detect and characterize 3-phenylpropanoylcarnitine, a metabolite of 3-phenylpropanoic acid, in the urine of a patient with MCAD deficiency [124].

Although both types of LC/MS interface have yielded valuable data and proven reliable, they require a degree of mass spectrometric expertise that is not available in most clinical laboratories. This may also apply to the new and potentially more sensitive electrospray ionization (ESI) source, which is ideally suited to LC/MS. Whilst a brief communication has described the electrospray mass spectra of acylcarnitines, which consist almost entirely of abundant $[M + H]^+$ ions [75], as yet there are no formal publications on the application of LC/ESIMS to the analysis of acylcarnitines in biological samples. The existing liquid chromatographic methods have now been largely superseded by tandem mass spectrometry, which avoids the practical difficulties of combined LC/MS.

7. *Tandem Mass spectrometry*

Currently, the most successful ionization method for acylcarnitines is fast atom bombardment. Since FABMS is a special case of secondary ion mass spectrometry in the liquid state (so-called "liquid secondary ion mass spectrometry" or LSIMS), in which a beam of either atoms or ions of high translational energy is directed at the surface of a solution of the sample in a suitable liquid matrix, the more generic term LSIMS is used in the remainder of the review.

In general, the types of molecule that are most amenable to LSIMS are those with the highest surface activity in hydrophilic solvents, in other words, surfactants. Such molecules contain a strongly basic or acidic functional group attached to a lipophilic moiety, usually an alkyl group. For example, the tetrabutylammonium ion is a pre-formed cation with high surface activity, requiring much less energy to become a gas-phase ion under LSIMS than most other molecules. Thus, the presence of a fraction of a per cent of tetrabutylammonium chloride in a sample solution will almost always result in a very large signal at m/z 242, corresponding to the tetrabutylammonium ion, to the virtual exclusion of all other species.

The presence of a quaternary ammonium functional group (also a pre-formed cation) in carnitine and its derivatives accounts for the sensitivity of these compounds to LSIMS. In the first detailed publication [118], the mass spectra of free carnitine, acetylcarnitine and propanoylcarnitine, showing abundant $[M + H]^+$ ions, were reported. Signals corresponding to these species were also detected in the urine of patients with propionic acidemia and Reyes syndrome after oral L-carnitine supplementation. In these complex mixtures, the presence of a signal with the correct mass is obviously insufficient for characterization. Even exact mass measurement using peak matching at high resolution, that confirmed the correct elemental composition for the putative acylcarnitine $[M + H]^+$ ions, was insufficient.

The most compelling evidence that signals observed at the correct mass for acylcarnitines were in fact due to acylcarnitines in urine was provided by using the technique of linked-field scanning at constant B/E ratio on a double-focusing mass spectrometer, a type of tandem mass spectrometer (*i.e.* one having two or more mass analysers). In this technique, an ion produced in the source can be isolated from the others and its fragmentation pattern studied independently. In the example shown in Figure 8.8 (top panel), the signal at m/z 218 in the LSIMS spectrum of a patient's urine extract is the putative $[M+H]^+$ ion for propanoylcarnitine. The fragmentation pattern of this ion by the linked-scan experiment (Figure 8.8, centre) exactly matched that of an authentic sample of propanoylcarnitine (Figure 8.8, bottom). This type of experiment is referred to as a product-ion scan, one of several types of useful tandem mass spectrometry, or MS/MS, experiments (the principles of which were introduced in the previous chapter). Using the same method, acylcarnitines were characterized in the urine of patients with isovaleric acidemia [141], methylmalonic aciduria [139] and MCAD deficiency [142]. This method was limited to urine samples from patients on carnitine supplement after partial purification of the carnitine esters by ion-exchange chromatography.

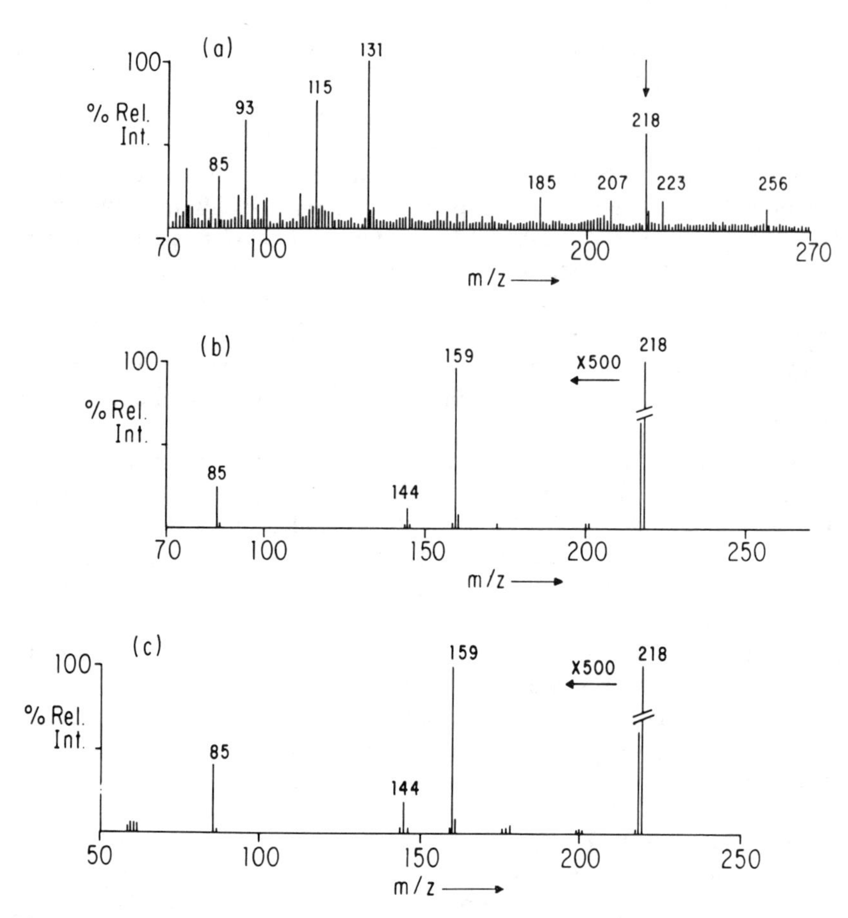

Fig. 8.8. Top: LSIMS spectrum of urine from a patient with propionic acidemia. The peak arrowed represents putative $[M + H]^+$ ions of propanoylcarnitine. Centre: fragmentation pattern of *m/z* 218 ions in the same clinical sample by B/E linked scan. Bottom: fragmentation pattern of *m/z* 218 ions from authentic propanoylcarnitine by B/E linked scan.

A considerable improvement in both sensitivity and selectivity was gained by esterification of the carboxylic acid function of acylcarnitines [111,140]. This is probably the combined result of increasing the surface activity of acylcarnitines and reducing the association of alkali metals with underivatized carboxylic acids. Esterification has enabled the detection of acylcarnitines in the urine of even non-carnitine supplemented patients down to levels of about 50 nmol/ml. Limitations of the method are lack of sensitivity for application to blood samples and the tedium of selecting ions one at a time for product ion analysis. These limitations were overcome by further improvements in technique and by using a tandem quadrupole mass spectrometer.

The triple quadrupole mass spectrometer is ideally suited to the analysis of small molecules ($M_r < 1,000$) by MS/MS [178]. This instrument

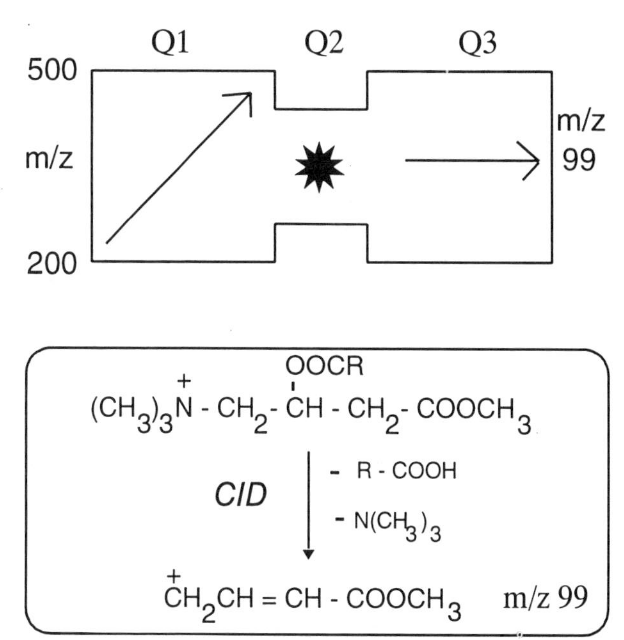

Fig. 8.9. Top: schematic representation of "precursors of *m/z* 99" scan function performed on a triple quadrupole mass spectrometer. Bottom: fragmentation of acylcarnitine methyl ester M^+ ions under collision-induced dissociation (CID) producing common ion at *m/z* 99.

consists of two quadrupole mass analysers, Q_1 and Q_3, separated by an RF-only quadrupole, Q_2, that serves the function of a reaction chamber (Figure 8.9). In Q_2, controlled fragmentation of ions selected by Q_1 is induced simply by inelastic collisions with inert gas or air molecules (collision-induced dissociation, CID). The fragment ions are then analysed according to *m/z* value by quadrupole Q_3. Because the functions of the instrument are fully controlled by a computer, several different types of MS/MS experiment can be carried out very rapidly. The most useful experiment from the point of view of acylcarnitine analysis is known as the "precursor ion scan", which enables all acylcarnitine species to be identified simultaneously in a complex biological sample. The basis of the method is that all acylcarnitine methyl esters exhibit a common fragment at *m/z* 99 under CID [114]. By scanning Q_1 and using Q_3 to allow only ions of *m/z* 99 to reach the final detector, a signal is produced only when Q_1 transmits a genuine precursor of *m/z* 99, such as an acylcarnitine methyl ester (Figure 8.9).

The biological acylcarnitines form a homologous series from acetyl to octadecenoyl (C_2-$C_{18:1}$), having M_r in the range *m/z* 200-500. The spectrum shown in Figure 8.10 is from a prepared sample of normal human plasma, applying the precursors of *m/z* 99 scan function on a triple quadrupole instrument. Before commenting on the spectrum, it should be noted that every ion in this spectrum actually has *m/z* 99, but the

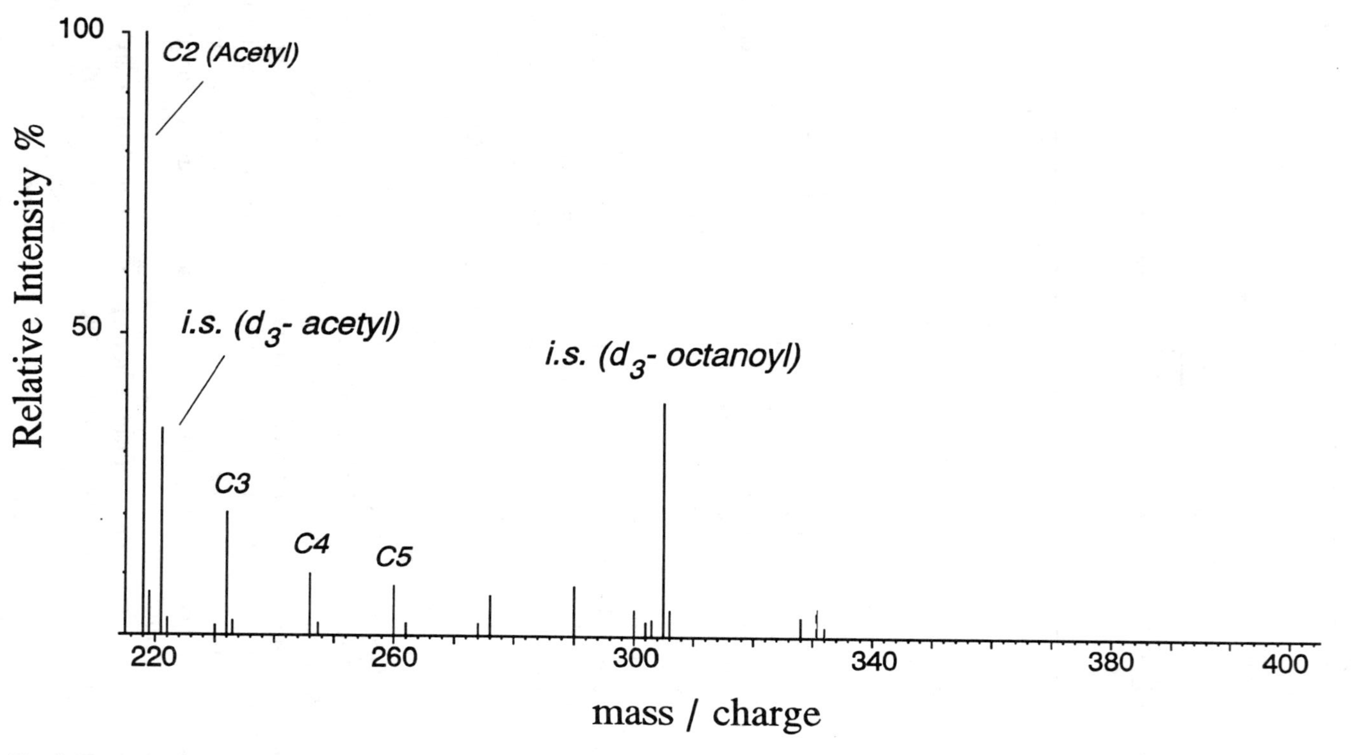

Fig. 8.10. Analysis of acylcarnitines in normal human plasma by LSIMS/MS showing detection of acetylcarnitine plus higher homologues (C_3-C_5) and two added internal standards. Note the lack of chemical interference.

computer has translated the time of the Q_1 scan to correspond to the mass range being scanned by that first quadrupole analyser, so that the masses displayed represent those of each of the *precursors* of *m/z* 99, and their relative abundances are related to the concentrations of the precursors in the sample. Thus, a homologous series of signals is observed, representing the masses of the intact molecular cations of acetyl (C_2), propanoyl (C_3), butanoyl (C_4) and isovaleryl (C_5) carnitine methyl esters respectively (*m/z* 218, 232, 246 and 260).

The result shown in Figure 8.10 is noteworthy for several reasons. The instrumental part of the analysis of this sample took only 1 minute, which of course is much faster than any chromatographic separation. The separation here is based on mass rather than retention time. The sensitivity and specificity for acylcarnitines is remarkable, and exceeds the capability of most chromatographic methods. Acylcarnitines are the only significant components detected in this sample, which was a simple ethanol extract of a 100 μl plasma aliquot, evaporated and derivatized with methanolic HCl, then added to a matrix of 1:1 methanol:glycerol containing 1% sodium octyl sulphate. The latter component is essential for acylcarnitine analysis [114]. The anion is a strongly acidic surfactant, and presumably forms ion pairs with acylcarnitines at or very near the surface layer [89]. Being negatively charged, it is transparent to the analysis of positive ions. The analysis is made quantitative simply by adding isotopically labelled analogues. In this example, 0.5 nmol acetyl-$^2H_3$-carnitine and 0.1 nmol octanoyl-$^2H_3$-carnitine were added to the plasma before extraction. The amount of acetylcarnitine in this sample was therefore about 1.5 nmol/100 μl (15μM), with lesser amounts of the other homologues.

Notwithstanding the obvious limitation that this method cannot distinguish between isomeric acylcarnitines, such as butanoyl- and 2-methylpropanoylcarnitine, the ability to analyse acylcarnitines in human blood at physiologically normal concentration [116] with high speed and accuracy has enabled numerous interesting applications.

An example of the diagnosis of metabolic disease is the analysis shown in Figure 8.10. In reality, this is a metabolic profile of acylcarnitines derived from the intramitochondrial acyl-CoA pool. In routine use in a clinical diagnostic environment, this method reveals a remarkable simplicity and consistency in the acylcarnitine profile both within and between normal individuals, in marked contrast to the metabolic profiles of urine (both of organic acids and acylcarnitines). In any metabolic disorder that leads to the intramitochondrial accumulation of abnormally elevated concentrations of intermediate acyl-CoA, there is also elevation of the corresponding acylcarnitine species. The diagnosis of at least 15 specific disorders of fatty acid and branched-chain amino acid catabolism is now practicable by acylcarnitine analysis [113], including some that cannot be diagnosed by analysis of urine. Specific examples include the long-chain acyl-CoA dehydrogenase defects. Long-chain acylcarnitines are protein-bound and are seldom excreted in the urine, but are readily detected in the plasma. The interpretation of disease profiles is straightforward in most cases.

The MS/MS method can also shed light on metabolism *in vitro* and *in vivo*. That is, the acylcarnitine profile can be used as a tool to investigate metabolic pathways and turnover rates of intermediates. By infusing $^2H_3$-labelled carnitine intravenously, the plasma carnitine/acylcarnitine system was shown to be in very rapid dynamic equilibrium [115]. Stable isotope studies have also supported the biochemical rationale for carnitine therapy in patients with some metabolic diseases. By investigating the metabolism of patients cells *in vitro*, it is hoped that the profiling technique can help to define new metabolic defects and improve further the diagnostic capabilities of MS/MS.

One of the most exciting possibilities for this method is the ability to perform large-scale neonatal screening of blood spots collected by the Guthrie method. Currently, this method of collection is almost universally employed to obtain specimens for screening for phenylketonuria and a few other inborn errors. The current analytical methods are non-specific and result in a large number of false positive results. The detection of PKU with a very low rate of false positive results using a MS/MS method to analyse amino acids in the blood spots has already been reported [35]. The ability to obtain amino acid and acylcarnitine profiles from the same sample has been demonstrated, and the automation of methods of sample preparation and analysis to achieve high sample throughput (several hundred analyses per instrument per day) is at an advanced stage of development. Tandem mass spectrometry could eventually replace the current technology and, in doing so, expand the capabilities of early detection of metabolic diseases to the benefit of many more families.

Finally, as noted in Section 4 of Part B, isotope-dilution tandem mass spectrometry has been reported for the routine analysis of free carnitine in urine [81]. A narrow mass range precursor scan is employed with the butyl ester derivatives which fragment to m/z 103. By repeating the assay after mild alkaline hydrolysis, both the free carnitine and short-chain total acylcarnitine fractions are determined. Recently, a semi-automated assay for free carnitine, acetylcarnitine and propanoylcarnitine has been developed and validated for use in clinical trials of acetyl-L-carnitine and propanoyl-L-carnitine.

D. SUMMARY

During the past 10 years, interest in carnitine metabolism has increased dramatically and numerous procedures for analysis of specific acylcarnitines have been developed in response to the demand for new methods. However, none of the methods described here could be considered a routine clinical diagnostic method. In fact, only a handful of specialist laboratories are attempting to analyse acylcarnitines in a clinical environment and most of these are unwilling or unable to consider mass spectrometric methods owing to the high capital cost of equipment. In comparing methods, therefore, the scientific merit or quality is not any more important than the purpose of its application.

In reviewing the suitability of methods for diagnosis of inborn errors of metabolism, it is now clear that urine is not the sample of choice whatever the method of analysis, and many of the methods are only applicable to urine. First, the urinary acylcarnitine profiles of normal individuals can be very complex, especially if carnitine flux is low. Numerous unusual species, especially dicarboxylated and unsaturated mono- and dicarboxylated acylcarnitines of medium chain length may be observed, that are never seen in the plasma. These components are of no clinical diagnostic importance and could even mask the presence of diagnostically useful markers. The most likely explanation of their formation is that the kidney is an active local metabolic system, part of the function of which is to conserve carnitine. In many urine samples, free carnitine and acetylcarnitine are minor components. In the early reports, urine samples were analysed by LSIMS from patients after carnitine supplement, in order to enhance the excretion of acylcarnitines to measurable levels. Under conditions of increased flux, the acylcarnitine profile more closely resembles that of plasma for normal individuals and those with metabolic defects in which short- to medium-chain acyl-CoA metabolites accumulate. It is true that, regardless of the method used, diagnostically useful profiles are more likely to be observed in patients who have received a hefty bolus of carnitine within 4-8 hours of the sample collection or who are being supplemented with carnitine in their diet. This effect is less important only in diseases such as propionic and isovaleric acidemias that produce large amounts of diagnostic metabolites, but these diseases are in any case easily diagnosed by analysis of organic acids. It must also be appreciated that monobasic acylcarnitines of chain length $>C_{10}$ are seldom seen in urine, and therefore defects of long-chain fatty acid catabolism are difficult, but not impossible [126], to recognize from urine analysis.

Thus, the most reliable methods for diagnosis are those that have sufficient sensitivity for detection of diagnostic acylcarnitines in plasma and blood spots. The concentration of the most abundant species, acetylcarnitine, ranges normally from about 4-10 μM, with others less than 1 μM. The only methods published to date with such sensitivity are LSIMS/MS and radioisotope exchange-HPLC. Two recent reports on the application of these methods to plasma and blood spots for diagnosis of MCAD deficiency show that the methods give comparable results [148,163]. It is not clear whether or not the isotope exchange-HPLC method can detect long-chain species in plasma. In addition, methods for the diagnosis of metabolic diseases through analysis of blood spots by electrospray MS/MS, and by GC/MS using a sensitive bench-top ion trap mass spectrometer following derivatization of acylcarnitines to lactones, will soon be reported. At present, LSIMS/MS must be considered the method of choice, especially in view of its speed and accuracy. The arguments against MS/MS most often cited are that it is too expensive, so complex that it requires PhD level scientists to operate, and that it does not separate mixtures of isomers. It certainly makes sense to have LSIMS/MS only in regional centres that have both the expertise and the sample volume to justify the expense. The throughput of samples that these instruments are capable of far exceeds the volume available in most hospitals. Contrary to the argument above, the MS/MS facility at Duke Medical Center (which has perhaps the most comprehensive diagnostic facility of its type and which has completed more than 7,000 specialist metabolic tests since its inception in 1982) is operated by personnel who are not qualified to PhD level.

In cases where unusual acylcarnitines have been observed using LSIMS/MS or when isomers need to be pinpointed, their identities have been confirmed by a chromatographic/mass spectrometric method, usually GC/MS analysis of the acyl groups liberated by mild alkaline hydrolysis of a partially purified fraction using ion-exchange or reversed-phase chromatography to remove potential free acids or other conjugates that could be hydrolysed. Since most of the interesting species not yet characterized will be found in low concentration in the plasma compartment, the methods used have to be sensitive. None of the HPLC methods is really suitable because of limited column resolution compared with capillary GC columns. The characterization of intact acylcarnitines by the two GC/MS methods that require chemical conversion to suitable derivatives are attractive alternatives to hydrolysis followed by GC/MS.

Finally, it would seem that any of the published methods could be developed and applied to follow the excretion of a particular acylcarnitine in a patient after an independent diagnosis is made, in order to monitor the efficacy of treatment, for example. They could also be applied for studies *in vitro* where conditions can be controlled to favour the formation of acylcarnitines and simpler methods of sample preparation would be appropriate.

ABBREVIATIONS

B/E, ratio of magnetic flux to electric sector potential in a magnetic-sector mass spectrometer; BIPM, *N*-[*p*-(2-benzimidazolyl)phenyl]-maleimide; CAT, carnitine acetyl transferase; CI, chemical ionization (mass spectrometry); CID, collision-induced dissociation; CoA, coenzyme A; CPT, carnitine palmitoyl transferase; DTNB, 5,5'-dithiobis-2-nitrobenzoic acid; EI, electron ionization (mass spectrometry); ESI, electrospray ionization (mass spectrometry); FAB, fast atom bombardment; GC, gas chromatography; GC/MS, gas chromatography/mass spectrometry; HEPES, [4-(2-hydroxyethyl)-1-piperazineethanesulphonic acid (buffer); HPLC, high-performance liquid chromatography; LCAD, long-chain acyl-coenzyme A dehydrogenase; LC/ESIMS, liquid chromatography/electrospray mass spectrometry; LC/MS, (high-performance) liquid chromatography/mass spectrometry; LSIMS, secondary ion mass spectrometry in the liquid phase; LSIMS/MS, tandem mass spectrometry in the LSIMS mode of operation; MADD, multiple acyl-coenzyme A dehydrogenation deficiency; MCAD, medium-chain acyl-coenzyme A dehydrogenase; M_r, relative molecular mass; MS/MS, tandem mass spectrometry; NMR, nuclear magnetic resonance; PKU, phenylketonuria; Q, quadrupole mass filter; REA, radioisotopic exchange assay; SCAD, short-chain acyl-coenzyme A dehydrogenase; SIM, selected ion monitoring; TLC, thin-layer chromatography; TLC/MS, combined thin-layer chromatography/mass spectrometry; TML, ε-*N*-trimethyllysine; TRIS, tris(hydroxymethyl)aminomethane (buffer).

REFERENCES

1. Alonso Diaz,R. and Herrera Carranza,J., *Farm. Clin.*, **5**, 702-714 (1988).
2. Arakawa,N., Ha,T.Y. and Otsuka,M., *J. Nutr. Sci. Vitaminol.*, **35**, 475-479 (1989).
3. Bachmann,C., in *Carnitine and Medicine*, pp. 61-63 (1987) (edited by R. Gitzelmann, K. Baerlocher and B. Steinmann, Schattauer, Stuttgart).
4. Bahl,J.J. and Bressler,R., *Ann. Rev. Pharmacol. Toxicol.*, **27**, 257-277 (1987).
5. Baker,H., DeAngelis,B., Baker,E.R., Reddi,A.S., Khalil,M. and Frank,O., *Food Chem.*, **43**, 141-146 (1992).
6. Barany,M., Spigos,D.G., Mok,E., Venkatasubramanian,P.N., Wilbur,A.C. and Langer,B.G., *Magn. Reson. Imaging*, **5**, 393-398 (1987).
7. Bennett,M.J., Worthy,E. and Pollitt,R.J., *Anal. Proc.*, **24**, 322-323 (1987).
8. Bhuiyan,A.K.M.J. and Bartlett,K., *Biochem. Soc. Trans.*, **16**, 796-797 (1988).
9. Bhuiyan,A.K.M.J., Causey,A.G. and Bartlett,K., *Biochem. Soc. Trans.*, **14**, 1072-1073 (1986).
10. Bhuiyan,A.K.M.J., Jackson,S., Turnbull,D.M., Aynsley-Green,A. and Leonard,J.V., *Clin. Chim. Acta*, **207**, 185-204 (1992).
11. Bhuiyan,A.K.M.J., Watmough,N.J., Turnbull,D.M., Aynsley-Green,A., Leonard,J.V. and Bartlett,K., *Clin. Chim. Acta*, **165**, 39-44 (1987).
12. Bieber,L.L. and Choi,Y.R., *Proc. Natl. Acad. Sci. U.S.A.*, **74**, 2795-2798 (1977).
13. Bieber,L.L.and Kerner,J., *Methods Enzymol.*, **123**, 264-276 (1986).
14. Bieber,L.L. and Lewin,L.M., *Methods Enzymol.*, **72**, 276-287 (1981).
15. Bieber,L.L., Markwell,M.A.K., Blair,M. and Helmrath,T.A., *Biochim. Biophys. Acta*, **326**, 145-154 (1973).
16. Blakeley,C.R. and Vestal,M.L., *Anal: Chem.*, **55**, 750-752 (1983).
17. Bogdarin,Y.A., *Khim.-Farm. Zh.*, **22**, 1516-1518 (1988).
18. Borum,P.R., *Ann. Rev. Nutr.*, **3**, 233-339 (1983).
19. Borum,P.R., *J. Nutr. Biochem.*, **1**, 111-114 (1990).
20. Borum,P.R., York,C.M. and Bennett,S.G., *Am. J. Clin. Nutr.*, **41**, 603-606 (1985).
21. Bosi,G. and Refrigeri,M.S., *Rass. Chim.*, **35**, 421-426 (1983).
22. Bounoure,J. and Souppe,J., *Analyst*, **113**, 1143-1144 (1988).
23. Brass,E.P. and Hoppel,C.L., *Clin. Chem.*, **31**, 491-492 (1985).
24. Bremer,J., *Biochim. Biophys. Acta.*, **57**, 327-335 (1962).
25. Bremer,J., *J. Biol. Chem.*, **237**, 2228-2231, (1962).
26. Bremer,J., *Physiol. Rev.*, **63**, 1420-1480 (1983).
27. Brown,J.C.C., Mills,G.A., Sadler,P.J. and Walker,V., *Magn. Reson. Med.*, **11**, 193-201 (1989).

28. Bruno,G., Gasparrini,F., Misiti,D., Arrigoni-Martelli,E. and Bronzetti,M., *J. Chromatogr.*, **504**, 319-333 (1990).
29. Caprioli,R., Fan,T. and Cottrell,J.S., *Anal. Chem.*, **58**, 2949-2954 (1986).
30. Carter,H.E., Bhattacharyya,P.K., Weidman,K.R. and Fraenkel,G., *Arch. Biochem. Biophys.*, **38**, 405-416 (1952).
31. Cederblad,G., Carlin,J.I., Constantin-Teodosiu,D., Harper,P. and Hultman,E., *Anal. Biochem.*, **185**, 274-278 (1990).
32. Cederblad,G. and Harper,P., *Clin. Chem.*, **32**, 2117-2118 (1986).
33. Cederblad,G., Harper,P. and Lindgren,K., *Clin. Chem.*, **32**, 342-346 (1986).
34. Cederblad,G. and Lindstedt,S., *Clin. Chim. Acta.*, **37**, 235-243 (1972).
35. Chace,D.H., Millington,D.S., Terada,N., Kahler,S.G., Roe,C.R. and Hofman,L.F., *Clin. Chem.*, **39**, 66-71 (1993).
36. Choi,Y.R. and Bieber,L.L., *Anal. Biochem.*, **79**, 413-418 (1977).
37. Choi,Y.R., Fogle,P.J., Clarke,P.R.H. and Bieber,L.L., *J. Biol. Chem.*, **252**, 7930-7931 (1977).
38. Colucci,W.J., Gandour,R.D. and Mooberry,E.A., *J. Am. Chem. Soc.*, **108**, 7141-7147 (1986).
39. Cox,R.A. and Hoppel,C.L., *J. Biochem.*, **136**, 1083-1090 (1973).
40. De Palo,E., Gatti,R., Dalla,P.F. and Erle,G., *G. Ital. Chim. Clin.*, **7**, 109-125 (1982).
41. De Sousa,C., English,N.R., Stacey,T.E. and Chalmers,R.A., *Clin. Chim. Acta*, **187**, 317-328 (1990).
42. Deufel,T., *J. Clin. Chem. Clin. Biochem.*, **28**, 307-311 (1990).
43. Dugan,R.E., Schmidt,M.J., Hoganson,G.E., Steele,J., Gilles,B.A. and Shug,A.L., *Anal. Biochem.*, **160**, 275-280 (1987).
44. Duran,M., Ketting,D., Beckeringh,T.E., Leupold,D. and Wadman,S.K., *J. Inherited Metab. Dis.*, 9, 202-207 (1986).
45. Duran,M., Ketting,D., Dorland,L. and Wadman,S.K., *J. Inherited Metab. Dis.*, **8**(Suppl. 2), 143-144 (1985).
46. Farquharson,J., Jamieson,E.C., Muir,J., Cockburn,F. and Logan,R.W., *Clin. Chim. Acta*, **205**, 233-240 (1992).
47. Fishlock,R.C., Bieber,L.L. and Snoswell,A.M., *Clin. Chem.*, **30**, 316-318 (1984).
48. Fraenkel,G., *Bio. Bull.*, **104**, 359-371 (1953).
49. Fraenkel,G. and Friedman,S., in *Vitamins and Hormones*, **15**, pp. 73-110 (1957) (edited by R.S. Harris, G.F. Marrian and K.V. Thimann, Acaedmic Press, New York.).
50. Fritz,I.B., *Adv. Lip. Res.*, **1**, 285-334 (1963).
51. Fritz,I.B. and Yue,K.T.N., *J. Lipid. Res.*, **4**, 279-288 (1963).
52. Gamcsik,M.P., Constantinidis,I. and Glickson,J.D., *Cancer Res.*, **51**, 3378-3383 (1991).
53. Garami,M. and Mincsovics,E., *Proc. Int. Conf. Biochem. Sep.*, **2**, 121-132 (1988).
54. Garami,M. and Mincsovics,E., *J. Planar Chromatogr.- Mod. TLC*, **2**, 438-441 (1989).
55. Gennaro,M.C., *J. Chromatogr. Sci.*, **29**, 410-415 (1991).
56. Glish,G.L., Todd,P.J., Busch,K.L. and Cooks,R.G., *Int. J. Mass Spectrom. Ion Processes*, **56**, 177-192 (1984).
57. Golan,R. and Lewin,L.M., *Anal. Biochem.*, **105**, 264-267 (1980).
58. Gulewitsch,V.S. and Krimberg,R., *Z. Physiol. Chem.*, **45**, 326-329 (1905).
59. Hamamoto,M., Shimoda,K., Matsuura,N. and Matsuura,H., *Nippon Eiyo*, **41**, 389-395 (1988).
60. Hamamoto,M., Shimoda,K., Matsuura,N. and Matsuura,H., *Nippon Eiyo*, **41**, 397-404 (1988).
61. Hand,O.W., Hsu,B.H. and Cooks,R.G., *Org. Mass Spectrom.*, **23**, 16-25 (1988).
62. Hayes,J.S., Alizade,M.A. and Brendel,K., *Anal. Chim. Acta*, **80**, 361-367 (1975).
63. Hoppel,C.L., Brass,E.P., Gibbons,A.P. and Turkaly,J.S., *Anal. Biochem.*, **156**, 111-117 (1986).
64. Horne,D.W., Tanphaichitir,V. and Broquist,H.P., *J. Biol. Chem.*, **248**, 2170-2175 (1973).
65. Houldsworth,P.E., personal communication and submitted for publication.
66. Huang,Z.H., Gage,D.A., Bieber,L.L. and Sweeley,C.C., *Anal. Biochem.*, **199**, 98-105 (1991).
67. Hvistendahl,G., Undheim,K. and Bremer,J., *Org. Mass Spectrom.*, **3**, 1433-1438 (1970).
68. Iles,R.A., Chalmers,R.A. and Hind,A.J., *Clin. Chim. Acta*, **161**, 173-189 (1986).
69. Iles,R.A., Hind,A.J. and Chalmers,R.A., *Clin. Chem.*, **31**, 1795-1801 (1985).
70. Iles,R.A., Jago,J.R., Williams,S.R., Stacey,T.E., De Sousa,C. and Chalmers,R.A.,

Biochem. Soc. Trans., **14**, 702-704 (1986).

71. Irias,J.J., in *Clinical Aspects of Human Carnitine Deficiency*, pp. 108-119 (1986) (edited by P.R. Borum, Pergamon Press, New York).
72. Ito,Y., Takeuchi,D., Ishi,D. and Goto,M., *J. Chromatogr.*, **346**, 161-163 (1985).
73. Kelly,B.M., Lowes,S. and Rose,M.E., in preparation.
74. Kelly,B.M. and Rose,M.E., in preparation.
75. Kelly,B.M., Rose,M.E., Wycherley,D. and Preece,S.W., *Org. Mass Spectrom.*, **27**, 924-926 (1992).
76. Kerner,J. and Bieber,L.L., *Anal. Biochem.*, **134**, 459-466 (1983).
77. Kidouchi,K., Niwa,T., Nohara,D., Asai,K., Sugiyama,N., Morishita,H., Kobayashi,M. and Wada,Y., *Clin. Chim. Acta*, **173**, 263-272 (1988).
78. Kidouchi,K., Sugiyama,N., Morishita,H., Kobayashi,M., Wada,Y. and Nohara,D., *Clin. Chim. Acta*, **164**, 261-266 (1987).
79. Kidouchi,K., Sugiyama,N., Morishita,H., Wada,Y., Nagai,S. and Sakakibara,J., *J. Chromatogr.*, **423**, 297-303 (1987).
80. Kler,R.S., Jackson,S., Bartlett,K., Bindoff,L.A., Eaton,S., Pourfarzam,M., Frerman,F.E., Goodman,S.I., Watmough,N.J. and Turnbull,D.M., *J. Biol. Chem.*, **266**, 22932-22938 (1991).
81. Kodo,N., Millington,D.S., Norwood,D. and Roe,C.R., *Clin. Chim. Acta*, **186**, 383-390 (1989).
82. Lehnert,W. and Hunkler,D., *Eur. J. Pediatr.*, **145**, 260-266 (1986).
83. Lehotay,D.C., *Biomed. Chromatogr.*, **5**, 113-121 (1991).
84. Lever,M., Bason,L., Leaver,C., Hayman,C.M. and Chambers,S.T., *Anal. Biochem.*, **205**, 14-21 (1992).
85. Lewin,L.M. and Bieber,L.L., *Anal. Biochem.*, **96**, 322-325 (1979).
86. Lewin,L.M., Peshin,A. and Sklarz,B., *Anal. Biochem.*, **68**, 531-536 (1975).
87. Liberato,D.J., Fenselau,C.C., Vestal,M.L. and Yergey,A.L., *Anal. Chem.*, **55**, 1741-1742 (1983).
88. Liberato,D.J., Millington,D.S. and Yergey,A.L., in *Mass Spectrometry in the Health and Life Sciences*, pp. 333-348 (1985) (edited by A.L. Burlingame and N. Castagnoli, Elsevier, Amsterdam).
89. Ligon,W.V. in *Biological Mass Spectrometry*, pp. 61-75 (1990) (edited by A.L. Burlingame and J.A. McCloskey, Elsevier, Amsterdam).
90. Liguori,A., Sindona,G. and Uccella,N., *J. Am. Chem. Soc.*, **108**, 7488-7491 (1986).
91. Liu,C.H., Yoshino,M. and Yuge,K., *Kurume Igakkai Zasshi*, **50**, 395-401 (1987).
92. Loester,H. and Mueller,D.M., *Wiss. Z. - Karl-Marx-Univ. Leipzig, Math.-Naturwiss. Reihe*, **34**, 212-223 (1985).
93. Loester,H., Seim,H. and Herzschuh,R., *Pharmazie*, **38**, 844-847 (1983).
94. Long,C.S., Haller,R.G., Foster,D.W. and McGarry,J.D., *Neurology*, **32**,663-672 (1982).
95. Lowes,S. and Rose,M.E., *Trends Anal. Chem.*, **8**, 184-188 (1989).
96. Lowes,S. and Rose,M.E., *Analyst*, **115**, 511-516 (1990).
97. Lowes,S. and Rose,M.E., *Philos. Trans. R. Soc. London, Ser. A*, **333**, 169-170 (1990).
98. Lowes,S., Rose,M.E., Mills,G.A. and Pollitt,R.J., *J. Chromatogr.*, **577**, 205-214 (1992).
99. McGarry,J.D. and Foster,D.W., *J. Lipid Res.*, **17**, 277-281 (1976).
100. McGarry,J.D. and Foster,D.W., in *Methods of Enzymic Analysis, 3rd edition*, **8**, pp. 474-481 (1985) (edited by H.U. Bergmeyer, VCH, Germany).
101. Maebashi,M., Kawamura,N., Sato,M., Yoshinaga,K. and Suzuki,M., *Tohoku J. Exp. Med.*, **116**, 203-204 (1975).
102. Maeda,J., Yamakawa,M., Mimura,Y., Furuya,K., Oohara,T. and Dudrick,S.J., *Nippon Jomyaku, Keicho Eiyo Kenkyukaishi*, **4**, 316-318 (1989).
103. Maehara,M., Kinoshita,S. and Watanabe,K., *Clin. Chim. Acta*, **171**, 311-316 (1988).
104. Marquis,N.R. and Fritz,I.B., *J. Lipid Res.*, **5**, 184-187 (1964).
105. Marzo,A., Cardace,G. and Arrigoni Martelli,E., *Chirality*, **4**, 247-251 (1992).
106. Marzo,A., Cardace,G., Monti,N., Muck,S. and Arrigoni Martelli,E., *J. Chromatogr.*, **527**, 247-258 (1990).
107. Marzo,A., Monti,N., Ripamonti,M. and Arrigoni Martelli,E., *J. Chromatogr.*, **459**, 313-317 (1988).
108. Matsumoto,K., Yamada,Y., Takahashi,M., Todoroki,T., Mizoguchi,K., Misaki,H. and Yuki,H., *Clin. Chem.*, **36**, 2072-2076 (1990).
109. Matsuyama,H., Morikawa,K., Hasegawa,T. and Kuroda,M., *Rinsho Byori*, **36**, 1296-

1302 (1988).
110. Melegh,B., Kerner,J. and Bieber,L.L., *Biochem. Pharmacol.*, **36**, 3405-3409 (1987).
111. Millington,D.S., in *Mass Spectrometry in Biomedical Research*, pp. 97-113 (1986) (edited by S.J. Gaskell, John Wiley and Sons, Chichester).
112. Millington,D.S., Bohan,T.P., Roe,C.R., Yergey,A.L. and Liberato,D.J., *Clin. Chim. Acta*, **145**, 69-76 (1985).
113. Millington,D.S. and Chace,D.H., in *Mass Spectrometry: Clinical and Biomedical Applications, Volume* 1, pp. 299-318 (1992) (edited by D.M. Desiderio, Plenum Press, New York).
114. Millington,D.S., Kodo,N., Terada,N., Roe,D. and Chace,D.H., *Int. J. Mass Spectrom. Ion Processes*, **111**, 211-228 (1991).
115. Millington,D.S., Maltby,D.A., Gale,D.S. and Roe,C.R., in *Synthesis and Applications of Isotopically Labelled Compounds*, pp. 189-194 (1989) (edited by T.A. Baillie and J.R. Jones, Elsevier, Amsterdam).
116. Millington,D.S., Norwood,D.L., Kodo,N. and Roe,C.R., *Anal. Biochem.*, **180**, 331-339 (1989).
117. Millington,D.S., Norwood,D.L., Kodo,N., Roe,C.R. and Inoue,F., *Anal. Biochem.*, **180**, 331-339 (1989).
118. Millington,D.S., Roe,C.R. and Maltby,D.A., *Biomed. Mass Spectrom.*, **11**, 236-241 (1984).
119. Millington, D.S., Roe, C.R. and Maltby, D.A., *Biomed. Environm. Mass Spectrom.*, **14**, 711-716 (1987).
120. Minkler,P.E. and Hoppel,C.L., *Clin. Chim. Acta*, **212**, 55-64 (1992).
121. Minkler,P.E., Ingalls,S.T. and Hoppel,C.L., *J. Chromatogr.*, **420**, 385-393 (1987).
122. Minkler,P.E., Ingalls,S.T. and Hoppel,C.L., *Anal. Biochem.*, **185**, 29-35 (1990).
123. Minkler,P.E., Ingalls,S.T., Kormos,L.S., Weir,D.E. and Hoppel,C.L., *J. Chromatogr.*, **336**, 271-283 (1984).
124. Moore,R., Millington,D.S., Norwood,D.L., Kodo,N., Robinson,P. and Glasgow,J.F.T., *J. Inher. Metab. Dis.*, **13**, 325-329 (1990).
125. Moretti,G.P. and Celletti,P., *Boll. Chim. Farm.*, **112**, 843-845 (1973).
126. Morrow,R.J. and Rose,M.E., *Clin. Chim. Acta*, **211**, 73-81 (1992).
127. Norwood,D.L., Kodo,N. and Millington,D.S., *Rapid Commun. Mass Spectrom.*, **2**, 269-272 (1988).
128. Norwood,D.L. and Millington,D.S., in *Continuous-Flow Fast Atom Bombardment Mass Spectrometry*, pp. 175-180 (1990) (edited by R.M. Caprioli, Wiley, Chichester).
129. Odessey,R., *Methods Enzymol.*, **123**, 284-290 (1986).
130. Odessey,R. and Chace,K.V., *Anal. Biochem.*, **122**, 41-46 (1982).
131. Ohtani,Y., Endo,F. and Matsuda,I., *J. Pediatr.*, **101**, 782-785 (1982).
132. Pace,J.A., Wannemacher,R.W. and Neufeld,H.A., *Clin. Chem.*, **24**, 32-35 (1978).
133. Peter,G. and Haubitz,I., *Laboratoriumsmedizin*, **11**, 134-141 (1987).
134. Pourfarzam,M. and Bartlett,K., *J. Chromatogr.*, **570**, 253-276 (1991).
135. Rebouche,C.J., in *Clinical Aspects of Human Carnitine Deficiency*, pp. 1-15 (1986) (edited by Borum, P.R. Pergamon, New York).
136. Rebouche,C.J. and Engel,A.G., *Biochim. Biophys. Acta.*, **630**, 22-29 (1980).
137. Richards,T., Tscholakoff,D. and Higgins,C.B., *Magn. Reson. Med.*, **4**, 555-566 (1987).
138. Rodriguez-Segade,S., De La Pena,C., Paz,M. and Del Rio,R., *Clin. Chem.*, **31**, 754-757 (1985).
139. Roe,C R., Hoppel,C.L., Stacey,T.E., Chalmers,R.A., Tracey,B.M. and Millington,D.S., *Arch. Dis. Child.*, **58**, 916-920 (1983).
140. Roe,C.R., Millington,D.S. and Maltby,D.A., in *Clinical Aspects of Human Carnitine Deficiency*, pp. 108-119 (1986) (edited by P.R. Borum, Pergamon Press, New York).
141. Roe,C.R., Millington,D.S., Maltby,D.A., Bohan,T.P. and Hoppel,C.L., *J. Clin. Invest.*, **74**, 2290-2295 (1984).
142. Roe,C.R., Millington,D.S., Maltby,D.A., Bohan,T.P., Kahler,S.G. and Chalmers,R.A., *Pediatr. Res.*, **19**, 459-466 (1985).
143. Roe,D.S., Terada,N. and Millington,D.S., *Clin. Chem.*, **38**, 2215-2220 (1992).
144. Roessle,C., Kohse,K.P., Franz,H.E. and Fuerst,P., *Clin. Chim. Acta*, **149**, 263-268 (1985).
145. Sandor,A., *Eur. J. Clin. Chem. Clin. Biochem.*, **29**, 347-349 (1991).
146. Sandor,A., Cseko,J. and Alkonyi,I., *J. Chromatogr.*, **497**, 250-257 (1989).
147. Schaefer,J. and Reichmann,H., *Clin. Chim. Acta*, **182**, 87-93 (1989).

148. Schmidt-Sommerfeld,E., Penn,D., Duran,M., Benett,M.J., Santer,R. and Stanley,C.A., *J. Pediatr.*, **122**, 708-714 (1993).
149. Schmidt-Sommerfeld,E., Penn,D., Kerner,J. and Bieber,L.L., *Clin. Chim. Acta*, **181**, 231-238 (1989).
150. Schmidt-Sommerfeld,E., Penn,D., Kerner,J., Bieber,L.L., Rossi,T.M. and Lebenthal,E., *J. Pediatr.*, **115**, 577-582 (1989).
151. Schmidt-Sommerfeld,E., Penn,D., Rinaldo,P., Kossak,B.D., Li,B.U.K., Huang,Z.H. and Gage,D.A., *Pediatr. Res.*, **31**, 545-551 (1992).
152. Seccombe,D.W., Dodek,P., Frohlich,J., Hahn,P., Skala,J.P. and Campbell,D.J., *Clin. Chem.*, **22**, 1589-1592 (1976).
153. Sekas,G. and Paul,H.S., *Anal. Biochem.*, **179**, 262-267 (1989).
154. Shihabi,Z.K., Oles,K.S., McCormick,C.P. and Penry,J.K., *Clin. Chem.*, **38**, 1414-1417 (1992).
155. Shinka,T., Matsumoto,M. and Matsumoto,I., *Iyo Masu Kenkyukai Koenshu*, **13**, 169-172 (1988).
156. Takahashi,M., Nakayama,K., Uehara,S., Kamata,K., Hagiwara,T. and Akiyama,K., *Kenkyu Nenpo - Tokyo-toritsu Eisei Kenkyusho*, 51-54 (1990).
157. Takeyama,N., Takagi,D., Adachi,K. and Tanaka,T., *Anal. Biochem.*, **158**, 346-354 (1986).
158. Tegelaers,F.P.W., Pickkers,M.M.G. and Seelen,P.J., *J. Clin. Chem. Clin. Biochem.*, **27**, 967-972 (1989).
159. Tomita,M. and Sendju,Y., *Z. Physiol. Chem.*, **169**, 263-277 (1927).
160. Tracey,B.M., Chalmers,R.A., Rosankiewicz,J.R., De Sousa,C. and Stacey,T.E., *Biochem. Soc. Trans.*, **14**, 700-701 (1986).
161. Travassos,L.R. and Sales,C.O., *Anal. Biochem*, **58**, 485-499 (1974).
162. Tyihak,E., Mincsovics,E. and Szekely,T.J., *J. Chromatogr.*, **471**, 375-387 (1989).
163. Van Hove,J.L.K., Zhang,W., Kahler,S.G., Roe,C.R., Chen,Y.-T., Terada,N., Chace,D.H., Iafolla,A.K., Ding,J.-H. and Millington,D.S., *Am. J. Hum. Genet.*, **52**, 958-966 (1993).
164. Vianey-Liaud,C., Divry,P., Gregersen,N. and Mathieu,M., *J. Inher. Metab. Dis.*, **10**, 159-198 (1987).
165. Voeffray,R., Perlberger,J.-C., Tenud,L. and Gosteli,J., *Helv. Chim. Acta*, **70**, 2058-2064 (1987).
166. Voet,D. and Voet,J.G., *Biochemistry*, Wiley and Sons, New York, 618-645 (1990).
167. Watanabe,F., Ishino,M., Hirose,Y. and Morimoto,I., *Eisei Kagaku*, **37**, 53-57 (1991).
168. Watmough,N.J., Bhuiyan,A.K.M.J., Bartlett,K., Sherratt,H.S.A. and Turnbull,D.M., *Biochem. Soc. Trans.*, **15**, 633-634 (1987).
169. Weavind,G.P., Mills,G.A. and Walker,V., *Ann. Clin. Biochem.*, **25**, 233s-234s (1988).
170. Wieland,O.H., Deufel,T. and Paetzke-Brunner,I., in *Methods of Enzymic Analysis, 3rd edition*, pp. 481-488 (1985) (edited by H.U. Bergmeyer, VCH, Germany).
171. Wittels,B. and Bressler,R., *J. Lipid Res.*, **6**, 313-314 (1965).
172. Wolf,G. and Berger,C.R.A., *Arch. Biochem. Biophys.*, **92**, 360-365 (1961).
173. Xia,L.J. and Folkers,K., *Biochem. Biophys. Res. Commun.*, **176**, 1617-1623 (1991).
174. Yamamoto,S., Kakinuma,H., Nishimuta,T. and Mori,K., *Masu Kenkyukai Koenshu*, **11**, 151-156 (1986).
175. Yasuda,T., Nakashima,K., Kawata,K. and Achi,T., *Iyakuhin Kenkyu*, **23**, 149-153 (1992).
176. Yergey,A.L., Liberato,D.J. and Millington,D.S., *Anal. Biochem.*, **139**, 278-283 (1984).
177. Yoshida,T., Aetake,A., Yamaguchi,H., Nimura,N. and Kinoshita,T., *J. Chromatogr.*, **445**, 175-82 (1988).
178. Yost,R.A. and Enke,C.G., in *Tandem Mass Spectrometry*, pp. 175-195 (1983) (edited by F.W. McLafferty, John Wiley and Sons, New York).

APPENDIX

Some Important References in Lipid Methodology - 1991

William W. Christie

The Scottish Crop Research Institute, Invergowrie, Dundee (DD2 5DA), Scotland

A. INTRODUCTION

When the *Journal of Lipid Research* ceased its current awareness service for lipid methodology, it left a gap which it is intended that this series will fill. As for our first volume, the search of the literature has been done mainly to keep my own research up to date, and may therefore be rather subjective in the selection. I have tried to list references with something new to say about lipid methodology rather than those that use tested methods, however competent or important these may be. Some papers may have been included simply because the title seemed apposite, although I may not have had personal access to the journal to check this. Others may have been omitted quite unjustly, because it is impracticable for one person to read every paper that deals with lipids in a comprehensive manner.

In "Advances in Lipid methodology - One", the years 1989 and 1990 were covered so papers for 1991 and 1992 are now listed. They are grouped in sections that correspond broadly to chapter headings in an earlier book ("Lipid Analysis (second edition)", Pergamon Press, 1982). Often this has caused difficulties, as methods in papers can be relevant to many chapters, especially with review articles. Occasionally, papers have been listed twice but usually I have selected the single section that appeared most appropriate. The literature on prostaglandins and clinical chemistry (especially lipoprotein analysis) is not represented comprehensively, but these areas are covered by current awareness services in the journals *Prostaglandins, Leukotrienes and Essential Fatty acids* and *Current Topics in Lipidology* respectively.

Note that the titles of papers listed below may not be literal transcriptions of the originals. In particular, a number of abbreviations have been introduced. References are listed alphabetically according to the surname of the first author in each section,

B. THE STRUCTURE, CHEMISTRY AND OCCURRENCE OF LIPIDS

Douglas,D.E., Higher aliphatic 2,4-diketones: a ubiquitous lipid class with chelating properties, in search of a physiological function. *J. Lipid Res.*, **32**, 553-558 (1991).

C. THE ISOLATION OF LIPIDS FROM TISSUES

Fidelio,G.D., Ariga,T. and Maggio,B., Molecular parameters of gangliosides in monolayers: comparative evaluation of suitable purification procedures. *J. Biochem. (Tokyo)*, **110**, 12-16 (1991).

Nelson,G.J., Isolation and purification of lipids from biological matrices. In *Analyses of Oils and Fats*, pp. 20-59 (ed. E.G. Perkins, American Oil Chemists' Society, Champaign, U.S.A.) (1991).

Sattler,W., Puhl,W., Hayn,M., Kostner,G.M. and Esterbauer,H., Determination of fatty acids in the main lipoprotein classes by capillary GC:BF_3/methanol transesterification of lyophilized samples instead of Folch extraction gives higher yields. *Anal. Biochem.*, **198**, 184-190 (1991).

Takeuchi,T., Ackman,R.G. and Lall,S.P., Differences in fatty acid composition of fish faeces as determined by two extraction methods. *J. Sci. Food Agric.*, **56**, 259-264 (1991).

D. CHROMATOGRAPHIC AND SPECTROSCOPIC ANALYSIS OF LIPIDS. GENERAL PRINCIPLES.

Wasiak,W., Chemically bonded chelates as selective complexing sorbents for gas chromatography. *J. Chromatogr.*, **547**, 259-268 (1991).

E. THE ANALYSIS OF FATTY ACIDS

This section corresponds to Chapters 4 and 5 in *Lipid Analysis* and deals with derivatization and the chromatographic analysis of fatty acids, for example by gas chromatography (GC) and high performance liquid chromatography (HPLC), together with spectrometric methods, especially mass spectrometry (MS). Papers dealing with analysis of a free fatty acid fraction of tissue lipids are listed in the next section mainly.

Ackman,R.G., Application of GLC to lipid separation and analysis: qualitative and quantitative analysis. In *Analyses of Oils and Fats*, pp. 270-300 (ed. E.G. Perkins, American Oil Chemists' Society, Champaign, U.S.A.) (1991).

Adkisson,H.D., Risener,F.S., Zarrinkar,P.P., Walla,M.D., Christie,W.W. and Wuthier,R.E., Unique fatty acid composition of normal cartilage: discovery of high levels of n-9 eicosatrienoic acid and low levels of n-6 polyunsaturated fatty acids. *FASEB J.*, **5**, 344-353 (1991).

Apps,P.J. and Willemse,C., Fast high precision fingerprinting of fatty acids in milk. *J. High Resolut. Chromatogr.*, **14**, 802-807 (1991).

Augustyn,O.P.H., Ferreira,D. and Kock,J.L.F., Differentiation between yeast species, and strains within a species, by cellular fatty acid analysis. 4. Saccharomyces-sensu-stricto, Hanseniaspora, Saccharomycodes and Wickerhamiella. *System. Appl. Microbiol.*, **14**, 324-334 (1991).

Baer,A.N., Costello,P.B. and Green,F.A., Stereospecificity of the products of the fatty acid oxygenases derived from psoriatic scales. *J. Lipid Res.*, **32**, 341-347 (1991).

Calvey,E.M., McDonald,R.E., Page,S.W., Mossoba,M.M. and Taylor,L.T., Evaluation of SFC/FT-IR for examination of hydrogenated soybean oil. *J. Agric. Food Chem.*, **39**, 542-548 (1991).

Cardillo,R., Fronza,G., Fuganti,C., Grasselli,P., Mele,A., Pizzi,D., Allegrone,G., Barbeni,M. and Pisciotta,A., Stereochemistry of the microbial generation of δ-decanolide, γ-dodecanolide, and γ-nonanolide from C_{18} 13-hydroxy, C_{18} 10-hydroxy and C_{19} 14-hydroxy unsaturated fatty acids. *J. Org. Chem.*, **56**, 5237-5239 (1991).

Caruso,U., Fowler,B., Erceg,M. and Romano,C., Determination of very-long-chain fatty acids in plasma by a simplified GC-MS procedure. *J. Chromatogr.*, **562**, 147-152 (1991).

Christie,W.W., Brechany,E.Y., Lie Ken Jie,M.S.F. and Bakare,O., MS characterization of picolinyl and methyl ester derivatives of isomeric thia fatty acids. *Biol. Mass Spectrom.*, **20**, 629-635 (1991).

Cocks,S. and Smith,R.M., Analysis for fatty acid methyl esters by using supercritical fluid chromatography with mass evaporative light scattering detection. *Anal. Proc. (London)*, **28**, 11-12 (1991).

Cohen,Z. and Cohen,S., Preparation of eicosapentaenoic acid (EPA) concentrate from *Porphyridium cruentum*. *J. Am. Oil Chem. Soc.*, **68**, 16-19 (1991).

Contado,M.J. and Adams,J., Collision-induced dissociations and B/E linked scans for structural determination of modified fatty acid esters. *Anal. Chim. Acta*, **246**, 187-197 (1991).

Contini,M., De Santis,D. and Anelli,G., Preparation of methyl esters for GC analysis of free fatty acids in vegetable oils. *Ital. J. Food Sci.*, **3**, 273-280 (1991).

Couderc,F., Berjeaud,J.M. and Prome,J.C., The use of remote-site fragmentation for the characterization of 2,3 unsaturated fatty acids. *Rapid Commun. Mass Spectrom.*, **5**, 92 (1991).

Cramer,G.L., Miller,J.F., Pendleton,R.B. and Lands,W.E.M., Iodometric measurement of lipid hydroperoxides in human plasma. *Anal. Biochem.*, **193**, 204-211 (1991).

Dirven,H.A.A.M., de Bruijn,A.A.G.M., Sessink,P.J.M. and Jongeneelen,F.J., Determination of the cytochrome P-450 IV marker, ω-hydroxylauric acid, by HPLC and fluorimetric detection. *J. Chromatogr.*, **564**, 266-271 (1991).

Dorland,L., Ketting,D., Bruinvis,L. and Duran,M., Medium-chain and long-chain 3-hydroxymonocarboxylic acids - analysis by GC combined with MS. *Biomed. Chromatogr.*, **5**, 161-164 (1991).

Eder,K., Reichlmayr-Lais,A.M. and Kirchgessner,M., GC analysis of fatty acid methyl esters - avoiding discrimination by programmed temperature vaporizing injection. *J. Chromatogr.*, **588**, 265-272 (1991).

Fay,L. and Richli,U., Location of double bonds in polyunsaturated fatty acids by GC-MS after 4,4-dimethyloxazoline derivatization. *J. Chromatogr.*, **541**, 89-98 (1991).

Frank,H.G. and Graf,R., Determination of free unsaturated C_{20} fatty acids in rat placentae using HPTLC. *J. Planar Chromatogr. - Mod. TLC*, **4**, 134-137 (1991).

Franotovic,M., Todoric,A. and Petrovic,J., GC analysis of fatty acids and identification of tuberculostearic acid in *Mycobacterium tuberculosis* by using GC-MS. *Period. Biol.*, **93**, 397-403 (1991).

Gaiday,N.V., Imbs,A.B., Kuklev,D.V. and Latyshev,N.A., Separation of natural polyunsaturated fatty acids by means of iodo-lactonization. *J. Am. Oil Chem. Soc.*, **68**, 230-233 (1991).

Gunstone,F.D., The ^{13}C-NMR spectra of six oils containing petroselinic acid and of aquilegia oil and meadowfoam oil which contain Δ-5 acids. *Chem. Phys. Lipids*, **58**, 159-167 (1991).

Gunstone,F.D., High resolution NMR studies of fish oils. *Chem. Phys. Lipids*, **59**, 83-89 (1991).

Gunstone,F.D., The composition of hydrogenated fats determined by high resolution ^{13}C NMR spectroscopy. *Chem. Ind. (London)*, 802-803 (1991).

Gutnikov,G. and Streng,J.R., Rapid HPLC determination of fatty acid profiles of lipids by conversion to their hydroxamic acids. *J. Chromatogr.*, **587**, 292-296 (1991).

Hamberg,M., Regio- and stereochemical analysis of trihydroxyoctadecenoic acids derived from linoleic acid 9- and 13-hydroperoxides. *Lipids*, **26**, 407-415 (1991).

Hamberg,M., Trihydroxyoctadecenoic acids in beer: qualitative and quantitative analysis. *J. Agric. Food Sci.*, **39**, 1568-1572 (1991).

Harvey,D.J., Lipids from the guinea pig harderian gland - use of picolinyl and other pyridine-containing derivatives to investigate the structures of novel branched-chain fatty acids and glycerol ethers. *Biol. Mass Spectrom.*, **20**, 61-69 (1991).

Harvey,D.J., Identification by GC and MS of lipids from the rat Harderian gland. *J. Chromatogr.*, **565**, 27-34 (1991).

Harvey,D.J., Identification and quantification of lipids from rabbit harderian glands by GC-MS. *Biomed. Chromatogr.*, **5**, 143-147 (1991).

Holman,R.T., Pusch,F., Svingen,B. and Dutton,H.J., Unusual isomeric polyunsaturated fatty acids in liver phospholipids of rats fed hydrogenated oils. *Proc. Natl. Acad. Sci. USA*, **88**, 4830-4834 (1991).

Husek,P., Derivatization and GC determination of hydroxycarboxylic acids treated with chloroformates. *J. Chromatogr.*, **547**, 307-314 (1991).

Imbs,A.B., Kuklev,D.B., Vereshchagin,A.D. and Latyshev,N.A., Application of an analytical modification of the iodolactonization reaction to selective detection of Δ5 (Δ4) unsaturated fatty acids. *Chem. Phys. Lipids*, **60**, 71-76 (1991).

Iwahashi,H., Parker,C.E., Mason,R.P. and Tomer,K.B., Radical adducts of nitrosobenzene and 2-methyl-2-nitrosopropane with 12,13-epoxylinoleic acid radical, 12,13-epoxylinolenic acid radical and 14,15-epoxyarachidonic acid radical. *Biochem. J.*, **276**, 447-453 (1991).

Jiang,Z.-Y., Woollard,A.C.S. and Wolff,S.P., Lipid hydroperoxide measurement by oxidation of Fe^{2+} in the presence of xylenol orange. Comparison of the TBA assay and an iodometric method. *Lipids*, **26**, 853-856 (1991).

Kamata,T., Tanno,H., Ohrui,H. and Meguro,H., A sensitive enzymatic determination of fatty acid hydroperoxide by glutathione peroxidase with *N*-(9-acridinyl)maleimide fluorometry. *Agric. Biol. Chem.*, **55**, 1985-1988 (1991).

Kase,B.F., Olund,J. and Sisfontes,L., Separation of pristanic and phytanic acid by HPLC: application of the method. *Anal. Biochem.*, **196**, 95-98 (1991).

Le Quere,J.L., Sebedio,J.L., Henry,R., Couderc,F., Demont,N. and Prome,J.C., GC-MS and GC-tandem MS of cyclic fatty acid monomers isolated from heated fats. *J. Chromatogr.*, **562**, 659-672 (1991).

Lie Ken Jie,M.S.F. and Choi,C.Y.C., GC-MS of saturated and unsaturated even and odd-number long chain fatty acids by analysis of the picolinyl esters. Application to a Chinese medicinial seed (*Cuscuta semen*) seed oil. *J. Int. Fed. Clin Chem.*, **3**, 121-128 (1991).

Lie Ken Jie,M.S.F., Choi,C.Y.C., Berger,A. and Berger,R.G., Re-examination of the fatty acid composition of *Biota orientalis* seed oil by GC MS of the picolinyl ester derivatives. *J. Chromatogr.*, **543**, 257-261 (1991).

Lin,C., Blank,E.W., Ceriani,R.L. and Baker,N., Evidence of extensive phospholipid fatty acid methylation during the assumed selective methylation of plasma free fatty acids by diazomethane. *Lipids*, **26**, 548-552 (1991).

Miwa,H., Yamamoto,M. and Asano,T., HPLC analysis of fatty acid compositions of platelet phospholipids as their 2-nitrophenylhydrazides. *J. Chromatogr.*, **568**, 25-34 (1991).

Moffat,C.F., McGill,A.S. and Anderson,R.S., The production of artifacts during preparation of fatty acid methyl esters from fish oils, food products and pathological samples. *J. High Resolut. Chromatogr.*, **14**, 322-326 (1991).

Mossoba,M.M., McDonald,R.E., Armstrong,D.J. and Page,S.W., Hydrogenation of soybean oil: a TLC and GC/matrix isolation/Fourier transform infrared study. *J. Agric. Food Chem.*, **39**, 695-699 (1991).

Mossoba,M.M., McDonald,R.E., Armstrong,D.J. and Page,S.W., Identification of minor C_{18} triene and conjugated diene isomers in hydrogenated soybean oil and margarine by GC-Mi-FT-IR spectroscopy. *J. Chromatogr. Sci.*, **29**, 324-330 (1991).

Nakamura,T. and Maeda,H., A simple assay for lipid hydroperoxides based on triphenylphosphine oxidation and HPLC. *Lipids*, **26**, 765-768 (1991).

Negri,A.P., Cornell,H.J. and Rivett,D.S., The nature of covalently bound fatty acids in wool fibres. *Aust. J. Agric. Res.*, **42**, 1285-1292 (1991).

Nilsson,R. and Liljenberg,C., The determination of double bond positions in polyunsaturated fatty acids - GC/MS of the diethylamide derivatives. *Phytochem. Anal.*, **2**, 253-259 (1991).

Petrzika,M., Engst,W. and Macholz,R., Mass spectrometric structural analysis of fatty acid mixtures from biological material after capillary GC. *Nahrung*, **35**, 491-502 (1991).

Pina,M., Montet,D., Graille,J., Ozenne,C. and Lamberet,G., Contribution of Grignard reagents in the analysis of short-chain fatty acids. *Rev. Franc. Corps Gras*, **38**, 7-8 (1991).

Quinn,R.J. and Tucker,D.J., Further acetylenic acids from the marine sponge *Xestospongia testudinaria*. *J. Nat. Products*, **54**, 290-294 (1991).

Rezanka,T., Zlatkin,I.V., Viden,I., Slabova,O.I. and Nikitin,D.I., Capillary GC-MS of unusual and very-long-chain fatty acids from soil Oligotrophic bacteria. *J. Chromatogr.*, **558**, 215-221 (1991).

Roemen,T.H.M. and van der Vusse,G.J., Assessment of fatty acids in cardiac tissue as 9-anthryldiazomethane esters by HPLC. *J. Chromatogr.*, **570**, 243-251 (1991).

Rojo,J.A. and Perkins,E.G., Isomer identification and gas chromatographic retention studies of monomeric cyclic fatty acid methyl esters. *J. Chromatogr.*, **537**, 329-344 (1991).

Sato,T., Kawana,S. and Iwamoto,M., Near infrared spectral patterns of fatty acid analysis from fats and oils. *J. Am. Oil Chem. Soc.*, **68**, 827-833 (1991).

Sattler,W., Puhl,W., Hayn,M., Kostner,G.M. and Esterbauer,H., Determination of fatty acids in the main lipoprotein classes by capillary GC:BF_3/methanol transesterification of lyophilized samples instead of Folch extraction gives higher yields. *Anal. Biochem.*, **198**, 184-190 (1991).

Shantha,N.C. and Ackman,R.G., Fish oil tetracosenoic acid isomers and GLC analyses of polyunsaturated fatty acids. *Lipids*, **26**, 237-239 (1991).

Shantha,N.C. and Ackman,R.G., Silica gel TLC method for concentration of longer-chain polyunsaturated fatty acids from food and marine lipids. *Canad. Inst. Food Sci. Technol. J.*, **24**, 156-160 (1991).

Shantha,N.C. and Ackman,R.G., Behaviour of a common phthalate plasticizer (dioctyl phthalate) during the alkali- and/or acid catalysed steps in an AOCS method for the preparation of methyl esters. *J. Chromatogr.*, **587**, 263-267 (1991).

Sidisky,L.M. and Ridley,H.S., Temperature dependence of equivalent chain length values on capillary columns of different polarity. *J. High Resolut. Chromatogr.*, **14**, 191-195 (1991).

Smuts,C.M. and Tichelaar,H.Y., Simple TLC purification procedure for the determination of cholesterol ester fatty acid compositions. *J. Chromatogr.*, **564**, 272-277 (1991).

Sonnet,P.E., Dudley,R.L., Osman,S., Pfeffer,P.E. and Schwartz,D., Configuration analysis of unsaturated hydroxy fatty acids. *J. Chromatogr.*, **586**, 255-258 (1991).

Spitzer,V., GC-MS (chemical ionization and electron impact modes) characterization of the methyl esters and oxazoline derivatives of cyclopropene fatty acids. *J. Am. Oil Chem. Soc.*, **68**, 963-969 (1991).

Spitzer,V., Matx,F., Maia,J.G.S. and Pfeilsticker,K., *Curupira tefeensis* II. Occurrence of acetylenic fatty acids. *Fat Sci. Technol.*, **93**, 169-174 (1991).

Thomas,D.W., van Kuijk,F.J.G.M., Dratz,E.A. and Stephens,R.J., Quantitative determination of hydroxy fatty acids as an indicator of *in vivo* lipid peroxidation: GC-MS methods. *Anal. Biochem.*, **198**, 104-111 (1991).

Wadano,T., Ikeda,T., Matumoto,M. and Himeno,M., Immobilized catalyst for detecting chemiluminescence in lipid hydroperoxide. *Agric. Biol. Chem.*, **55**, 1217-1223 (1991).

Wood,R., Sample preparation, derivatization and analysis. In *Analyses of Oils and Fats*, pp.

236-269 (ed. E.G. Perkins, American Oil Chemists' Society, Champaign, U.S.A.) (1991).
Yamamoto,K., Shibahara,A., Nakayama,T. and Kajimoto,G., Determination of double bond positions in methylene-interrupted dienoic fatty acids by GC-MS as their dimethyl disulphide adducts. *Chem. Phys. Lipids*, **60**, 39-50 (1991).
Yamamoto,K., Shibahara,A., Nakayama,T. and Kajimoto,G., Double bond localization in heneicosapentaenoic acid by a GC/MS method. *Lipids*, **26**, 948-950 (1991).
Zamir,I., Derivatization of saturated long-chain fatty acids with phenacyl bromide in non-ionic micelles. *J. Chromatogr.*, **586**, 347-350 (1991).

F. THE ANALYSIS OF SIMPLE LIPID CLASSES

This section corresponds to Chapter 6 in *Lipid Analysis* and deals mainly with chromatographic methods, especially TLC and HPLC, for the isolation and analysis of simple lipid classes. Separations of molecular species of simple lipids are listed in Section H below.

Ackman,R.G., Application of TLC to lipid separation: neutral lipids. In *Analyses of Oils and Fats*, pp. 60-82 (ed. E.G. Perkins, American Oil Chemists' Society, Champaign, U.S.A.) (1991).
Ackman,R.G., Application of TLC to lipid separation: detection methods. In *Analyses of Oils and Fats*, pp. 97-121 (ed. E.G. Perkins, American Oil Chemists' Society, Champaign, U.S.A.) (1991).
Aiken,J.H. and Huie,C.W., Use of hematoporphyrin as a fluorescent stain for detection of lipids in HP TLC. *J. Chromatogr.*, **588**, 295-301 (1991).
Artiss,J.D., Feldbruegge,D.H., Kroll,M.H., McQueen,M.J., Pry,T., Zak,B. and Ziegenhorn,J., Measurement of cholesterol concentration. In *Lipid and Lipoprotein Risk Factors*, pp. 33-49 (edited by N. Rifai and G.R. Warnick, AACC Press) (1991).
Beckman,J.K., Morley,S.A. and Greene,H.L., Analysis of aldehydic peroxidation products by TLC/densitometry. *Lipids*, **26**, 155-161 (1991).
Behrens,W.A. and Madere,R., Malonaldehyde determination in tissues and biological fluids by ion-pairing HPLC. *Lipids*, **26**, 232-236 (1991).
Bergheim,S., Malterud,K.E. and Anthonsen,T., Preparative scale separation of neutral lipids and phospholipids by centrifugally accelerated TLC. *J. Lipid Res.*, **32**, 877-879 (1991).
Bernert,J.T., Akins,J.R., Cooper,G.R., Poulose,A.K., Myers,G.L. and Sampson,E.J., Factors influencing the accuracy of the national reference system total cholesterol reference method. *Clin Chem.*, **37**, 2053-2061 (1991).
Bilyk,A., Piazza,G.J., Bistline,R.G. and Haas,M.J., Separation of cholesterol, and fatty acylglycerols, acids and amides by TLC. *Lipids*, **26**, 405-406 (1991).
Bruns,A., Applications of preparative HPLC in oleochemical research. *J. Chromatogr.*, **536**, 75-84 (1991).
Cooper,G.R., Myers,G.L. and Henderson,L.O., Establishment of reference methods for lipids, lipoproteins and apolipoproteins. *Eur. J. Clin Chem. Clin. Biochem.*, **29**, 269-275 (1991).
Eckfeldt,J.H., Lewis,L.A., Belcher,J.D., Singh,J. and Frantz,I.D., Determination of serum cholesterol by isotope dilution MS with a benchtop capillary GC-MS - comparison with the national reference system's definitive and reference methods. *Clin Chem.*, **37**, 1161-1165 (1991).
Fillion,L., Zee,J.A. and Gossellin,C., Determination of a cholesterol oxide mixture by a single-run HPLC analysis using benzoylation. *J. Chromatogr.*, **547**, 105-112 (1991).
France,J.E., Snyder,J.M. and Wing,J.W., Packed microbore supercritical fluid chromatography with flame ionization detection of abused vegetable oils. *J. Chromatogr.*, **540**, 271-278 (1991).
Fried,B., Lipids. (review of TLC of) *Chromatogr. Sci.*, **55**, 593-623 (1991).
Grob,K., Artho,A. and Mariani,C., Coupled LC-GC for the analysis of olive oils. *Fat Sci. Technol.*, **93**, 494-500 (1991).
Gu,Y., Hirano,C. and Horiike,M., Fuzzy classification analysis of continuously scanned mass spectra of binary mixtures of positionally isomeric tetradecenols. *Rapid Commun. Mass Spectrom.*, **5**, 622-623 (1991).

Gunstone,F.D., ^{13}C-NMR studies of mono-, di- and triacylglycerols leading to qualitative and semiquantitative information about mixtures of these glycerol esters. *Chem. Phys. Lipids*, **58**, 219-224 (1991).
Hannan,R.M. and Hill,H.H., Analysis of lipids in aging seeds using capillary supercritical fluid chromatography. *J. Chromatogr.*, **547**, 393-401 (1991).
Harvey,D.J., Identification by GC and MS of lipids from the rat Harderian gland. *J. Chromatogr.*, **565**, 27-34 (1991).
Horiike,M., Yuan,G. and Hirano,C., Fuzzy classification of location of double bonds in tetradecenyl acetates by electron impact mass spectrometry. *Agric. Biol. Chem.*, **55**, 2521-2526 (1991).
Indrasena,W.M., Paulson,A.T., Parrish,C.C. and Ackman,R.G., A comparison of alumina and silica gel chromarods for the separation and characterization of lipid classes by Iatroscan TLC-FID. *J. Planar Chromatogr.*, **4**, 182-188 (1991).
Kamata,T., Tanno,H., Ohrui,H. and Meguro,H., A sensitive enzymatic determination of acylglycerol hydroperoxide by lipase and glutathione peroxidase with *N*-(9-acridinyl)maleimide fluorometry. *Agric. Biol. Chem.*, **55**, 1989-1992 (1991).
Kaphalia,B.S. and Ansari,G.A.S., Rapid chromatographic analysis of fatty acid anilides suspected of causing toxic oil syndrome. *J. Anal. Toxicol.*, **15**, 90-94 (1991).
Korytowski,W., Bachowski,G.J. and Girotti,A.W., Chromatographic separation and electrochemical determination of cholesterol hydroperoxides generated by photodynamic action. *Anal. Biochem.*, **197**, 149-156 (1991).
Kramer,J.K.G., Fouchard,R.C., Sauer,F.D., Farnworth,E.R. and Wolynetz,M.S., Quantitating total and specific lipids in a small amount of biological sample by TLC-FID. *J. Planar. Chromatogr. - Mod. TLC*, **4**, 42-45 (1991).
Kuksis,A., Marai,L. and Myher,J.J., Plasma lipid profiling by liquid chromatography with chloride-attachment MS. *Lipids*, **26**, 240-246 (1991).
Kurantz,M.J., Maxwell,R.J., Kwoczak,R. and Taylor,F., Rapid and sensitive method for the quantitation of non-polar lipids by high-performance TLC and fluorodensitometry. *J. Chromatogr.*, **549**, 387-399 (1991).
Lee,T.W., Bobik,E. and Malone,W., Quantitative determination of monoglycerides and diglycerides with and without derivatization by capillary supercritical fluid chromatography. *J. Assn. Offic. Anal. Chem.*, **74**, 533-537 (1991).
Leitch,C.A. and Jones,P.J.H., Measurement of triglyceride synthesis in humans using deuterium oxide and isotope ratio mass spectrometry. *Biol. Mass Spectrom.*, **20**, 392-396 (1991).
Mariani,C., Fedeli,E. and Grob,K., Determination of free and esterified minor components in fatty materials. *Riv. Ital. Sostanze Grasse*, **68**, 233-242 (1991).
Markello,T.C., Guo,J. and Gahl,W.A., HPLC of lipids for the identification of human metabolic diseases. *Anal. Biochem.*, **198**, 368-374 (1991).
Ohshima,T. and Ackman,R.G., New developments in Chromarod/Iatroscan TLC-FID: analysis of lipid class composition. *J. Planar. Chromatogr. - Mod. TLC*, **4**, 27-34 (1991).
Pahuja,S.L., Zielinski,J.E., Giordano,G., McMurray,W.J. and Hochberg,R.B., The biosynthesis of D-ring fatty acid esters of estriol. *J. Biol. Chem.*, **266**, 7410-7416 (1991).
Pick,J. and Kovacs,L., Lipid analysis by HP-TLC II. Multidevelopment technique. *J. Planar. Chromatogr. - Mod. TLC*, 4, 91-92 (1991).
Redden,P.R. and Huang,Y.S., Automated separation and quantitation of lipid fractions by HPLC and mass detection. *J. Chromatogr.*, **567**, 21-27 (1991).
Sacchi,R., Paolillo,L., Giudicianni,I. and Addeo,F., Rapid ^{1}H-NMR determination of 1,2- and 1,3-diglycerides in virgin olive oils. *Ital. J. Food Sci.*, **3**, 253-262 (1991).
Sadeghi-Jorabchi,J.D., Wilson,R.H., Belton,P.S., Edwards-Webb,J.D. and Coxon,D.T., Quantitative analysis of oils and fats by Fourier transform Raman spectroscopy. *Spectrochim. Acta,* Part A, **47A**, 1449-1558 (1991).
Sebedio,J.L. and Juaneda,P., Quantitative lipid analyses using the new Iatroscan TLC-FID system. *J. Planar. Chromatogr. - Mod. TLC*, **4**, 35-41 (1991).
Shukla,V.K.S. and Perkins,E.G., The presence of oxidative polymeric materials in encapsulated fish oils. *Lipids*, **26**, 23-26 (1991).
Stein,J.M., Smith,G.A, and Luzio,J.P., An acetylation method for the quantification of membrane lipids, including phospholipids, polyphosphoinositides and cholesterol. *Biochem. J.*, **274**, 375-379 (1991).
Szucs,R., Vindovogel,J. and Sandra,P., Micellar electrokinetic chromatography of aliphatic

compounds with indirect UV detection. *J. High Resolut. Chromatogr.*, **14**, 694-695 (1991).
Volkman,J.K. and Nichols,P.D., Application of TLC-flame ionization detection to the analysis of lipids and pollutants in marine and environmental samples. *J. Planar Chromatogr. - Mod. TLC*, **4**, 19-26 (1991).
Warne,T.R. and Robinson,M., A method for the simultaneous determination of alkylacylglycerol, diacylglycerol, monoalkylglycerol, monoacylglycerol and cholesterol by HPLC. *Anal. Biochem.*, **198**, 302-307 (1991).
Wong,W.W., Hachey,D.L., Feste,A., Leggitt,J., Clarke,L.L., Pond,W.G. and Klein,P.D., Measurement of *in vivo* cholesterol synthesis from ^2H_2O: a rapid procedure for the isolation, combustion, and isotopic assay of erythrocyte cholesterol. *J. Lipid Res.*, **32**, 1049-1056 (1991).
Yang,B. and Chen,J., Analysis of neutral lipids and glycerolysis products from olive oil by liquid chromatography. *J. Am. Oil Chem. Soc.*, **68**, 980-982 (1991).
Zamir,I., Grushka,E. and Cividalli,G., HPLC analysis of free palmitic and stearic acids in cerebrospinal fluid. *J. Chromatogr.*, **565**, 424-429 (1991).
Zamir,I., Grushka,E. and Chemke,J., Separation and determination of saturated very-long-chain free fatty acids in plasma of patients with adrenoleukodystrophy using solid-phase extraction and HPLC. *J. Chromatogr.*, **567**, 319-330 (1991).

G. THE ANALYSIS OF COMPLEX LIPIDS

This section corresponds to Chapter 7 in *Lipid Analysis* and deals mainly with chromatographic methods, especially TLC and HPLC, for the isolation and analysis of complex lipid classes including both phospholipids and glycolipids. Degradative procedures for the identification of polar moieties and spectrometric methods for intact lipids are also listed here. Separations of molecular species of complex lipids are listed in the next section.

Abidi,S.L., HPLC of phosphatidic acids and related polar lipids. *J. Chromatogr.*, **587**, 193-203 (1991).
Aloisi,J., Fried,B. and Sherma,J., Comparison of mobile phases for separation of phospholipids on preadsorbent TLC plates. *J. Liqu. Chromatogr.*, **14**, 3269-3275 (1991).
Artiss,J.D., Amniotic fluid phospholipid analysis. *Clin Chem.*, **37**, 1321-1322 (1991).
Balazy,M., Braquet,P. and Bazan,N.G., Determination of platelet-activating factor and alkyl-ether phospholipids by GC-MS via direct derivatization. *Anal. Biochem.*, **196**, 1-10 (1991).
Bergheim,S., Malterud,K.E. and Anthonsen,T., Preparative scale separation of neutral lipids and phospholipids by centrifugally accelerated TLC. *J. Lipid Res.*, **32**, 877-879 (1991).
Burgers,J.A. and Akkerman,J.W.N., Measurement of platelet-activating factor in biological specimens. *J. Lipid Mediators*, **3**, 241-248 (1991).
Cecconi,O., Ruggieri,S. and Mugnai,G., Use of *N*-acetylpsychosine as internal standard for quantitative HPLC analysis of glycosphingolipids. *J. Chromatogr.*, **555**, 267-271 (1991).
Chai,W., Cashmore,G.C., Stoll,M.S., Gaskell,S.J., Orkiszewski,R.S. and Lawson,A.M., Oligosaccharide sequence determination using B/E linked field scanning or tandem MS of phosphatidylethanolamine derivatives. *Biol. Mass Spectrom.*, **20**, 313-323 (1991).
Christman,B.W., Gay,J.C., Christman,J.W., Prakash,C. and Blair,I.A., Analysis of effector cell-derived lyso platelet activating factor by electron capture negative ion MS. *Biol. Mass Spectrom.*, **20**, 545-552 (1991).
Coene,J., Ghijs,M., Van den Eeckhout,E., Van den Bosche,W., and Sandra,P., Evaluation of different packings for HPLC analysis of alkyl lysophospholipids. *J. Chromatogr.*, **553**, 285-297 (1991).
Cole,M.J. and Enke,C.G., FAB tandem MS employing ion-molecule reactions for the differentiation of phospholipid classes. *J. Am. Soc. Mass Spectrom.*, **2**, 470-475 (1991).
Dandurand,D.M., Kiechle,F.L., Strandbergh,D.R., Zak,B. and Artiss,J.D., Fluorometric determination of phosphatidylcholine as a measure of phospholipid methylation. *Anal. Biochem.*, 196, 356-359 (1991).

de la Vigne,U. and Janchen,D.E., Determination of phospholipids by gradient HPTLC using AMD. *Int. Lab.*, (Nov./Dec.), 22-29 (1991).

De Schrijver,R. and Vermeulen,D., Separation and quantitation of phospholipids in animal tissues by Iatroscan TLC/FID. *Lipids*, **26**, 74-76 (1991).

Ellin,A., Vandenberg,S. and Strandvik,B., A simplified analysis of fatty acids in serum phospholipids using Sep-Pak cartridges. *Clin. Chim. Acta*, 200, 59-61 (1991).

Fischer,W., One-step purification of bacterial lipid macroamphiphiles by hydrophobic interaction chromatography. *Anal. Biochem.*, **194**, 353-358 (1991).

Fried,B., Lipids. (review of TLC of). *Chromatogr. Sci.*, **55**, 593-623 (1991).

Gallant,J. and LeBlanc,R.M., Purification of galactolipids by HPLC for Langmuir-Blodgett film studies. *J. Chromatogr.*, **542**, 307-316 (1991).

Gornati,R., Rapelli,S., Montorfano,G., Cattaneo,C. and Berra,B., A new procedure for gangliosidic *N*-acetylneuraminic acid analysis in serum. *Int. J. Biol. Markers*, **6**, 91-98 (1991).

Grit,M., Crommelin,D.J.A. and Lang,J., Determination of phosphatidylcholine, phosphatidylglycerol and their lyso forms from liposome dispersions by HPLC using high-sensitivity refractive index detection. *J. Chromatogr.*, **585,** 239-246 (1991).

Gross,S.K., Daniel,P.F., Evans,J.E. and McCluer,R.H., Lipid composition of lysosomal multilamellar bodies of male mouse urine. *J. Lipid Res.*, **32**, 157-164 (1991).

Hanras,C. and Perrin,J.L., Gram-scale preparative HPLC of phospholipids from soybean lecithins. *J. Am. Oil Chem. Soc.*, 68, 804-808 (1991).

Hara-Hotta,H., Miyazaki,Y., Yano,I., Matsuyama,T. and Cotter,R.J., Mass spectrometry with soft ionization techniques for structural analysis of lipids in *Serratia* sp. *Anal. Chim. Acta*, 247, 283-293 (1991).

Harvey,D.J., Lipids from the guinea pig harderian gland - use of picolinyl and other pyridine-containing derivatives to investigate the structures of novel branched-chain fatty acids and glycerol ethers. *Biol. Mass Spectrom.*, **20**, 61-69 (1991).

Harvey,D.J., Application of nicotinate derivatives to the structural determination of glycerol ethers. *Biol. Mass Spectrom.*, **20**, 303-307 (1991).

Hasegawa,Y., Kunow,E., Shindou,J. and Yuki,H., Determination of platelet-activating factor by a chemiluminescence method and its application to stimulated guinea pig neutrophils. *Lipids*, **26**, 1117-1121 (1991).

Hedrick,D.B., Guckert,J.B. and White,D.C., Archaebacterial ether lipid diversity analyzed by supercritical fluid chromatography: integration with a bacterial lipid protocol. *J. Lipid Res.*, **32**, 659-666 (1991).

Hirata,H., Higuchi,K., Yamashina,T. and Sugiura,M., Chemistry of succinimido esters. XXI. Determination of sphingosine bases as *N*-arylacetyl derivatives by normal phase HPLC. *Yukagaku*, **40**, 1088-1094 (1991).

Holley,A.E. and Slater,T.F., Measurement of lipid hydroperoxides in normal human blood plasma using HPLC chemiluminescence linked to a diode array detector for measuring conjugated dienes. *Free Radical Res. Commun.*, **15**, 51-63 (1991).

Itonori,S., Kamemura,K., Narushima,K., Sonku,N., Itasaka,O., Hori,T. and Sugita,M., Characterization of a new phosphonocerebroside, *N*-methyl-2-aminoethyl phosphonylglucosylceramide, from the Antarctic krill, *Euphasia superba*. *Biochim. Biophys. Acta*, **1081**, 321-327 (1991).

Karara,A., Dishman,E., Falck,J.R. and Capdevila,J.H., Endogenous epoxyeicosatrienoyl-phospholipids. A novel class of cellular glycerolipids containing epoxidized arachidonate moieties. *J. Biol. Chem.*, **266**, 7561-7569 (1991).

Kasama,T. and Handa,S., Structural studies of gangliosides by fast atom bombardment ionization, low-energy collision-activated dissociation and tandem mass spectrometry. *Biochemistry*, **30**, 5621-5624 (1991).

Kato,N., Gasa,S., Makita,A. and Oguchi,H., Improved separation of lysoglycolipids from solvolyzates by reversed-phase HPLC. *J. Chromatogr.*, **549**, 133-139 (1991).

Kilby,P.M., Bolas,N.M. and Radda,G.K., ^{31}P-NMR study of brain phospholipid structures *in vivo*. *Biochim. Biophys. Acta*, **1085**, 257-264 (1991).

Kloppel.K.-D. and Frederickson,H.L., FAB MS as a rapid means of screening mixtures of ether-linked polar lipids from extremely halophilic archaebacteria for the presence of novel chemical structures. *J. Chromatogr.*, **562**, 369-376 (1991).

Klyashchitsky,B.A., Mezhova,I.V., Krasnopolsky,Y.M. and Shvets,V.I., Preparative isolation of polyphosphoinositides and other anionic phospholipids from natural sources using chromatography on adsorbents containing primary amino groups. *Biotechol. Appl. Biochem.*, **14**, 284-295 (1991).

Kobayashi,T. and Goto,I., A sensitive assay of lysogangliosides using HPLC. *Biochim. Biophys. Acta*, **1081**, 159-166 (1991).

Kojima,M., Seki,T., Ohnishi,M. and Ito,S., Two digalactosyldiacylglycerols with different anomeric configuration in leguminous seeds. *J. Sci. Food Agric.*, **54**, 35-41 (1991).

Koul,O., Prada-Maluf,M., McCluer,R.H. and Ullman,M.D., Rapid isolation of monosialogangliosides from bovine brain gangliosides by selective-overload chromatography. *J. Lipid Res.*, **32**, 1712-1715 (1991).

Kuksis,A., Marai,L. and Myher,J.J., Plasma lipid profiling by liquid chromatography with chloride-attachment MS. *Lipids*, **26**, 240-246 (1991).

Kuksis,A., Myher,J.J., Geher,K., Breckenridge,W.C., Feather,T., McGuire,V. and Little,J.A., Gas chromatographic profiles of plasma total lipids as indicators of dietary history - correlation with fat intake based on 24-H dietary recall. *J. Chromatogr.*, **564**, 11-26 (1991).

Lendrath,G., Bonekamp-Nasner,A. and Kraus,L., Analytical possibilities of qualitative and quantitative determination of phospholipids of different sources. *Fat Sci. Technol.*, **93**, 53-61 (1991).

Lendrath,G., Bonekamp,A. and Kraus,L., Quantitative planar chromatography of phospholipid*s* with different fatty acid compositions. *J. Chromatogr.*, **588**, 303-305 (1991).

Linard,A., Guesnet,P. and Durand,G., Separation of phospholipid classes by overpressure layer chromatography (OPLC). *Rev. Franc. Corps Gras*, **38**, 377-380 (1991).

Liu,Y. and Chan,K.-F.J., High-performance capillary electrophoresis of gangliosides. *Electrophoresis (Weinheim)*, **12**, 402-408 (1991).

Manuguerra,J.-C., DuBois,C. and Hannoun,C., Analytical detection of 9(4)-*O*-acetylated sialoglycoproteins and gangliosides using influenza C virus. *Anal. Biochem.*, **194**, 425-432 (1991).

Markello,T.C., Guo,J. and Gahl,W.A., HPLC of lipids for the identification of human metabolic diseases. *Anal. Biochem.*, **198**, 368-374 (1991).

Matsubara,T. and Hayashi,A., Fragmentation pathways of *O*-trimethylsilyl ethers of dihydroxy long-chain bases analysed by linked-scan MS. *J. Chromatogr.*, **562**, 119-124 (1991).

Matsubara,T. and Hayashi,A., FAB/MS of lipids. *Prog. Lipid Res.*, **30**, 301-322 (1991).

Matsuki,N., Tamura,S., Ono,K., Watari,T., Goitsuka,R., Takahi,S. and Hasegawa,A., The HPLC analysis for the peroxidized phospholipids in equine erythrocytes and skeletal muscle. *J. Vet. Med. Sci.*, **53**, 717-719 (1991).

Merchant,T.E., Meneses,P., Gierke,L.W., Otter,W., Den,W. and Glonek,T. P-31 NMR analysis of phospholipid profiles of neoplastic human breast tissues. *Brit. J. Cancer*, **63**, 693-698 (1991).

Merritt,M.V., Sheeley,D.M. and Reinhold,V.N., Characterization of glycosphingolipids by supercritical fluid chromatography-mass spectrometry. *Anal. Biochem.*, **193**, 24-34 (1991).

Myers,R.L., Ullmann,M.D., Ventura,R.F. and Yates,A.J., A HPLC method for the analysis of glycosphingolipid*s* using galactose oxidase/NaB^3H_4 labelling of intact cells and synaptosomes. *Anal. Biochem.*, **192**, 156-164 (1991).

Nanjee,M.N., Gebre,A.K. and Miller,N.E., Enzymatic fluorometric procedure for phospholipid quantification with an automated mictrotiter plate fluorometer. *Clin Chem.*, **37**, 868-874 (1991).

Nishimura,K., Suzuki,A. and Kino,H., Sphingolipids of a cestode *Metrioliasthes coturnix*. *Biochim. Biophys. Acta*, **1086**, 141-150 (1991).

Ohara,K., Sano,M., Kondo,A. and Kato,I., Two-dimensional mapping by HPLC of pyridylamino oligosaccharides from various glycolipids. *J. Chromatogr.*, **586**, 35-41 (1991).

Ohashi,Y., and Nagai,Y., Fast-atom-bombardment chemistry of sulfatide (3-sulfogalactosylceramide). *Carbohydr. Res.*, **221**, 235-243 (1991).

Piretti,M.V. and Pagliuca,G., Regeneration of column activity after the GC separation of membrane lipids on thermostable SE-52 phase. *J. Chromatogr.*, **585**, 342-344 (1991).

Ranny,M., Sedlacek,J. and Michalec,C., Resolution of phospholipids on chromarods impregnated with salts of some divalent metals. *J. Planar Chromatogr. - Mod. TLC*, **4**, 15-18 (1991).

Redden,P.R. and Huang,Y.S., Automated separation and quantitation of lipid fractions by

HPLC and mass detection. *J. Chromatogr.*, **567**, 21-27 (1991).
Rivera,C.J., Plishker,G.A., Morrisett,J.D., Moore,J. and Percy,A.K., C-13 NMR spectroscopic analysis of phospholipid metabolism in adrenal chromaffin cells. *NMR in Biomedicine*, **4**, 133-136 (1991).
Salyan,M.E.K., Stroud,M.R. and Levery,S.B., Differentiation of type 1 and type 2 chain linkages of native glycosphingolipids by positive-ion fast-atom bombardment MS with collision-induced dissociation and linked scanning. *Rapid Commun. Mass Spectrom.*, **5**, 456-462 (1991).
Shaffiq-Ur-Rehman, Rapid isocratic method for the separation and quantification of major phospholipid classes by HPLC. *J. Chromatogr.*, **567**, 29-37 (1991).
Shimizu,A., Ashida,Y. and Fujiwara,F., Measurement of the ratio of lecithin to sphingomyelin in amniotic fluid by fast atom bombardment MS. *Clin Chem.*, **37**, 1370-1374 (1991).
Song,Y., Kitajima,K., Inoue,S. and Inoue,Y., Isolation and structure elucidation of a novel type of ganglioside, deaminated neuraminic acid (KDN)-containing glycosphingolipid, from rainbow trout sperm. *J. Biol. Chem.*, **266**, 21929-21935 (1991).
Suzuki,M., Yamakawa,T. and Suzuki,A., A micromethod involving micro HPLC-MS for the structural characterization of neutral glycosphingolipids and monosialogangliosides. *J. Biochem. (Tokyo)*, **109**, 503-506 (1991).
Triolo,A., Bertini,J., Mannucci,C., Perico,A. and Pestellini,V., Analysis of platelet-activating factor by GC-MS/low energy electron impact of the corresponding 3-acetyl-2-*tert*-butyldimethylsilyl derivative. *J. Chromatogr.*, **568**, 281-290 (1991).
Weintraub,S.T., Pinckard,R.N. and Hail,M., Electrospray ionisation for analysis of platelet activating factor. *Rapid Commun. Mass Spectrom.*, **5**, 309-311 (1991).
Yamamoto,H., Nakamura,K., Nakatani,M. and Terada,H., Determination of phospholipids on 2-dimensional TLC plates by imaging densitometry. *J. Chromatogr.*, **543**, 201-210 (1991).
Yasugi,E., Kasama,T. and Seyama,Y., Composition of long chain bases in ceramide of the guinea pig Harderian gland. *J. Biochem. (Tokyo)*, **110**, 202-206 (1991).
Yokoyama,Y., Hashimoto,M., Tsuchiya,M. and Yabe,R., Phospholipids in middle ear effusion and serum analysed by liquid ionization mass spectrometry. *Int. J. Mass Spectrom. Ion Processes*, **111**, 263-272 (1991).
Zhu,X., Hara,A. and Taketomi,T., The existence of cerebroside I_3-sulphate in serums of various mammals and its anticoagulant activity. *J. Biochem. (Tokyo)*, **110**, 241-245 (1991).

H. THE ANALYSIS OF MOLECULAR SPECIES OF LIPIDS

This section corresponds to Chapter 8 in *Lipid Analysis* and deals mainly with chromatographic methods for the isolation and analysis of molecular species of lipid classes, including simple lipids, phospholipids and glycolipids. Many of the references in the next section are relevant here also and *vice versa*.

Abidi,S.L., HPLC of phosphatidic acids and related polar lipids. *J. Chromatogr.*, **587**, 193-203 (1991).
Abidi,S.L., Mounts,T.L. and Rennick,K.A., Reversed-phase ion-pair HPLC of phosphatidylinositols. *J. Liqu. Chromatogr.*, **14**, 573-588 (1991).
Adlof,R., Fractionation of egg and soybean phosphatidylcholines by silver resin chromatography. *J. Chromatogr.*, **538**, 469-473 (1991).
Aitzetmuller,K., The use of HPLC in modern fat analysis. *Fat Sci. Technol.*, **93**, 501-510 (1991).
Artz,W.E., Application of GLC to lipid separation and analysis: qualitative and quantitative analysis. In *Analyses of Oils and Fats*, pp. 301-318 (ed. E.G. Perkins, American Oil Chemists' Society, Champaign, U.S.A.) (1991).
Bouteillier,J.C. and Maurin,R., Relation between the triacylglycerol structure and their retention data in HPLC. *Rev. Franc. Corps Gras*, **38**, 297-304 (1991).
Bruschweiler,H. and Dieffenbacher,A., Determination of mono- and diglycerides by capillary GC. *Pure Appl. Chem.*, **63**, 1153-62 (1991).

Bryant,D.K., Orlando,R.C., Fenselau,C., Sowder,R.C. and Henderson,L.E., Four sector tandem MS analysis of complex mixtures of phosphatidylcholines in a human immunodeficiency virus preparation. *Anal. Chem.*, **63**, 1110-1114 (1991).

Burkow,I.C. and Henderson,R.J., Analysis of polymers from autoxidized marine oils by gel permeation HPLC using a light-scattering detector. *Lipids*, **26**, 227-231 (1991).

Burkow,I.C. and Henderson,R.J., Isolation and quantification of polymers from oxidized fish oils by high-performance size-exclusion chromatography with an evaporative mass detector. *J. Chromatogr.*, **552**, 501-506 (1991).

Chilton,F.H., Assays for measuring arachidonic acid release from phospholipids. *Methods Enzymol.*, **197**, 166-182 (1991).

Chobanov,D., Amidzhin,B. and Nikolova-Damyanova,B., Densitometric determination of triglycerides separated by reversed phase TLC. Application to natural oils. *Riv. Ital. Sostanze Grasse*, **68**, 357-362 (1991).

Christie,W.W., Fractionation of the triacylglycerols of evening primrose oil by HPLC in the silver ion mode. *Fat Sci. Technol.*, **93**, 65-66 (1991).

Coene,J., Herdewijn,P., Vandeneeckhout,E., Vandenbossche,W. and Sandra,P., Capillary GC analysis of alkyl lysophospholipids after derivatization with trimethylsilylbromide. *J. High Resolut. Chromatogr.*, **14**, 699-701 (1991).

Cole,M.J. and Enke,C.G., Direct determination of phospholipid structure in microorganisms by fast atom bombardment triple quadrupole mass spectrometry. *Anal. Chem.*, **63**, 1032-1038 (1991).

DaTorre,S.D. and Creer,M.H., Differential turnover of polyunsaturated fatty acids in plasmalogen and diacylglycerophospholipids of isolated cardiac myocytes. *J. Lipid Res.*, **32**, 1159-1172 (1991).

Demirbuker,M and Blomberg,L.G., Separation of triacylglycerols by supercritical-fluid argentation chromatography. *J. Chromatogr.*, **550**, 765-769 (1991).

Duffin,K.L., Henion,J.D. and Shieh,J.J., Electrospray and tandem MS characterization of acylglycerol mixtures that are dissolved in non-polar solvents. *Anal. Chem.*, **63**, 1781-1788 (1991).

Evans,C., Traldi,P., Bambagiotti-Alberti,M., Giannellini,V., Coran,S.A. and Vincieri,F.F., Positive and negative fast atom bombardment MS and collision spectroscopy in the structural characterization of mono-, di- and triglycerides. *Biol. Mass Spectrom.*, **20**, 351-356 (1991).

Evershed,R.P. and Goad,L.J., Chromatographic and mass-spectrometric methods for the analysis of steryl esters. *Stud. Nat. Prod. Chem.*, **9**, 447-486 (1991).

Fang,X., Sheikh,S.U. and Touchstone,J.C., HPLC of cholesteryl esters. *J. Liqu. Chromatogr.*, **14**, 589-598 (1991).

Foglia,T.A. and Maeda,K., HPLC separation of enantiomeric alkyl glycerol ethers. *Lipids*, **26**, 769-773 (1991).

Gasser,H., Strohmaier,W., Schlag,G., Schmid,E.R. and Allmaier,G., Characterization of phosphatidylcholines in rabbit lung lavage fluid by positive and negative FAB MS. *J. Chromatogr.*, **562**, 257-266 (1991).

Glass,R.L., Semipreparative HPLC separation of phosphatidylcholine molecular species from soybean leaves. *J. Liqu. Chromatogr.*, **14**, 339-349 (1991).

Graber,L. and Rohrbasser,C., Study on composition of plasma triacylglycerols using supercritical fluid chromatography. *Chimia*, **45**, 342-345 (1991).

Harvey,D.J., Nicotinylidene derivatives for the structural elucidation of glycerol mono-ethers and monoesters by GC-MS. *Biol. Mass Spectrom.*, **20**, 87-93 (1991).

Hedrick,D.B., Guckert,J.B. and White,D.C., Archaebacterial ether lipid diversity analyzed by supercritical fluid chromatography: integration with a bacterial lipid protocol. *J. Lipid Res.*, **32**, 659-666 (1991).

Hogge,L.R., Taylor,D.C., Reed,D.W. and Underhill,E.W., Characterization of castor bean neutral lipids by MS/MS. *J. Am. Oil Chem. Soc.*, **68**, 863-868 (1991).

Hori,M., Sugiura,K., Sahashi,Y. and Koike,S., Determination of molecular species of triacylglycerols by reversed phase liquid chromatography/double focussing MS with a frit-CI interface. *Shitsuryo Bunseki*, **39**, 133-140 (1991).

Hoving,E.B., Muskiet,F.A. and Christie,W.W., Separation of cholesterol esters by silver ion chromatography using HPLC or solid-phase extraction columns packed with a bonded sulphonic acid phase. *J. Chromatogr.*, **565**, 103-110 (1991).

Ishinaga,M., Tanimoto,M., Sugiyama,S., Kumamoto,R. and Yokoro,K., Molecular species of phospholipids in rats in primary and transplanted fibrosarcomas induced by soybean oil containing tocopherol acetate. *Biochem. Cell Biol.*, **69**, 655-660 (1991).

Itabashi,Y., Marai,L. and Kuksis,A., Identification of natural diacylglycerols as the 3,5-dinitrophenylurethanes by chiral phase liquid chromatography with MS. *Lipids*, **26**, 951-956 (1991).

Jeffrey,B.S.J., Silver-complexation liquid chromatography for fast, high-resolution separation of triacylglycerols. *J. Am. Oil Chem. Soc.*, 68, 289-293 (1991).

Jeong,Y., Ohshima,T. and Koizumi,C., Changes in molecular species compositions of glycerophospholipids in Japanese oyster *Crassostrea gigas* (Thunberg) during frozen storage. *Comp. Biochem. Physiol.*, **100B**, 99-105 (1991).

Kallio,H., Vauhkonen,T. and Linko,R.R., Thin-layer silver ion chromatography and supercritical fluid chromatography of Baltic herring (*Clupea harrengus membras*) triacylglycerols. *J. Agric. Food Chem.*, **39**, 1573-1577 (1991).

Kallio,H., Johansson,A. and Oksman,P., Composition and development of turnip rapeseed (*Brassica campestris*) oil triacylglycerols at different stages of maturation. *J. Agric. Food Chem.*, **39**, 1752-1756 (1991).

Karlsson,K.A., Lanne,B., Pimlott,W. and Teneberg,S., The resolution into molecular species on desorption of glycolipids from TLC chromatograms, using combined TLC and fast-atom-bombardment MS. *Carbohydrate Res.*, **221**, 49-61 (1991).

Kaufmann,P. and Herslof,B.G., A multivariate identification of natural triglyceride oils. *Fat Sci. Technol.*, **93**, 179-183 (1991).

Kennerley,D.A., Quantitative analysis of water-soluble products of cell-associated phospholipase C- and phospholipase D-catalyzed hydrolysis of phosphatidylcholine. *Methods Enzymol.*, **197**, 191-197 (1991).

Kitsos,M., Gandini,C., Massolimi,G., De Lorenzi,E. and Caccialanza,G., HPLC post-column derivatization with fluorescence detection to study the influence of ambroxol on dipalmitoylphosphatidylcholine levels in rabbit Eustachian tube washings. *J. Chromatogr.*, **553**, 1-6 (1991).

Kojima,M., Ohnishi,M. and Ito,S., Composition and molecular species of ceramide and cerebroside in scarlet runner beans (*Phaseolus coccineus* L.) and kidney beans (*Phaseolus vulgaris* L.). *J. Agric. Food Chem.*, **39**, 1709-1714 (1991).

Kuksis,A., Marai,L. and Myher,J.J., Reversed-phase liquid chromatography mass spectrometry of complex mixtures of natural triacylglycerols with chloride attachment negative chemical ionization. *J. Chromatogr.*, **588**, 73-87 (1991).

Kuksis,A., Marai,L., Myher,J.J., Itabashi,Y. and Pind,S., Qualitative and quantitative analysis of molecular species of glycerolipids by HPLC. In *Analyses of Oils and Fats*, pp. 214-232 (ed. E.G. Perkins, American Oil Chemists' Society, Champaign, U.S.A.) (1991).

Kuksis,A., Marai,L., Myher,J.J., Itabashi,Y. and Pind,S., Applications of GC/MS, LC/MS and FAB/MS to determination of molecular species of glycerolipids. In *Analyses of Oils and Fats*, pp. 464-495 (ed. E.G. Perkins, American Oil Chemists' Society, Champaign, U.S.A.) (1991).

Kuypers,F.A., Butikofer,P. and Shackleton,C.H., Application of liquid chromatography-thermospray MS in the analysis of glycerophospholipid molecular species. *J. Chromatogr.*, **562**, 191-206 (1991).

Laakso,P. and Christie,W.W., Combination of silver ion and reversed-phase HPLC in the fractionation of herring oil triacylglycerols. *J. Am. Oil Chem. Soc.*, **68**, 213-223 (1991).

Lee,C., Fisher,S.K., Agranoff,B.W. and Hajra,A.K., Quantitative analysis of molecular species of diacylglycerol and phosphatidate formed upon muscarinic receptor activation of human SK-N-SH neuroblastoma cells. *J. Biol. Chem.*, **266**, 22837-22846 (1991).

Lee,C. and Hajra,A.K., Molecular species of diacylglycerols and phosphoglycerides and the postmortem changes in the molecular species of diacylglycerols in rat brain. *J. Neurochem.*, **56**, 370-379 (1991).

Maniongui,C., Gresti,J., Bugaut,M., Gautier,S. and Bezard,J., Determination of bovine butter fat triacylglycerols by reversed-phase liquid chromatography and gas chromatography. *J. Chromatogr.*, **543**, 81-103 (1991).

Mares,P., Rezanka,T. and Novak,M., Analysis of human blood plasma triacylglycerols using capillary GC, silver ion TLC fractionation and desorption CI-MS. *J. Chromatogr.*, **568**, 1-10 (1991).

Mariani,C., Fedeli,E. and Grob,K., Determination of free and esterified minor components in fatty materials. *Riv. Ital. Sostanze Grasse*, **68**, 233-242 (1991).

Matsubara,T. and Hayashi,A., FAB/MS of lipids. *Prog. Lipid Res.*, **30**, 301-322 (1991).
Nikolova-Damyanova,B. and Amidzhin,B., Densitometric quantitation of triglycerides. *J. Planar Chromatogr. - Mod. TLC.*, **4**, 397-401 (1991).
Nishimura,K., Suzuki,A. and Kino,H., Sphingolipids of a cestode *Metrioliasthes coturnix*. *Biochim. Biophys. Acta*, **1086**, 141-150 (1991).
Norman,H.A., Pillai,P. and St John,J.B., *In vitro* desaturation of monogalactosyldiacylglycerol and phosphatidylcholine molecular species by chloroplast homogenates. *Phytochem.*, **30**, 2217-2222 (1991).
Ohshima,T. and Koizumi,C., Selected ion monitoring GC/MS of 1,2-diacylglycerol tert-butyldimethylsilyl ethers derived from glycerophospholipids. *Lipids*, **26**, 940-947 (1991).
Pchelkin,V.P. and Vereshchagin,A.G., Identification of individual diacylglycerols by adsorption TLC of their coordination complexes. *J. Chromatogr.*, **538**, 373-383 (1991).
Rezanka,T. and Mares,P., Determination of plant triacylglycerols using capillary GC, HPLC and MS. *J. Chromatogr.*, **542**, 145-159 (1991).
Schaller,H., High temperature stationary phases for triglyceride analysis with capillary columns. *Fat Sci. Technol.*, **93**, 510-515 (1991).
Schlame,M., Beyer,K., Hayer-Hartl,M. and Klingenberg,M., Molecular species of cardiolipin in relation to other mitochondrial phospholipids. Is there an acyl specificity of the interaction between cardiolipin and the ADP/ATP carrier. *Eur. J. Biochem.*, **199**, 459-466 (1991).
Schlame,M. and Otten,D., Analysis of cardiolipin molecular species by HPLC of its derivative 1,3-bisphosphatidyl-2-benzoyl-*sn*-glycerol. *Anal. Biochem.*, **195**, 290-295 (1991).
Sebedio,J.L., Chromatographic techniques applied to heated fats and oils. *Chromatography and Analysis*, (Oct. Issue), 9-11 (1991).
Sempore,B.G. and Bezard,J.A., Analysis and fractionation of natural source diacylglycerols as urethane derivatives by reversed-phase HPLC. *J. Chromatogr.*, **547**, 89-103 (1991).
Sempore,B.G. and Bezard,J.A., Enantiomer separation by chiral-phase liquid chromatography of urethane derivatives of natural diacylglycerols previously fractionated by reversed-phase LC. *J. Chromatogr.*, **557**, 227-240 (1991).
Sempore,G. and Bezard,J., Determination of molecular species of oil triacylglycerols by reversed-phase and chiral-phase HPLC. *J. Am. Oil Chem. Soc.*, **68**, 702-709 (1991).
Stinson,A.M., Wiegand,R.D. and Anderson,R.E., Recycling of docosahexaenoic acid in rat retinas during *n*-3 fatty acid deficiency. *J. Lipid Res.*, **32**, 2009-2017 (1991).
Sugatani,J., Fujimura,K., Miwa,M., Satouchi,K. and Saito,K., Molecular heterogeneity of platelet-activating factor (PAF) in rat glandular stomach determined by GC/MS. PAF molecular species changes upon water-immersion stress. *Lipids*, **26**, 1347-1353 (1991).
Takahashi,K. and Hirano,T., Chromatographic behaviour of glycerolipid with respect to 1,2-;2,1- and 1,2-;1,3- isomers on reverse phase mode. *Yukagaku*, **40**, 277-282 (1991).
Takahashi,K. and Hirano,T., Theoretical aspects of the resolution of lipid molecular species containing unsaturated fatty acid on reverse phase HPLC. *Yukagaku*, **40**, 300-305 (1991).
Takamura,H. and Kito,M., A highly sensitive method for quantitative analysis of phospholipid molecular species by HPLC. *J. Biochem. (Tokyo)*, **109**, 436-439 (1991).
van Oosten,H.J., Klooster,J.R., Vandeginste,B.G.M. and de Galan,L., Capillary supercritical fluid chromatography for the analysis of oils and fats. *Fat Sci. Technol.*, **93**, 481-487 (1991).
Weintraub,S.T., Pinckard,R.N., Heath,T.G. and Gage,D.A., Novel mass spectral fragmentation of heptafluorobutyryl derivatives of acyl analogs of platelet-activating factor. *J. Am. Soc. Mass Spectrom.*, **2**, 476-482 (1991).
Wolff,J.P., Mordret,F.X. and Dieffenbacher,A., Determination of polymerized triglycerides in oils and fats by HPLC. Results of a collaborative study and the standardized method. *Pure Appl. Chem.*, **63**, 1163-71 (1991).
Wolff,J.P., Mordret,F.X. and Dieffenbacher,A., Determination of triglycerides in vegetable oils in terms of their partition numbers by HPLC. Results of a collaborative study and the standardized method. *Pure Appl. Chem.*, **63**, 1173-82 (1991).
Yang,L.-Y. and Kuksis,A., Apparent convergence (at 2-monoacylglycerol level) of phosphatidic acid and 2-monoacylglycerol pathways of synthesis of chylomicron triacylglycerols. *J. Lipid Res.*, **32**, 1173-1186 (1991).
Yang,L.-Y., Kuksis,A. and Myher,J.J., Similarities in surface lipids of chylomicrons from glycerol and alkyl ester feeding: major components. *Lipids*, **26**, 806-818 (1991).
Zeitoun,M.A.M., Neff,W.E., Selke,E. and Mounts,T.L., Analyses of vegetable oil

triglyceride molecular species by reversed phase HPLC. *J. Liqu. Chromatogr.*, **14**, 2685-2698 (1991).
Zirrolli,J.A., Clay,K.L. and Murphy,R.C., Tandem MS of negative ions from choline phospholipid molecular species related to platelet activating factor. *Lipids*, **26**, 1112-1116 (1991).

I. STRUCTURAL ANALYSIS OF LIPIDS BY MEANS OF ENZYMATIC HYDROLYSIS

This section corresponds to Chapter 9 in *Lipid Analysis* and relates to simple lipids, phospholipids and glycolipids. Many of the references in the last section are relevant here also and *vice versa*. Some methods for the resolution of chiral lipids are listed here when they deal with methods for determining positional distributions of fatty acids within lipid classes.

Blank,M.L. and Snyder,F., Chromatographic analysis of phospholipase reaction products. *Methods Enzymol.*, **197**, 158-165 (1991).
Christie,W.W., Nikolova-Damyanova,B., Laakso,P. and Herslof,B., Stereospecific analysis of triacyl-*sn*-glycerols via resolution of diastereomeric diacylglycerol derivatives by HPLC on silica. *J. Am. Oil Chem. Soc.*, **68**, 695-701 (1991).
Ghosh,S., Lee,S., Brown,T.A., Basu,M., Hawes,J.W., Davidson,D. and Basu,S., Use of exoglycosidases from *Mercenaria mercenaria* (hard shelled clam) as a tool for structural studies of glycosphingolipids and glycoproteins. *Anal. Biochem.*, **196**, 252-261 (1991).
Itabashi,Y., Marai,L. and Kuksis,A., Identification of natural diacylglycerols as the 3,5-dinitrophenylurethanes by chiral phase liquid chromatography with MS. *Lipids*, **26**, 951-956 (1991).
Mazur,A.W., Hiler,G.D., Lee,S.S.C., Armstrong,M.P. and Wendel,J.D., Regio- and stereoselective enzymatic esterification of glycerol and its derivatives. *Chem. Phys. Lipids*, 60, 189-199 (1991).
Murakami,N., Imamura,H., Morimoto,T., Ueda,T., Nagai,S., Sakakibara,J. and Yamada,N., Studies on glycolipids 2. Selective preparation of *sn*-1 and *sn*-2 lysogalactolipids by enzymatic hydrolysis using lipase (from *Rhizopus arrizhus*). *Tetrahedron Letts*, **32**, 1331-1334 (1991).
Ohara,K., Sano,M., Kondo,A. and Kato,I., Two-dimensional mapping by HPLC of pyridylamino oligosaccharides from various glycolipids. *J. Chromatogr.*, **586**, 35-41 (1991).
Sempore,B.G. and Bezard,J.A., Enantiomer separation by chiral-phase liquid chromatography of urethane derivatives of natural diacylglycerols previously fractionated by reversed-phase LC. *J. Chromatogr.*, **557**, 227-240 (1991).
Sempore,G. and Bezard,J., Determination of molecular species of oil triacylglycerols by reversed-phase and chiral-phase HPLC. *J. Am. Oil Chem. Soc.*, **68**, 702-709 (1991).
Takagi,T., Chromatographic resolution of chiral lipid derivatives. *Prog. Lipid Res.*, **29**, 277-298 (1991).
Takagi,T. and Ando,Y., Stereospecific analysis of triacylglycerols by chiral-phase HPLC. Direct derivatization of partially hydrolyzed products. *Yukagaku*, **40**, 288-292 (1991).
Takagi,T. and Ando,Y., Stereospecific analysis of triacyl-*sn*-glycerols by chiral HPLC. *Lipids*, **26**, 542-547 (1991).
Wolff,R.L. and Entressangles,B., Compositional changes of fatty acids in the 1(1")- and 2(2")- positions of cardiolipin from liver, heart, and kidney mitochondria of rats fed a low-fat diet. *Biochim. Biophys. Acta*, **1082**, 136-142 (1991).

J. THE ANALYSIS AND RADIOASSAY OF ISOTOPICALLY-LABELLED LIPIDS

This section corresponds to Chapter 10 in *Lipid Analysis*. Only papers in which the radioactivity of the sample appeared to be central to the analysis are listed.

Haroldsen,P.E., Gaskell,S.J., Weintraub,S.T. and Pinckard,R.N., Isotopic exchange during derivatization of platelet activating factor for GC-MS. *J. Lipid Res.*, **32**, 723-729 (1991).
Myers,R.L., Ullmann,M.D., Ventura,R.F. and Yates,A.J., A HPLC method for the analysis of glycosphingolipids using galactose oxidase/NaB^3H_4 labelling of intact cells and synaptosomes. *Anal. Biochem.*, **192**, 156-164 (1991).
Stein,J.M., Smith,G.A, and Luzio,J.P., An acetylation method for the quantification of membrane lipids, including phospholipids, polyphosphoinositides and cholesterol. *Biochem. J.*, **274**, 375-379 (1991).

K. THE SEPARATION OF PLASMA LIPOPROTEINS

This section corresponds to Chapter 11 in *Lipid Analysis*, and a only few key papers of particular interest are listed.

Bauer,J.E., Single-spin density gradient systems and micropreparative ultracentrifugation. In *Analyses of Oils and Fats*, pp. 555-572 (ed. E.G. Perkins, American Oil Chemists' Society, Champaign, U.S.A.) (1991).
Borghini,I., James,R.W., Blatter,M.-C. and Pometta,D., Distribution of apoprotein E between free and A-II complexed forms in very-low- and high-density lipoproteins: functional implications. *Biochim. Biophys. Acta*, **1083**, 139-146 (1991).
Cooper,G.R., Myers,G.L. and Henderson,L.O., Establishment of reference methods for lipids, lipoproteins and apolipoproteins. *Eur. J. Clin Chem. Clin. Biochem.*, **29**, 269-275 (1991).
Fless,G.M., General preparative ultracentrifugation and considerations for lipoprotein isolation. In *Analyses of Oils and Fats*, pp. 512-523 (ed. E.G. Perkins, American Oil Chemists' Society, Champaign, U.S.A.) (1991).
France,D.S., Weinstein,D.B., Quinby,R.E., Babiak,J., Lapen,D.C., Murdoch,M.K. and Paterniti,J.R., Clinical chemistry applications of robotic systems: cloning of a microplate management system and interface to a FPLC system for online compositional analysis of serum lipoproteins. *Adv. Lab. Autom. Rob.*, **7**, 645-682 (1991).
Keift,K.A., Bocan,T.M.A. and Krause,B.R., Rapid on-line determination of cholesterol distribution among plasma lipoproteins after high-performance gel filtration chromatography. *J. Lipid Res.*, **32**, 859-866 (1991).
Matsumoto,U., Nakayama,H., Shibusawa,Y. and Niimura,T., Separation of human serum lipoproteins into three major classes by hydroxyapatite chromatography. *J. Chromatogr.*, **566**, 67-76 (1991).
Nyyssonen,K. and Salonen,J.T., Comparison of gel permeation chromatography, density gradient ultracentrifugation and precipitation methods for quantitation of VLDL, LDL and HDL cholesterol. *J. Chromatogr.*, **570**, 782-789 (1991).
Orr,J.R., Adamson,G.L. and Lindgren,F.T., Preparative ultracentrifugation and analytical ultracentrifugation of plasma lipoproteins. In *Analyses of Oils and Fats*, pp. 524-554 (ed. E.G. Perkins, American Oil Chemists' Society, Champaign, U.S.A.) (1991).
Otvos,J.D., Jeyarajah,E.J. and Bennett,D.W., Quantitation of plasma lipoproteins by proton NMR. *Clin Chem.*, **37**, 377-386 (1991).
Shimano,H., Yamada,N., Ishibashi,S., Mokuno,H., Mori,N., Gotoda,T., Harada,K., Akanuma,Y., Murase,T., Yazaki,Y. and Takaku,F., Oxidation-labile subfraction of human plasma LDL isolated by ion-exchange chromatography. *J. Lipid Res.*, **32**, 763-773 (1991).
Shore,V.G., Non-denaturing electrophoresis of lipoproteins in agarose and polyacrylamide gradient gels. In *Analyses of Oils and Fats*, pp. 573-598 (ed. E.G. Perkins, American Oil Chemists' Society, Champaign, U.S.A.) (1991).
Vedie,B., Myara,I., Pech,M.A., Maziere,J.C., Maziere,C., Caprania,A. and Moatti,N., Fractionation of charge-modified LDL by fast protein liquid chromatography. *J. Lipid Res.*, **32**, 1359-1359 (1991).

L. SOME MISCELLANEOUS SEPARATIONS

Analyses of lipids such as prostaglandins, acylcarnitines, coenzyme A esters and so forth that do not fit conveniently into other sections are listed here. More complete listings for prostaglandins are available elsewhere (*Prostaglandins, Leukotrienes and Essential Fatty acids*). The decision on whether to list some papers here or in Section E was sometimes arbitrary.

Abian,J., Pages,M. and Gelpi,E., Use of methyl oxime derivatives to enhance structural information in thermospray HPLC-MS analysis of linoleic acid lipoxygenase metabolites in maize embryos. *J. Chromatogr.*, **554**, 155-174 (1991).

Baer,A.N., Costello,P.B. and Brash,F.A., Stereospecificity of the hydroxyeicosatetraenoic and hydroxyoctadecadienoic acids produced by cultured bovine endothelial cells. *Biochim. Biophys. Acta*, **1085**, 45-52 (1991).

Bernstrom,K., Kayganich,K. and Murphy,R.C., Collisionally induced dissociation of epoxyeicosatrienoic acids and epoxyeicosatrienoic acid-phospholipid molecular species. *Anal. Biochem.*, **198**, 203-211 (1991).

Capdevila,J.H., Dishman,E., Karara,A. and Falck,J.R., Cytochrome P450 arachidonic acid expoxygenase - stereochemical characterization of epoxyeicosatrienoic acids. *Methods Enzymol.*, **206**, 441-453 (1991).

Chadwick,R.R. and Hsieh,J.C., Separation of *cis* and *trans* double-bond isomers using capillary zone electrophoresis. *Anal. Chem.*, **63**, 2380-2383 (1991).

Clare,R.A., Huang,S., Doig,M.V. and Gibson,G.G., GC-MS characterization of some novel hydroxyeicosatetraenoic acids formed on incubation of arachidonic acid with microsomes from induced rat livers. *J. Chromatogr.*, **562**, 237-247 (1991).

Coors,U., Utilization of the tocopherol pattern for recognition of fat and oil adulteration. *Fat Sci. Technol.*, **93**, 519-526 (1991).

Crabtree,D.V. and Adler,A.J., Derivatization of hydroxyeicosatetraenoic fatty acid esters with pentafluorobenzoic anhydride and analysis with supercritical fluid chromatography CI-MS. *J. Chromatogr.*, **543**, 405-412 (1991).

Huang,Z.-H., Gage,D.A., Bieber,L.L. and Sweeley,C.C., Analysis of acylcarnitines as their *N*-demethylated ester derivatives by GC-CI/MS. *Anal. Biochem.*, **199**, 98-105 (1991).

Hurst,J.S., Balazy,M., Bazan,H.E.P. and Bazan,N.G., The epithelium, endothelium and stroma of the rat cornea generate (12*S*)-hydroxyeicosatetraenoic acid as the main lipoxygenase metabolite in response to injury. *J. Biol. Chem.*, **266**, 6726-6730 (1991).

Kler,R.S., Jackson,S., Bartlett,K., Bindoff,L.A., Eaton,S., Pourfarzam,M., Frerman,F.E., Goodman,S.I., Watmough,N.J. and Turnbull,D.M., Quantitation of acyl-CoA and acylcarnitine esters accumulated during abnormal mitochondrial fatty acid oxidation. *J. Biol. Chem.*, **266**, 22932-22938 (1991).

Kuhn,H., Heydeck,D. and Sprecher,H., On the mechanistic reason for the dual positional specificity of the reticulocyte lipoxygenase. *Biochim. Biophys. Acta*, **1081**, 129-134 (1991).

Lohmus,M., Vamets,A., Jarning,I., Samel,N., Lille,U. and Pehk,T., Preparative separation of natural prostaglandins E. *Prep. Chromatogr.*, **1**, 279-300 (1991).

Lysz,T.W., Wu,Y., Brash,A.R., Keeting,P.A., Lin,C. and Fu,S.C.J., Identification of 12(*S*)-hydroxyeicosatetraenoic acid in the young rat lens. *Current Eye Res.*, 10, 331-337 (1991).

Okita,R.T., Clark,J.E., Okita,J.R. and Masters,B.S.S., ω-Hydroxylation and (ω-1)-hydroxylation of eicosanoids and fatty acids by HPLC. *Methods Enzymol.*, **206**, 432-441 (1991).

Pourfarzam,M. and Bartlett,K., Synthesis, characterisation and HPLC of C_6-C_{16} dicarboxylyl-mono-coenzyme A and -mono-carnitine esters. *J. Chromatogr.*, **570**, 253-276 (1991).

Rackham,D.M. and Harvey,G.A., Stereochemical analysis of a leukotriene related hydroxypentadecadiene using a chiral HPLC column and diode array detection. *J. Chromatogr.*, **542**, 189-192 (1991).

Squire,R.S.T., Synthesis and purification of radioactive fatty acylcarnitines of high specific activity. *Anal. Biochem.*, **197**, 104-107 (1991).

Stein,T.A., Bailey,B., Auguste,L.-J. and Wise,L., Measurement of prostaglandin G/H synthase and lipoxygenase activity in the stomach wall by HPLC. *BioChromatogr.*, **10**, 222-225 (1991).

Weber,C., Holler,M., Beetens,J., de Clerck,F. and Tegtmeier,F., Determination of 6-keto-PGF_{1a}, 2,3-dinor-6-ketoPGF_{1a}, thromboxane B_2, 2,3-dinor-thromboxane B_2, PGE_2, PGD_2 and PGF_{2a} in human urine by GC-chemical ion MS. *J. Chromatogr.*, **562**, 599-611 (1991).

Acknowledgement

This paper is published as part of a programme funded by the Scottish Office Agriculture and Fisheries Dept.

APPENDIX

Some Important References in Lipid Methodology - 1992

William W. Christie

The Scottish Crop Research Institute, Invergowrie, Dundee (DD2 5DA), Scotland

A. Introduction
B. The Structure, Chemistry and Occurrence of Lipids
C. The Isolation of Lipids from Tissues
D. Chromatographic and Spectroscopic Analysis of Lipids - General Principles.
E. The Analysis of Fatty Acids
F. The Analysis of Simple Lipid Classes
G. The Analysis of Complex Lipids
H. The Analysis of Molecular Species of Lipids
I. Structural Analysis of Lipids by means of Enzymatic Hydrolysis
J. The Separation of Plasma Lipoproteins
K. Some Miscellaneous Separations

A. INTRODUCTION

The purpose of this chapter is the same as the last, except that the year 1992 is covered. It has been compiled in the same way with sections corresponding to Chapters in *Lipid Analysis* (Second Edition, Pergamon Press, 1982) by the author, and the strengths and weaknesses are the same as in the previous listings. Again, note that the titles of papers listed below may not be literal transcriptions of the originals. In particular, a number of abbreviations have been introduced. References are listed alphabetically according to the surname of the first author in each section,

B. THE STRUCTURE, CHEMISTRY AND OCCURRENCE OF LIPIDS

Hetherington,A.M. and Drobak,B.K., Inositol-containing lipids in higher plants. *Prog. Lipid Res.*, **31**, 53-63 (1992).

Kaya,K., Chemistry and biochemistry of taurolipids. *Prog. Lipid Res.*, **31**, 87-108 (1992).

Kundu,S.K., Glycolipids - structure, synthesis, functions. In *Glycoconjugates. Composition, Structure and Function*, pp. 203-262 (ed. H.J. Allen & E.C. Kisalius, Marcel Dekker, N.Y.) (1992).

C. THE ISOLATION OF LIPIDS FROM TISSUES

Cabrini,L., Landi,L., Stefanelli,C., Barzanti,V. and Sechi,A.M., Extraction of lipids and lipophilic antioxidants from fish tissues - a comparison among different methods. *Comp. Biochem. Physiol.*, **101B**, 383-386 (1992).

Christie,W.W., Solid-phase extraction columns in the analysis of lipids. In *Advances in Lipid Methodology - One* (ed. W.W. Christie, Oily Press, Ayr), pp. 1-17 (1992).

Christie,W.W., Solid-phase extraction columns and lipid analysis. In *Contemporary Lipid Analysis, 2nd Symposium Proceedings*, pp. 23-29 (ed. N.U. Olsson & B.G. Herslof, LipidTeknik, Stockholm) (1992).

Elmer-Frolich,K. and Lachance,P.A., Faster and easier methods for quantitative lipid extraction and fractionation from miniature samples of animal tissues. *J. Am. Oil Chem. Soc.*, **69**, 243-245 (1992).

King,J.W., France,J.E. and Snyder,J.M., On-line supercritical fluid extraction-supercritical fluid reaction-capillary GC analysis of the fatty acid composition of oil seeds. *Fres. J. Anal. Chem.*, **344**, 474-478 (1992).

Miller-Podraza,H., Mansson,J.-E. and Svennerholm,L., Isolation of complex gangliosides from bovine brain. *Biochim. Biophys. Acta*, **1124**, 45-51 (1992).

Soares,M.G.C.B., Da Silva,K.M.O. and Guedes,L.S., Lipid extraction - a proposal of substitution of chloroform by dichloromethane in the method of Folch, Lees and Stanley. *Arquiv. Biol. Tecnol.*, **35**, 655-658 (1992).

Ulberth,F. and Henninger,M., One-step extraction/methylation method for determining the fatty acid composition of processed foods. *J. Am. Oil Chem. Soc.*, **69**, 174-177 (1992).

Whiteley,G.S.W., Fuller,B.J. and Hobbs,K.E.F., Lipid peroxidation in liver tissue specimens stored at subzero temperatures. *Cryo-Letters*, **13**, 83-86 (1992).

Whiteley,G.S.W., Fuller,B.J. and Hobbs,K.E.F., Deterioration of cold-stored tissue specimens due to lipid peroxidation - modulation by antioxidants at high subzero temperatures. *Cryobiology*, **26**, 668-673 (1992).

D. CHROMATOGRAPHIC AND SPECTROSCOPIC ANALYSIS OF LIPIDS. GENERAL PRINCIPLES.

Cole,L.A. and Dorsey,J.G., Temperature dependence of retention in reversed-phase liquid chromatography. 1. Stationary-phase considerations. *Anal. Chem.*, **64**, 1317-1323 (1992).

Cole,L.A., Dorsey,J.G. and Dill,K.A., Temperature dependence of retention in reversed-phase liquid chromatography. 2. Mobile-phase considerations. *Anal. Chem.*, **64**, 1324-1327 (1992).

Van der Meeren,P., Vanderdeelen,J. and Baert,L., Simulation of the mass response of the evaporative light scattering detector. *Anal. Chem.*, **64**, 1056-1062 (1992).

E. THE ANALYSIS OF FATTY ACIDS

This section corresponds to Chapters 4 and 5 in *Lipid Analysis* and deals with derivatization and the chromatographic analysis of fatty acids, for example by gas chromatography (GC) and high performance liquid chromatography (HPLC), together with spectrometric methods, especially mass spectrometry (MS). Papers dealing with analysis of a free fatty acid fraction of tissue lipids are listed in the next section mainly.

Ackman,R.G., Applications of GLC to lipid separation and analysis: qualitative and quantitative analysis. In *Fatty acids in foods and their health implications*, pp. 47-63 (ed. C.K. Chow, Marcel Dekker, NY) (1992).

Ackman,R.G., GLC and the arachidonic acid content of Atlantic food fish. *Food Res. Int.*, **25**, 453-456 (1992).

Antoine,M. and Adams,J., Implications of the charge site in charge-remote fragmentations. *J. Am. Soc. Mass Spectrom.*, **3**, 776-778 (1992).

Aursand,M. and Grasdalen,H., Interpretation of the ^{13}C-NMR spectra of omega-3 fatty acids and lipid extracted from the white muscle of Atlantic salmon (*Salmo salar*). *Chem. Phys. Lipids*, **62**, 239-251 (1992).

Banerjee,P., Dawson,G. and Dasgupta,A., Enrichment of saturated fatty acid containing phospholipids in sheep brain serotonin receptor preparations: use of microwave irradiation for rapid transesterification of phospholipids. *Biochim. Biophys. Acta*, **1110**, 65-74 (1992).

Barnathan,G., Miralles,J., Gaydou,E.M., Boury-Esnault,N. and Kornprobst,J.-M., New phospholipid fatty acids from the marine sponge *Cinachyrella alloclada* Uliczka. *Lipids*, **27**, 779-784 (1992).

Bhat,U.R. and Carlson,R.W., A new method for the analysis of amide-linked hydroxy fatty acids in lipid-As from gram negative bacteria. *Glycobiol.*, **2**, 535-539 (1992).

Biedermann,M., Grob,K., Froelich,D. and Meier,W., On-line coupled LC-GC and LC-LC-GC for detecting the irradiation of fat-containing foods. *Z. Lebensm.-Unters. Forsch.*, **195**, 409-416 (1992).

Bortolomeazzi,R., Pizzale,L. and Lercker,G., Chromatographic determination of the position and configuration of isomers of methyl oleate hydroperoxides. *J. Chromatogr.*, **626**, 109-116 (1992).

Brechany,E.Y. and Christie,W.W., Identification of the saturated oxo fatty acids in cheese. *J. Dairy Res.*, **59**, 57-64 (1992).

Brodowsky,I.D. and Oliw,E.H., Metabolism of 18:2(*n*-6), 18:3(*n*-3), 20:4(*n*-6) and 20:5(*n*-3) by the fungus *Gaeumannomyces graminis*: identification of metabolites formed by 8-hydroxylation and by ω-2 and ω-3 oxygenation. *Biochim. Biophys. Acta*, **1124**, 59-65 (1992).

Brondz,I. and Olsen,I., Intra-injector methylation of free fatty acids from aerobically and anaerobically cultured *Actinobacillus actinomycetemcomitans* and *Haemophilus aphrophilus*. *J. Chromatogr.*, **576**, 328-333 (1992).

Butler,W.R., Thibert,L. and Kilburn,J.O., Identification of *Mycobacterium avium* complex strains and some similar species by HPLC. *J. Clin. Microbiol.*, **30**, 2698-2704 (1992).

Carballeira,N.M., Negron,V. and Reyes,E.D., Novel naturally occurring alpha-methoxy acids from the phospholipids of Caribbean sponges. *Tetrahedron*, **48**, 1053-1058 (1992).

Carballeira,N.M. and Sepulveda,J.A., Two novel naturally occurring α-methoxy acids from the phospholipids of two Caribbean sponges. *Lipids*, **27**, 72-74 (1992).

Chang,Y.S. and Watson,J.T., Charge-remote fragmentation during FAB-CAD-b/e linked-scan mass spectrometry of (aminoethyl)-triphenylphosphonium derivatives of fatty acids. *J. Am. Soc. Mass Spectrom.*, **3**, 769-775 (1992).

Chen,H.W. and Anderson,R.E., Quantitation of phenacyl esters of retinal fatty acids by HPLC. *J. Chromatogr.*, **578**, 124-129 (1992).

Chin,S.F., Liu,W., Storkson,J.M., Ha,Y.L. and Pariza,M.W., Dietary sources of conjugated dienoic isomers of linoleic acid, a newly recognised class of anticarcinogens. *J. Food Compos. Anal.*, **5**, 185-197 (1992).

Chobanov,D., Tarandjiiska,R. and Nikolova-Damyanova,B., Quantitation of isomeric unsaturated fatty acids by argentation TLC. *J. Planar Chromatogr.-Mod. TLC.*, **5**, 157-163 (1992).

Christie,W.W., The chromatographic resolution of chiral lipids. In *Advances in Lipid Methodology - One* (ed. W.W. Christie, Oily Press, Ayr), pp. 121-148 (1992).

Christie,W.W., Brechany,E.Y., Lie Ken Jie,M.S.F. and Wong,C.F., Mass spectrometry of derivatives of isomeric allenic fatty acids. *Biol. Mass Spectrom.*, **21**, 267-270 (1992).

Christie,W.W., Brechany,E.Y., Stefanov,K. and Popov,S., The fatty acids of the sponge *Dysidea fragilis* from the Black Sea. *Lipids*, **27**, 640-644 (1992).

Dasgupta,A., Banerjee,P. and Malik,S., Use of microwave irradiation for rapid transesterification of lipids and accelerated synthesis of fatty acyl pyrrolidides for analysis by GC-MS: study of fatty acid profiles of olive oil, evening primrose oil, *Chem. Phys.*

Lipids, **62**, 281-291 (1992).

Demirbuker,M., Hagglund,I. and Blomberg,L.G., Separation of unsaturated fatty acid methyl esters by packed capillary supercritical fluid chromatography - comparison of different column packings. *J. Chromatogr.*, **605**, 263-268 (1992).

Eder,K., Reichlmayr-Lais,A.M. and Kirchgessner,M., GC analysis of fatty acid methyl esters: avoiding discrimination by programmed temperature vaporizing injection. *J. Chromatogr.*, **588**, 265-272 (1992).

Eder,K., Reichlmayr-Lais,A.M. and Kirchgessner,M., Studies on the methanolysis of small amounts of purified phospholipids for GC analysis of small amounts of fatty acid methyl esters. *J. Chromatogr.*, **607**, 55-67 (1992).

Firestone,D. and Sheppard,A., Determination of *trans* fatty acids. In *Advances in Lipid Methodology - One* (ed. W.W. Christie, Oily Press, Ayr), pp. 273-322 (1992).

Gardner,D.R., Sanders,R.A., Henry,D.E., Tallmadge,D.H. and Wharton,H.W., Characterization of used frying oils. Part. 1. Isolation and identification of compound classes. *J. Am. Oil Chem. Soc.*, **69**, 499-508 (1992).

Gerard,H.C., Moreau,R.A., Fett,W.F. and Osman,S.F., Separation and quantitation of hydroxy and epoxy fatty acids by HPLC with an evaporative light-scattering detector. *J. Am. Oil Chem. Soc.*, **69**, 301-304 (1992).

Gerard,H.C., Osman,S.F., Fett,W.F. and Moreau.R.A., Separation, identification and quantification of monomers from cutin polymers by HPLC with evaporative light scattering detection. *Phytochem. Anal.*, **3**, 139-144 (1992).

Griffiths,D.W., Robertson,G.W., Millam,S. and Holmes,A.C., The determination of the petroselinic acid content of Coriander (*Coriander sativum*) oil by capillary GC. *Phytochem. Anal.*, **3**, 250-253 (1992).

Guth,H. and Grosch,W., Furan fatty acids in butter and butter oil. *Z. Lebensm. Unters. Forsch.*, **194**, 360-362 (1992).

Hamberg,M., A method for the determination of the stereochemistry of α,β-epoxy alcohols derived from fatty acid hydroperoxides. *Lipids*, **27**, 1042-1046 (1992).

Hamberg,M., Gerwick,W.H. and Asen,P.A., Linoleic acid metabolism in the red alga *Lithothamnion coralloides*: biosynthesis of 11(*R*)-hydroxy-9(*Z*),12(*Z*)-octadecadienoic acid. *Lipids*, **27**, 487-493 (1992).

Harvey,D.J. Mass spectrometry of picolinyl and other nitrogen-containing derivatives of fatty acids. In *Advances in Lipid Methodology - One* (ed. W.W. Christie, Oily Press, Ayr), pp. 19-80 (1992).

Husek,P., Fast derivatization with chloroformates for GC analysis. *LC-GC International*, **5** (9), 43-49 (1992).

Ikeda,M. and Kusaka,T., Liquid chromatography-mass spectrometry of hydroxy and non-hydroxy fatty acids as amide derivatives. *J. Chromatogr.*, **575**, 197-205 (1992).

Ioneda,T. and Beaman,B.L., Molecular weight determination of methyl esters of mycolic acids using thermospray mass spectrometry. *Chem. Phys. Lipids*, **63**, 41-46 (1992).

Jansen,E.H.J.M. and Defluiter,P., Determination of lauric acid metabolites in peroxisome proliferation after derivatization and HPLC analysis with fluorometric detection. *J. Liqu. Chromatogr.*, **15**, 2247-2260 (1992).

Jensen,N.J., Haas,G.W. and Gross,M.L., Ion-neutral complex intermediate for loss of water from fatty acid carboxylates. *Org. Mass Spectrom.*, **27**, 423-427 (1992).

Jin,S.-J., Hoppel,C.L. and Tserng,K.-Y., Incomplete fatty acid oxidation. The production and epimerization of fatty acids. *J. Biol. Chem.*, **267**, 119-125 (1992).

Johnson,D.W., Beckman,K., Fellenberg,A.J., Robinson,B.S. and Poulos,A. Monoenoic fatty acids in human brain lipids: isomer identification and distribution. *Lipids*, **27**, 177-180 (1992).

Johnson,S.B. and Brown,R.E., Simplified derivatization for determining sphingolipid fatty acyl composition by GC-MS. *J. Chromatogr.*, **605**, 281-6 (1992).

Joseph,J.D. and Ackman,R.G., Capillary column GC method for analysis of encapsulated fish oils and fish oil ethyl esters - collaborative study. *J. Assoc. Off. Anal Chem. Int.*, **75**, 488-506 (1992).

Kaya,K., Sano,T. and Shiraishi,F., Tetrahydropyran ring-containing fatty acid-combined taurine (tetrathermoyltaurine) in the taurolipid fraction of *Tetrahymena thermophila*. *Biochim. Biophys. Acta*, **1127**, 22-27 (1992).

King,J.W., France,J.E. and Snyder,J.M., On-line supercritical fluid extraction-supercritical fluid reaction-capillary GC analysis of the fatty acid composition of oil seeds. *Fres. J.*

Anal. Chem., **344**, 474-478 (1992).

Kuhn,H., Belker,J., Wiesner,R., Schewe,T., Lankin,V.Z. and Tikhaze,A.K., Structure elucidation of oxygenated lipids in human atherosclerotic lesions. *Eicosanoids*, **5**, 17-22 (1992).

Kusters,E., Spondlin,C., Volken,C. and Eder,C., Direct resolution of β-hydroxy myristic acid enantiomers by chiral phase GC and HPLC. *Chromatographia*, **33**, 159-162 (1992).

Laakso,P., Supercritical fluid chromatography of lipids. In *Advances in Lipid Methodology - One* (ed. W.W. Christie, Oily Press, Ayr), pp. 81-119 (1992).

Lie Ken Jie,M.S.F. and Cheung,Y.K., Mass spectral studies on methyl and picolinyl ester derivatives of isomeric selena fatty acids. *Biol. Mass Spectrom.*, **21**, 505-508 (1992).

Lie Ken Jie,M.S.F., Cheung,Y.K., Chau,S.H., Christie,W.W. and Brechany,E.Y., Mass spectra of the picolinyl ester derivatives of some conjugated diacetylenic acids. *Chem. Phys. Lipids*, **63**, 65-68 (1992).

Lie Ken Jie,M.S.F. and Choi,Y.C., GC-MS of the picolinyl ester derivatives of deuterated acetylenic fatty acids. *J. Chromatogr.*, **625**, 271-276 (1992).

Lie Ken Jie,M.S.F. and Choi,Y.C., Mass spectral determination of deuterium-labelled picolinyl fatty esters in the determination of double-bond position. *J. Am. Oil Chem. Soc.*, **69**, 1245-1247 (1992).

Matucha,M., Jockisch,W., Verner,P. and Anders,G., Isotope effects in GLC of labelled compounds. *J. Chromatogr.*, **588**, 251-258 (1992).

Medina,I., Aubourg,S., Gallardo,J.M. and Perez-Martin,R., Comparison of six methylation methods for analysis of the fatty acid composition of albacore lipid. *Int. J. Food Sci. Technol.*, **27**, 597-601 (1992).

Mielniczuk,Z., Alugupalli,S., Mielniczuk,E. and Larsson,L., GC-MS of lipopolysaccharide 3-hydroxy fatty acids - comparison of pentafluorobenzyl and trimethylsilyl methyl ester derivatives. *J. Chromatogr.*, **623**, 115-122 (1992).

Morvai,M., Palyka,I. and Molnar-Perl,I., Flame ionization detector response factors using the effective carbon number concept in the quantitative analysis of esters. *J. Chromatogr. Sci.*, **30**, 448-452 (1992).

Moss,C.W. and Daneshvar,M.I., Identification of some uncommon monounsaturated fatty acids of bacteria. *J. Clin. Microbiol.*, **30**, 2511-2512 (1992).

Nikolova-Damyanova,B., Silver ion chromatography and lipids. In *Advances in Lipid Methodology - One* (ed. W.W. Christie, Oily Press, Ayr), pp. 181-237 (1992).

Nikolova-Damyanova,B., Herslof,B.G. and Christie,W.W., Silver ion high-performance liquid chromatography of derivatives of isomeric fatty acids. *J. Chromatogr.*, **609**, 133-140 (1992).

Oliw,E.H., Enantioselective separation of some polyunsaturated epoxy fatty acids by HPLC on a cellulose phenylcarbamate (Chiralcel OC) stationary phase. *J. Chromatogr.*, **583**, 231-235 (1992).

Painuly,P. and Grill,C.M., Purification of erucic acid by preparative HPLC and crystallization. *J. Chromatogr.*, **590**, 139-145 (1992).

Pawlosky,R.J., Sprecher,H.W. and Salem,N., High sensitivity negative ion GC-MS method for detection of desaturated and chain-elongated products of deuterated linoleic and linolenic acids. *J. Lipid Res.*, **33**, 1711-1717 (1992).

Pfeffer,E.P., Sonnet,P.E., Schwartz,D.P., Osman,S.F. and Weisleder,D., Effects of bis homoallylic and homoallylic substitution on the olefinic ^{13}C resonance shifts in fatty acid methyl esters. *Lipids*, **27**, 285-288 (1992).

Rakoff,H. and Rohwedder,W.K., Catalytic deuteration of alkynols and their tetrahydropyranyl esters. *Lipids*, **27**, 567-569 (1992).

Ratnayake,W.M.N., AOCS method Ce 1c-89 underestimates the *trans* octadecenoate content in favour of the *cis* isomers in partially hydrogenated vegetable oils. *J. Am. Oil Chem. Soc.*, **69**, 192 (1992).

Ratnayake,W.M.N. and Pelletier,G., Positional and geometrical isomers of linoleic acid in partially hydrogenated oils. *J. Am. Oil Chem. Soc.*, **69**, 95-105 (1992).

Rezanka,T., Identification of very-long-chain acids from peat and coals by capillary GC-MS. *J. Chromatogr.*, **627**, 241-245 (1992).

Robinson,B.S., Johnson,D.W. and Poulos,A., Novel molecular species of sphingomyelin containing 2-hydroxylated polyenoic very-long-chain fatty acids in mammalian testes and spermatozoa. *J. Biol. Chem.*, **267**, 1746-1751 (1992).

Rubino,F.M. and Zecca,L., Application of triple quadrupole tandem MS to the analysis of pyridine-containing derivatives of long-chain acids and alcohols. *J. Chromatogr.*, **579**, 1-12 (1992).

Rubino,F.M. and Zecca,L., Triple quadrupole tandem MS study of the 3-picolinyl esters of fatty acids. *Org. Mass Spectrom.*, **27**, 1240-1247 (1992).

Schwartz,D.P. and Rady,A.H., Quantitation and occurrence of hydroxy fatty acids in fats and oils. *J. Am. Oil Chem. Soc.*, **69**, 170-173 (1992).

Shantha,N.C. and Napolitano,G.E.,, Gas chromatography of fatty acids. *J. Chromatogr.*, **624**, 37-51 (1992).

Spitzer,V., Marx,F., Maia,J.G.S. and Pfeilsticker,K., Occurrence of α-eleostearic acid in the seed oil of *Parinari montana* (Chrysobalanaceae). *Fat Sci. Technol.*, **94**, 58-60 (1992).

Stein,J., Kulemeier,J., Lembcke,B. and Caspary,W.F., Simple and rapid method for determination of short-chain fatty acids in biological materials by HPLC with UV detection. *J. Chromatogr.*, **576**, 53-61 (1992).

Stransky,K., Jursik,T., Vitek,A. and Skorepa,J., An improved method of characterizing fatty acids by equivalent chain length values. *J. High Resolut. Chromatogr.*, **15**, 730-740 (1992).

Takadate,A., Masuda,T., Murata,C., Haratake,C., Isobe,A., Irikura,M. and Goya,S., 3-Bromoacetyl-6,7-methylenedioxycoumarin as a highly reactive and sensitive fluorescence labelling reagent for fatty acids. *Anal. Sci.*, **8**, 695-697 (1992).

ten Brink,H.J., Stellaard,F., van den Heuvel,C.M.M., Kok,R.M., Schor,D.S.M., Wanders,R.J.A. and Jakobs,C., Pristanic acid and phytanic acid in plasma from patients with peroxisomal disorders: stable isotope dilution analysis with electron capture negative ion mass fragmentography. *J. Lipid Res.*, **33**, 41-47 (1992).

Teng,J.I., Reversed-phase HPLC of underivatized fatty acids by fatty acid analysis column. *J. Liqu. Chromatogr.*, **15**, 1473-1485 (1992).

Teng,J.I., Chen,X. and Guerrero,S., Fatty acid group separation: polyunsaturated from saturated and monounsaturated by TLC. *J. Planar Chromatogr.*, **5**, 64-66 (1992).

Thomas,D.W., van Kuijk,F.J.G.M. and Stephens,R.J., Quantitative determination of hydroxy fatty acids as an indicator of *in vivo* lipid peroxidation: oxidation products of arachidonic and docosapentaenoic acids in rat liver after exposure to carbon tetrachloride. *Anal. Biochem.*, **206**, 353-358 (1992).

Toyo'oka,T., Ishibashi,M., Takeda,Y., Nakashima,K., Akiyama,S., Uzu,S. and Imai,K., Precolumn fluorescence tagging reagent for carboxilic acids in HPLC: 4-substituted-7-aminoalkylamino-2,1,3-benzoxadiazoles. *J. Chromatogr.*, **588**, 61-71 (1992).

Toyo'oka,T., Ishibashi,M. and Terao,T., Fluorescent chiral derivatization reagents for carboxilic acid enantiomers in HPLC. *Analyst (London)*, **117**, 727-733 (1992).

Traitler,H. and Wille,H.J., Isolation of pure fatty acids from fats and oils. *Fat Sci. Technol.*, **94**, 506-511 (1992).

Tsuyama,Y., Uchida,T. and Goto,T., Analysis of underivatized C_{12}-C_{18} fatty acids by reversed-phase ion-pair HPLC with conductivity detection. *J. Chromatogr.*, **596**, 181-184 (1992).

Ulberth,F. and Haider,H.J., Determination of low level *trans* unsaturation in fats by Fourier transform infrared spectroscopy. *J. Food Sci.*, **57**, 1444-1447 (1992).

Ulberth,F. and Henninger,M., On-column injection of fatty acid methyl esters on to polar capillary columns without distortion of early running peaks. *J. High Resolut. Chromatogr.*, **15**, 54-56 (1992).

Ulberth,F. and Henninger,M., Simplified method for the determination of *trans* monoenes in edible fats by TLC-GLC. *J. Am. Oil Chem. Soc.*, **69**, 829-831 (1992).

Ulberth,F. and Kampter,W., Artifact formation during transmethylation of lipid peroxides. *Anal. Biochem.*, **203**, 35-38 (1992).

Vorderwuhlbecke,T., Epple,M., Cammenga,H.K., Petersen,S. and Kieslich,K., Racemic lactones from butterfat: an advanced approach that includes stereodifferentiation. *J. Am. Oil Chem. Soc.*, **69**, 797-801 (1992).

Vreeken,R.J., Jager,M.E., Ghijsen,R.T. and Brinkman,U.A.T. The derivatization of fatty acids by (chloro)alkyl chloroformates in non-aqueous and aqueous media for GC analysis. *J. High Resolut. Chromatogr.*, **15**, 785-790 (1992).

Wang,T., Wu,W. and Powell,W.S., Formation of monohydroxy derivatives of arachidonic acid, linoleic acid, and oleic acid during oxidation of low density lipoprotein by copper ions and endothelial cells. *J. Lipid Res.*, **33**, 525-537 (1992).

Wesen,C., Mu,H.L., Kvernheim,A.L. and Larsson,P., Identification of chlorinated fatty acids in fish lipids by partitioning studies and by GC with Hall electrolytic conductivity

detection. *J. Chromatogr.*, **625**, 257-269 (1992).
Wilson,R. and Sargent,J.R., High-resolution separation of polyunsaturated fatty acids by argentation TLC. *J. Chromatogr.*, **623**, 403-407 (1992).
Wolff,R.L., Resolution of linolenic acid geometrical isomers by GLC on a capillary column coated with a 100% cyanopropyl polysiloxane film (CP Sil-88). *J. Chromatogr. Sci.*, **30**, 17-22 (1992).
Wolff,R.L., *trans*-Polyunsaturated fatty acids in French edible rapeseed and soybean oils. *J. Am. Oil Chem. Soc.*, **69**, 106-110 (1992).
Wolff,R.L. and Vandamme,F.F., Separation of petroselinic acid (*cis*-6 18:1) and oleic (*cis*-9 18:1) acids by GLC of their isopropyl esters. *J. Am. Oil Chem. Soc.*, **69**, 1228-1231 (1992).
Wu,M., Church,D.F., Mahler,T.J., Barker,S.A. and Pryor,W.A., Separation and spectral data of the six isomeric ozonides from methyl oleate. *Lipids*, **27**, 129-135 (1992).
Yamane,M. and Abe,A., HPLC-thermospray MS of hydroxy-polyunsaturated fatty acid acetyl derivatives. *J. Chromatogr.*, **575**, 1-18 (1992).
Yang,G.C., Detection of lipid hydroperoxides by HPLC coupled with post-column reaction. *Trends Food Sci. Tecnol.*, **3**, 15-18 (1992).
Yoo,J.S. and McGuffin,V.L., Determination of fatty acids in fish oil dietary supplements by capillary liquid chromatography with laser-induced fluorescence detection. *J. Chromatogr.*, **627**, 87-96 (1992).
Yoo,J.S., Watson,J.T. and McGuffin,V.L., Temperature-programmed microcolumn liquid chromatography/mass spectrometry. *J. Microcol. Sep.*, **4**, 349-362 (1992).

F. THE ANALYSIS OF SIMPLE LIPID CLASSES

This section corresponds to Chapter 6 in *Lipid Analysis* and deals mainly with chromatographic methods, especially TLC and HPLC, for the isolation and analysis of simple lipid classes. Separations of molecular species of simple lipids are listed in Section H below.

Agren,J.J., Julkunen,A. and Pentilla,I., Rapid separation of serum lipids for fatty acid analysis by a single aminopropyl column. *J. Lipid Res.*, **33**, 1871-1876 (1992).
Akasaka,K., Sasaki,I., Ohrui,H. and Meguro,H., A simple fluorometry of hydroperoxides in oils and foods. *Biosci. Biotech. Biochem.*, **56**, 605-607 (1992).
Alvarez,J.G. and Touchstone,J.C., Separation of acidic and neutral lipids by aminopropyl-bonded silica gel chromatography. *J. Chromatogr.*, **577**, 142-145 (1992).
Amelio,M., Rizzo,R. and Varazini,F., Determination of sterols, erythrodiol, uvaol and alkanols in olive oils using combined solid-phase extraction, HPLC and high-resolution GC techniques. *J. Chromatogr.*, **606**, 179-185 (1992).
Christie,W.W., Solid-phase extraction columns in the analysis of lipids. In *Advances in Lipid Methodology - One* (ed. W.W. Christie, Oily Press, Ayr), pp. 1-17 (1992).
Christie,W.W., Detectors for HPLC of lipids with special reference to evaporative light-scattering detection. In *Advances in Lipid Methodology - One* (ed. W.W. Christie, Oily Press, Ayr), pp. 239-271 (1992).
Contreras,J.A., Castro,M., Brocos,C., Herrera,E. and Lasuncion,M.A., Combination of an enzymatic method and HPLC for the quantitation of cholesterol in cultured cells. *J. Lipid Res.*, **33**, 931-936 (1992).
Danno,H., Jincho,Y., Budiyanto,S., Furukawa,Y. and Kimura,S., A simple enzymatic quantitative analysis of triglycerides in tissues. *J. Nutr. Sci. Vitamin.*, **38**, 517-521 (1992).
Debrauwer,L., Paris,A., Rao,D., Fournier,F. and Tabet,J.C., MS studies on 17β-estradiol-17-fatty acid esters: evidence for the formation of anion-dipole intermediates. *Org. Mass Spectrom.*, **27**, 707-719 (1992).
Fenton,M., Chromatographic separation of cholesterol in foods. *J. Chromatogr.*, **624**, 369-388 (1992).
Giron,D., Link,R. and Bouissel,S., Analysis of monoglycerides, diglycerides and triglycerides in pharmaceutical fluids by capillary supercritical fluid chromatography. *J. Pharm. Biomed. Anal.*, **10**, 821-830 (1992).
Hammad,S., Siegel,H.S., Marks,H.L. and Barbato,G.F., A fast HPLC analysis of cholesterol

and cholesterol esters in avian plasma. *J. Liqu. Chromatogr.*, **15**, 2005-2014 (1992).
Hopia,A.I., Piironen,V.I., Koivistoienen,P.E. and Hyvonen,L., Analysis of lipid classes by solid-phase extraction and high-performance size-exclusion chromatography. *J. Am. Oil Chem. Soc.*, **69**, 772-776 (1992).
Horiike,M., Yuan,G., Kim,C.S., Hirano,C. and Shibuya,K., Determination of the double bond position in hexadecenols by MS without prior chemical modification. *Org. Mass Spectrom.*, **27**, 944-948 (1992).
Iwata,T., Inoue,K., Nakamura,M. and Yamaguchi,M., Simple and highly sensitive determination of free fatty acids in human serum by HPLC with fluorescence detection. *Biomed. Chromatogr.*, **6**, 120-123 (1992).
Jandera,P. and Guiochon,G., Adsorption isotherms of cholesterol and related compounds in non-aqueous reversed-phase chromatographic systems. *J. Chromatogr.*, **605**, 1-17 (1992).
Kamido,H., Kuksis,A., Marai,L., Myher,J.J. and Pang,H., Preparation, chromatography and mass spectrometry of cholesteryl ester and glycerolipid-bound aldehydes. *Lipids*, **27**, 645-650 (1992).
Larner,J.M., Shackleton,C.H.L., Roitman,E., Schwartz,P.E. and Hochberg,R.B., Measurement of estradiol-17-fatty acid esters in human tissues. *J. Clin. Endocrinol. Metab.*, **75**, 195-200 (1992).
Lognay,G., Severin,M., Boenke,A. and Wagstaffe,P.J., Edible fats and oils reference materials for sterols analysis with particular reference to cholesterol. Part. 1. Investigation of some analytical aspects by experienced laboratories. *Analyst (London)*, **117**, 1093-1097 (1992).
Lovaas,E., A sensitive spectrophotometric method for hydroperoxide determination. *J. Am. Oil Chem. Soc.*, **69**, 777-783 (1992).
Marini,D., HPLC of lipids. In *Food Analysis by HPLC*, (ed. L.M.L. Nollet, Marcel Dekker, New York), pp. 169-240 (1992).
Murata,M. and Ide,T., Determination of cholesterol in subnanomolar quantities in biological fluids by HPLC. *J. Chromatogr.*, **579**, 329-333 (1992).
Oishi,M., Onishi,K., Nishijima,K., Nakagomi,K., Nakazawa,H., Uchiyama,S. and Suzuki,S., Rapid and simple coulimetric measurements of peroxide value in edible oils and fats. *J. Assoc. Off. Anal Chem. Int.*, **75**, 507-510 (1992).
Olsson,N.U., Advances in planar chromatography for the separation of food lipids. *J. Chromatogr.*, **624**, 11-19 (1992).
Pallavicini,M., Villa,L. and Cesarotti,E., HPLC determination of the enantiomeric excess of 1,3-glyceryl diethers obtained by stereoselective catalytic reduction. *J. Chromatogr.*, **604**, 197-202 (1992).
Parrish,C.C., Bodennec,G. and Gentien,P., Separation of polyunsaturated and saturated lipids from marine phytoplankton on silica gel-coated chromarods. *J. Chromatogr.*, **607**, 97-104 (1992).
Pioch,D., Lozano,P., Frater,C. and Graille,J., A quick method for cholesterol titration in complex media. *Fat Sci. Technol.*, **94**, 268-272 (1992).
Polette,A., Durand,P., Floccard,B. and Blache,D., A method for specific analysis of free fatty acids in biological samples by capillary GC. *Anal. Biochem.*, **206**, 241-245 (1992).
Prieto,J.A., Ebri,A. and Collar,C., Optimized separation of nonpolar and polar lipid classes from wheat flour by solid-phase extraction. *J. Am. Oil Chem. Soc.*, **69**, 387-391 (1992).
Rubino,F.M. and Zecca,L., Application of triple quadrupole tandem MS to the analysis of pyridine-containing derivatives of long-chain acids and alcohols. *J. Chromatogr.*, **579**, 1-12 (1992).
Sakura,S. and Terao,J., Determination of hydroperoxides by electroluminescence. *Anal. Chim. Acta*, **262**, 59-65 (1992).
Sakura,S. and Terao,J., Comparison of electrochemiluminescence and ampometric detection of lipid hydroperoxides. *Anal. Chim. Acta*, **262**, 217-223 (1992).
Serrano-Carreon,L., Hathout,Y., Bensoussan,M. and Belin,J.M., Quantitative separation of Trichoderma lipid classes on a bonded phase column. *J. Chromatogr.*, **584**, 129-133 (1992).
Shantha,N.C., Thin-layer chromatography flame ionization detection Iatroscan system. *J. Chromatogr.*, **624**, 21-35 (1992).
Stewart,G., Gosselin,C. and Pandian,S., Selected ion monitoring of *tert*-butyldimethylsilyl cholesterol ethers for determination of total cholesterol content in foods. *Food Chem.*, **44**, 377-380 (1992).

Toschi,T.G. and Caboni,M.F., Cholesterol oxides: biological behaviour and analytical determination. *Ital. J. Food Sci.*, **4**, 223-228 (1992).
Wasilchuk,B.A., Le Quesne,P.W. and Vouros,P., Monitoring cholesterol oxidation processes using multideuteriated cholesterol. *Anal. Chem.*, **64**, 1077-1087 (1992).
Wieland,E., Diedrich,F., Schletter,V., Schuffwerner,P. and Oellerich,M., Lipid hydroperoxide determination in serum by HPLC and iodometry. *Fres. J. Anal. Chem.*, **343**, 62-63 (1992).
Wieland,E., Schettler,V., Diedrich,F., Schuffwerner,P. and Oellerich,M., Determination of lipid hydroperoxides in serum. Iodometry and HPLC compared. *Eur. J. Clin. Chem. Clin. Biochem.*, **30**, 363-369 (1992).

G. THE ANALYSIS OF COMPLEX LIPIDS

This section corresponds to Chapter 7 in *Lipid Analysis* and deals mainly with chromatographic methods, especially TLC and HPLC, for the isolation and analysis of complex lipid classes including both phospholipids and glycolipids. Degradative procedures for the identification of polar moieties and spectrometric methods for intact lipids are also listed here. Separations of molecular species of complex lipids are listed in the next section.

Aiken,J.H. and Huie,C.W., Comparison with hematoporphyrin with other fluorogenic agents for the detection of cationic surfactants and lipids after separation by HPTLC. *J. Planar Chromatogr. Mod. TLC.*, **5**, 87-91 (1992).
Aluyi,H.S., Boote,V., Drucker,D.B., Wilson,J.M. and Ling,Y.H., Analysis of polar lipids from some representative enterobacteria, *Plesiomonas* and *Acinobacter* by fast atom bombardment-mass spectrometry. *J. Appl. Biol.*, **73**, 426-432 (1992).
Alvarez,J.G., Storey,B.T., Hemling,M.L. and Grob,R.L., Chromatographic and spectroscopic analysis of globotriaosyl ceramide from bovine spermatozoa. *J. Liqu. Chromatogr.*, **15**, 1621-1638 (1992).
Alvarez,J.G. and Touchstone,J.C., Separation of acidic and neutral lipids by aminopropyl-bonded silica gel chromatography. *J. Chromatogr.*, **577**, 142-145 (1992).
Anderson,L., Cummings,J. and Smyth,J.F., Rapid and selective isolation of radiolabelled inositol phosphates from cancer cells using solid phase extraction. *J. Chromatogr.*, **574**, 150-155 (1992).
Ann,Q. and Adams,J., Structure determination of ceramides and neutral glycosphingolipids by collisional activation of $[M + Li]^+$ ions. *J. Am. Soc. Mass Spectrom.*, **3**, 260-263 (1992).
Beare-Rogers,J.L., Bonekamp-Nasner,A. and Dieffenbacher,A., Determination of the phospholipid profile of lecithins by HPLC - results of a collaborative study and the standardized method. *Pure Appl. Chem.*, **64**, 447-454 (1992).
Benfenati,E., Perico,N., Peterlongo,F., Imberti,O., Schieppati,A. and Remuzzi,G., Analysis by fast atom bombardment MS of phospholipids from tubuli, glomeruli and urine of rats with acute renal failure. *Biochem. Med. Metab. Biol.*, **48**, 219-226 (1992).
Bioque,G., Tost,D., Closa,D., Rosello-Catafu, Ramis,I., Cabrer,F., Carcigano,G. and Gelpi,E., Concurrent C_{18}-solid phase extraction of platelet activating factor (PAF) and arachidonic acid metabolites. *J. Liqu. Chromatogr.*, **15**, 1249-1258 (1992).
Bonanno,L.M., Denizot,B.A., Tchoreloff,P.C., Puisieux,F. and Cardot,P.J., Determination of phospholipids from pulmonary surfactant using an on-line coupled silica/reversed-phase HPLC system. *Anal. Chem.*, **64**, 371-379 (1992).
Butikofer,P., Zollinger,M. and Brodbeck,U., Alkylacyl glycerophosphoinositol in human and bovine erythrocytes. Molecular species composition and comparison with glycosyl-inositolphospholipid anchors of erythrocyte acetylcholinesterases. *Eur. J. Biochem.*, **208**, 677-683 (1992).
Caldwell,K.A. and Gross,M.L., Structure determination of lipids: comparison of classical methods and new approaches involving charge-remote fragmentation. *NATO ASI Ser., Ser. C.*, **353**, 413-425 (1992).

Capomacchia,A.C., Cho,J.K., Do,N.H. and Bunce,O.R., Luminol chemiluminescence liquid chromatography analysis of rat plasma phosphoglyceride hydroperoxides. *Anal. Chim. Acta*, **266**, 287-294 (1992).

Capuani,G., Aureli,T., Miccheli,A., Di Cocco,M.E., Ramacci,M.T. and Delfini,M., Improved resolution of ^{31}P NMR spectra of phospholipids. *Lipids*, **27**, 389-391 (1992).

Chen,S., Menon,G. and Traldi,P., Identification of aminophospholipid stereomers by positive-ion fast atom bombardment combined with collisional activation mass-analyzed ion kinetic energy analysis and HPLC. *Org. Mass Spectrom.*, **27**, 215-218 (1992).

Christie,W.W., Detectors for HPLC of lipids with special reference to evaporative light-scattering detection. In *Advances in Lipid Methodology - One* (ed. W.W. Christie, Oily Press, Ayr), pp. 239-271 (1992).

Curtiss,J.M., Derrick,P.J., Holgersson,J., Samuelsson,B.E. and Breimer,M.E., Electron ionization tandem MS of glycosphingolipids. I. The identification of compound-specific sequence ions in the collision-induced dissociation spectra of the ammonium ions of 2 isomeric hexaglycosylceramides. *J. Am. Soc. Mass Spectrom.*, **3**, 353-359 (1992).

Davey,M.W. and Lambein,F., Quantitative derivatization and HPLC analysis of cyanobacterial heterocyst-type glycolipids. *Anal. Biochem.*, **206**, 323-327 (1992).

Dick,D., Pluskey,S., Sukumaran,D.K. and Lawrence,D.S., NMR spectral analysis of cytotoxic ether lipids. *J. Lipid Res.*, **33**, 605-609 (1992).

Duh,J.S. and Her,G.R., Analysis of permethylated glycosphingolipids by desorption chemical ionisation/triple quadrupole tandem MS. *Biol. Mass Spectrom.*, **21**, 391-396 (1992).

Eder,K., Reichlmayr-Lais,A.M. and Kirchgessner,M., Simultaneous determination of amounts of major phospholipid classes and their fatty acid composition in erythrocyte membranes using HPLC and GC. *J. Chromatogr.*, **598**, 33-42 (1992).

Edmonds,J.S., Shibata,Y., Francesconi,K.A., Yoshinaga,J. and Morita,M., Arsenic lipids in the digestive gland of the western Rock Lobster *Panulirus cygnus* - an investigation by HPLC ICP-MS. *Sci. Total Environm.*, **122**, 321-335 (1992).

Edzes,H.T., Teerlink,T., Van der Knaap,M.S. and Valk,J., Analysis of phospholipids in brain tissue by phosphorus-31 NMR at different compositions of the solvent system chloroform-methanol-water. *Magn. Res. Med.*, **26**, 46-59 (1992).

Gage,D.A., Huang,Z.-H. and Benning,C., Comparison of sulfoquinovosyl diacylglycerol from spinach and the purple bacterium *Rhodobacter sphaeroides* by fast atom bombardment tandem MS. *Lipids*, **27**, 632-636 (1992).

Han,X., Zupan,L.A., Hazen,S.L. and Gross,R.W., Semisynthesis and purification of homogeneous plasmenylcholine molecular species. *Anal. Biochem.*, **200**, 119-124 (1992).

Hetherington,A.M. and Drobak,B.K., Inositol-containing lipids in higher plants. *Prog. Lipid Res.*, **31**, 53-63 (1992).

Hofmann,D., Herzschuh,R. and Kertscher,H.P., FAB-MS investigation of phospholipids. *Int. J. Mass Spectrom. Ion Processes*, **115**, 21-32 (1992).

Isobe,R., Higuchi,R. and Komori,T., Negative-ion fast-atom-bombardment MS of native gangliosides using a high-polar matrix system. *Carbohydr. Res.*, **233**, 231-235 (1992).

Jiang,Z.-Y., Hunt,J.V. and Wolff,S.P., Ferrous oxidation in the presence of xylenol orange for detection of lipid hydroperoxide in low density lipoprotein. *Anal. Biochem.*, **202**, 384-389 (1992).

Juhasz,P. and Costello,C.E., Matrix-assisted laser desorption ionization time-of-flight MS of underivatized and permethylated gangliosides. *J. Am. Soc. Mass Spectrom.*, **3**, 785-793 (1992).

Kamido,H., Kuksis,A., Marai,L., Myher,J.J. and Pang,H., Preparation, chromatography and mass spectrometry of cholesteryl ester and glycerolipid-bound aldehydes. *Lipids*, **27**, 645-650 (1992).

Kaufmann,P., Chemometrics in lipid analysis. In *Advances in Lipid Methodology - One* (ed. W.W. Christie, Oily Press, Ayr), pp. 149-180 (1992).

Kaya,K., Chemistry and biochemistry of taurolipids. *Prog. Lipid Res.*, **31**, 87-108 (1992).

Khan,S.M., Khan,T.M., Wells,R.D., Maslin,D.J. and Connock,M.J., A sensitive-enzyme based colorimetric assay for choline-containing phospholipids. *J. Sci. Food Agric.*, **58**, 443-445 (1992).

Kondo,Y., Miyazawa,T. and Mizutani,J., Detection and time-course analysis of phospholipid hydroperoxide in soybean seedlings after treatment with fungal elicitor, by chemiluminescence-HPLC assay. *Biochim. Biophys. Acta*, **1127**, 227-232 (1992).

Kropp,J., Ambrose,K.R., Knapp,F.F., Nissen,H.P. and Biersack,H.J., Incorporation of radioiodinated IPPA and BMIPP fatty acid analogues into complex lipids from isolated rat hearts. *Nucl. Med. Biol.*, **19**, 283-288 (1992).

Lagana,A., Marino,A., Fago,G. and Miccheli,A., Determination of free sphingosine in biological systems by HPLC. *Annali di Chimica*, **81**, 721-734 (1992).

Lendrath,G., Bonekamp,A. and Kraus,Lj. Quantitative planar chromatography of phospholipids with different fatty acid compositions. *J. Chromatogr.*, **588**, 303-305 (1992).

Letter,W.S., A rapid method for phospholipid class separation by HPLC using an evaporative light-scattering detector. *J. Liqu. Chromatogr.*, **15**, 253-266 (1992).

Levery,S.B., Nudelman,E.D. and Hakomori,S.-I., Novel modification of glycosphingolipids by long-chain cyclic acetals: isolation and characterization of plasmalocerebroside from human brain. *Biochemistry*, **31**, 5335-5340 (1992).

Marini,D., HPLC of lipids. In *Food Analysis by HPLC* (ed. L.M.L. Nollet, Marcel Dekker, New York), pp. 169-240 (1992).

McMaster,C.R. and Choy,P.C., The determination of tissue ethanolamine levels by reverse-phase HPLC. *Lipids*, **27**, 560-563 (1992).

Melton,S.L., Analysis of soybean lecithins and beef phospholipids by HPLC with an evaporative light scattering detector. *J. Am. Oil Chem. Soc.*, **69**, 784-788 (1992).

Miller-Podraza,H., Mansson,J.-E. and Svennerholm,L., Isolation of complex gangliosides from bovine brain. *Biochim. Biophys. Acta*, **1124**, 45-51 (1992).

Miyazawa,T., Suzuki,T., Fujimoto,K. and Yasuda,K., Chemiluminescent simultaneous determination of phosphatidylcholine hydroperoxide and phosphatidylethanolamine hydroperoxide in the liver and brain of the rat. *J. Lipid Res.*, **33**, 1051-1058 (1992).

Morrow,J.D., Awad,J.A., Boss,H.J., Blair,I.A. and Roberts,L.J., Non-cyclooxygenase-derived prostanoids (F_2-isoprostanes) are formed *in situ* on phospholipids. *Proc. Natl. Acad. Sci.* USA, **89**, 10721-10725 (1992).

Mounts,T.L., Abidi,S.L. and Rennick,K.A., HPLC analysis of phospholipids by evaporative laser light-scattering detector. *J. Am. Oil Chem. Soc.*, **69**, 438-442 (1992).

Muething,J. and Unland,F., Detection of gangliosides with the fluorochrome NBD dihexadecylamine and its application for preparative HP TLC. *Biomed. Chromatogr.*, **6**, 227-230 (1992).

Newburg,D.S. and Chaturvedi,P., Neutral glycolipids of human and bovine milk. *Lipids*, **27**, 923-927 (1992).

Nudelman,E.D., Levery,S.B., Igarashi,Y. and Hakomori,S.-I., Plasmalopsychosine, a novel plasmal (fatty aldehyde) conjugate of psychosine with cyclic acetal linkage. Isolation and characterization from human brain white matter. *J. Biol. Chem.*, **267**, 11007-11016 (1992).

Olsson,N.U., Advances in planar chromatography for the separation of food lipids. *J. Chromatogr.*, **624**, 11-19 (1992).

Ouhaza,M. and Siouffi,A.M., Liquid chromatographic analysis of some phospholipids with fluorescence detection. *Analusis*, **20**, 185-188 (1992).

Pak,Y. and Larner,J., Identification and characterization of chiroinositol-containing phospholipids from bovine liver. *Biochem. Biophys. Res. Commun.*, **184**, 1042-1047 (1992).

Persat,F., Bouhours,J.-F., Mojon,M. and Petavy,A.-F., Glycosphingolipids with Galβ1-6Gal sequences in metacestodes of the parasite *Echinococcus multilocularis*. *J. Biol. Chem.*, **267**, 8764-8769 (1992).

Petit,J.-M., Maftah,M., Ratinaud,M.-H. and Julien,R., 10*N*-Nonyl acridine orange interacts with cardiolipin and allows the quantification of this phospholipid in isolated mitochondria. *Eur. J. Biochem.*, **209**, 267-273 (1992).

Previti,M., Dotta,F., Pontieri,G.M., Di Mario,U. and Lenti,L., Determination of gangliosides by HPLC with photodiode-array detection. *J. Chromatogr.*, **605**, 221-225 (1992).

Ren,S., Scarsdale,J.N., Ariga,T., Zhang,Y., Klein,R.A., Hartmann,R., Yasunori,K., Egge,H. and Yu,R.K., *O*-Acetylated gangliosides in bovine butter milk. Characterisation of 7-*O*-acetyl, 9-*O*-acetyl and 7,9-di-*O*-acetyl GD_3. *J. Biol. Chem.*, **267**, 12632-12638 (1992).

Riboni,L., Acquotti,D., Casellato,R., Ghidoni,R., Montagnolo,G., Benevento,A., Zecca,L., Rubino,F. and Sonnino,S., Changes of the human liver ganglioside molecular species during ageing. *Eur. J. Biochem.*, **203**, 107-113 (1992).

Rubino,F.M., Zecca,L. and Sonnino,S., Characterization of sphingosine long-chain bases by fast atom bombardment and high-energy collision-induced decomposition tandem MS. *Org. Mass Spectrom.*, **27**, 1357-1364 (1992).

Scherer,O.W., Budzikiewicz,H., Hartmann,R., Klein,R.A. and Egge,H., The structural elucidation of the two positional isomers of a mono-glucopyranosyl monoacylglycerol derivative from *Cystobacter fuscus* (Myxobacterales). *Biochim. Biophys. Acta*, **1117**, 42-46 (1992).

Singh,A.K., Quantitative analysis of inositol lipids and inositol phosphates in synaptosomes and microvessels by column chromatography - comparison of the mass analysis and the radiolabelling methods. *J. Chromatogr.*, **581**, 1-10 (1992).

Sugita,M., Fujii,H., Suzuki,M., Hayata,C. and Hori,T., Polar glycosphingolipids in Annelida. A novel series of glycosphingolipids containing choline phosphate from the earthworm, *Pheretima hilgendorfi*. *J. Biol. Chem.*, **267**, 22595-22598 (1992).

Svennerholm,L., Bostrom,K., Fredman,P., Jungbjer,B., Mansson,J.-E. and Rynmark,B.-M., Membrane lipids of human peripheral nerve and spinal cord. *Biochim. Biophys. Acta*, **1128**, 1-7 (1992).

Tao,L. and Li,W., Rapid and sensitive anion-exchange HPLC determination of radiolabeled inositol phosphates and inositol triphosphate isomers in cellular systems. *J. Chromatogr.*, **607**, 19-24 (1992).

Tokumura,A., Yotsumoto,T., Hoshikawa,T., Tanaka,T. and Tsukatani,H., Quantitative analysis of platelet-activating factor in rat brain. *Life Sci.*, **51**, 303-308 (1992).

Unland,F. and Muthing,J., An improved method for preparation of perbenzoylated ganglioside-derived sialic acids and nanogram detection of *N*-acetylneuraminic acid and *N*-glycolylneuraminic acid by HPLC. *Biomed. Chromatogr.*, **6**, 155-159 (1992).

Vajdi,M., Rapid isolation of GM_1 and GD_1 from bovine brain gangliosides by propylamine and Q-Sepharose chromatography. *J. Liqu. Chromatogr.*, **15**, 1869-1885 (1992).

Van Bremen,R.B., Structural analysis of phosphatidylinositol from carrot cell membranes by fast atom bombardment and tandem MS. *NATO ASI Ser., Ser. C.*, 353, 443-451 (1992).

Van der Meeren,P., Vanderdeelen,J. and Baert,L., Phospholipid analysis by HPLC. In *Food Analysis by HPLC*, (ed. L.M.L. Nollet, Marcel Dekker, New York), pp. 241-258 (1992).

Van der Meeren,P., Vanderdeelen,J., Huyghebaert,G. and Baert,L., Partial resolution of molecular species during liquid chromatography of soybean phospholipids and effect on quantitation by light-scattering. *Chromatographia*, **34**, 557-562 (1992).

Wait,R., The use of FAB MS of cellular lipids for the characterization of medically important bacteria. *NATO ASI Ser., Ser. C.*, **353**, 427-441 (1992).

Wang,W.Q. and Gustafson,A., One-dimensional TLC separation of phospholipids and lysophospholipids from tissue lipid extracts. *J. Chromatogr.*, **581**, 139-142 (1992).

Watanabe,M., Kudoh,S., Yamada,Y. Iguchi,K. and Minnikin,D.E., A new glycolipid from *Mycobacteria avium-M. intracellulare* complex. *Biochim. Biophys. Acta*, **1165**, 53-60 (1992).

Yang,L.Y., Kuksis,A., Myher,J.J. and Pang,H., Surface components of chylomicrons from rats fed glyceryl or alkyl esters of fatty acid: minor components. *Lipids*, **27**, 613-618 (1992).

Yongmanitchai,W. and Ward,O.P., Separation of lipid classes from *Phaedactylum tricornutum* using silica cartridges. *Phytochem.*, **31**, 3405-3408 (1992).

Zhou,X. and Arthur,G., Improved procedure for the determination of lipid phosphorus by malachite green. *J. Lipid Res.*, **33**, 1233-1236 (1992).

H. THE ANALYSIS OF MOLECULAR SPECIES OF LIPIDS

This section corresponds to Chapter 8 in *Lipid Analysis* and deals mainly with chromatographic methods for the isolation and analysis of molecular species of lipid classes, including simple lipids, phospholipids and glycolipids. Many of the references in the next section are relevant here also and *vice versa.*

Abidi,S.L. and Mounts,T.L., HPLC separation of molecular species of neutral phospholipids. *J. Chromatogr.*, **598**, 209-218 (1992).

Abidi,S.L. and Mounts,T.L., Separation of molecular species of phosphatidylserine by

reversed-phase ion-pair HPLC. *J. Liqu. Chromatogr.*, **15**, 2487-2502 (1992).
Aitzetmuller,K. and Grondheim,M., Separation of highly unsaturated triacylglycerols by reversed phase HPLC with short wavelength UV detection. *J. High Resolut. Chromatogr.*, **15**, 219-226 (1992).
Akasaka,K., Ijichi,S., Watanabe,K., Ohrui,H. and Meguro,H., HPLC and post-column derivatization with diphenyl-1-pyrenyl phosphine for fluorometric determination of triacylglycerol hydroperoxides. *J. Chromatogr.*, **596**, 197-202 (1992).
Amari,J.V., Brown,P.R., Pivarnik,P.E., Sehgal,R.K. and Turcotte,J.G., Isolation of experimental anti-AIDS glycerophospholipids by micro-preparative reversed-phase HPLC. *J. Chromatogr.*, **590**, 153-161 (1992).
Bartle,K.D. and Clifford,T.A., Supercitical fluid extraction and chromatography of lipids and related compounds. In *Advances in Applied Lipid Research*, Vol. 1, pp. 217-264 (ed. F.B. Padley, JAI Press, London) (1992).
Bergqvist,M.H.J. and Olsson,N.U., Characterisation of honeysuckle (*Lonicera caprifolium*) seed oil triacylglycerols by HPLC and light-scattering detection. *Phytochem. Anal.*, **3**, 215-217 (1992).
Bernstrom,K., Kayganich,K., Murphy,R.C. and Fitzpatrick,F.A., Incorporation and distribution of epoxyeicosatrienoic acids into cellular phospholipids. *J. Biol. Chem.*, **267**, 3686-3690 (1992).
Booker,M.L., LaMorte,W.W., Ahrendt,S.A., Lillemoe,K.D. and Pitt,H.A., Distribution of phosphatidylcholine molecular species between mixed micelles and phospholipid-cholesterol vesicles in human gallbladder bile: dependence on acyl chain length and unsaturation. *J. Lipid Res.*, **33**, 1485-1492 (1992).
Caboni,M.F., Conte,L.S. and Lercker,G., Rapid HPLC analysis of triacylglycerols by isocratic elution and light scattering detection. *Ital. J. Food Sci.*, **4**, 125-132 (1992).
Cagniant,D., Argentation chromatography - applications to the determination of olefins, lipids, and heteroatomic compounds. In *Complexation Chromatography*, pp. 149-195 (ed. D. Cagniant, Marcel Dekker, New York) (1992).
Cantafora,A. and Masella,R., Improved determination of individual molecular species of phosphatidylcholine in biological samples by HPLC with internal standards. *J. Chromatogr.*, **593**, 139-146 (1992).
Chen,S., Curcuruto,O., Catinella,S., Traldi,P. and Menon,G., Characterization of the molecular species of glycerophospholipids from rabbit kidney: an alternative approach to the determination of fatty acyl chain position by negative ion fast atom bombardment combined with mass-analysed ion kinetic energy analysis. *Biol. Mass Spectrom.*, **21**, 655-666 (1992).
Chen,S., Menon,G. and Traldi,P., Identification of aminophospholipid stereomers by positive-ion fast atom bombardment combined with collisional activation mass-analyzed ion kinetic energy analysis and HPLC. *Org. Mass Spectrom.*, **27**, 215-218 (1992).
Christie,W.W., Detectors for HPLC of lipids with special reference to evaporative light-scattering detection. In *Advances in Lipid Methodology - One* (ed. W.W. Christie, Oily Press, Ayr), pp. 239-271 (1992).
Cserhati,C. and Szogyi,M., Anomalous retention behaviour of some synthetic phospholipids in reversed phase chromatography. *J. High Resolut. Chromatogr.*, **15**, 277-278 (1992).
Davey,M.W. and Lambein,F., Semipreparative isolation of individual cyanobacterial heterocyst-type glycolipids by reverse-phase HPLC. *Anal. Biochem.*, **206**, 226-230 (1992).
Debrauwer,L., Paris,A., Rao,D., Fournier,F. and Tabet,J.C., Mass spectrometric studies on 17β-estradiol-17-fatty acid esters - evidence for the formation of anion dipole intermediates. *Org. Mass Spectrom.*, **27**, 709-719 (1992).
Demirbuker,M. and Blomberg,L.G., Permanganate-impregnated packed capillary columns for group separation of triacylglycerols using supercritical media as mobile phases. *J. Chromatogr.*, **600**, 358-363 (1992).
Demirbuker,M., Blomberg,L.G., Olsson,N.U., Bergqvist,M., Herslof,B.G. and Jacobs,F.A., Characterization of triacylglycerols in the seeds of *Aquilegia vulgaris* by chromatographic and mass spectrometric methods. *Lipids*, **27**, 436-441 (1992).
Demirbuker,M., Hagglund,I. and Blomberg,L.G., Separation of lipids by packed fused silica capillary SFC: selectivity of some stationary phases. In *Contemporary Lipid Analysis, 2nd Symposium Proceedings*, pp. 30-47 (ed. N.U. Olsson & B.G. Herslof, LipidTeknik, Stockholm) (1992).
Evershed,R.P., Prescott,M.C. and Goad,L.J., Deuteration as an aid to the HPLC-MS of

steryl fatty acyl esters. *J. Chromatogr.*, **590**, 305-313 (1992).

Foglia,T.A. and Maeda,K., HPLC separation of enantiomeric benzyl glycerides. *Lipids*, **27**, 396-399 (1992).

Gordon,M.H. and Griffith,R.E., Steryl ester identification as an aid to the identification of oils in blends. *Food Chem.*, **43**, 71-78 (1992).

Hierro,M.T.G., Najera,A.I. and Santa-Maria,G., Analysis of triglycerides by reversed-phase HPLC with gradient elution using a light-scattering detector. *Rev. Espan. Cienc. Tecnol. Aliment.*, **32**, 635-651 (1992).

Hierro,M.T.G., Tomas,M.C., Fernandez-Martin,F. and Santa-Maria,G., Determination of the triglyceride composition of avocado oil by HPLC using a light-scattering detector. *J. Chromatogr.*, **607**, 329-338 (1992).

Holbrook,P.G., Pannell,L.K., Murata,Y. and Daly,J.W., Bis(monoacylglycero)phosphate from PC12 cells, a phospholipid that can comigrate with phosphatidic acid: molecular species analysis by fast atom bombardment MS. *Biochim. Biophys. Acta*, **1125**, 330-334 (1992).

Holbrook,P.G., Pannell,L.K., Murata,Y. and Daly,J.W., Molecular species analysis of a product of phospholipase D activation. Phosphatidylethanol is formed from phosphatidylcholine in phorbol ester- and bradykin-stimulated PC12 cells. *J. Biol. Chem.*, **267**, 16834-16840 (1992).

Huang,Z.H., Gage,D.A. and Sweeley,C.C., Characterization of diacylglycerylphosphocholine molecular species by FAB-CAD-MS/MS - a general method not sensitive to the nature of the fatty acyl groups. *J. Am. Soc. Mass Spectrom.*, **3**, 71-78 (1992).

Kallio,H., Tandem mass spectrometric analysis of triacylglycerols of low unsaturation level of Baltic herring (*Clupea harengus membras*) flesh oil. In *Contemporary Lipid Analysis, 2nd Symposium Proceedings*, pp. 48-62 (ed. N.U. Olsson & B.G. Herslof, LipidTeknik, Stockholm) (1992).

Karrer,R. and Herberg,H., Analysis of sucrose fatty acid esters by high temperature GC. *J. High Resolut. Chromatogr.*, **15**, 785-790 (1992).

Kaufmann,P., Chemometrics in lipid analysis. In *Advances in Lipid Methodology - One* (ed. W.W. Christie, Oily Press, Ayr), pp. 149-180 (1992).

Kaufmann,P. and Olsson,N.U., Molecular species separation of polar lipids, utilizing a multivariate liquid chromatography optimization strategy. In *Contemporary Lipid Analysis, 2nd Symposium Proceedings*, pp. 72-83 (ed. N.U. Olsson & B.G. Herslof, LipidTeknik, Stockholm) (1992).

Kayganich,K.A. and Murphy,R.C., Fast atom bombardment tandem MS identification of diacyl, alkylacyl, and alk-1-enylacyl molecular species of glycerophosphoethanolamine in human polymorphonuclear leukocytes. *Anal. Chem.*, **64**, 2965-2971 (1992).

Kuksis,A., Lipids. In *J. Chromatogr. Library. Vol. 51. Part B.*, pp. B171-B227 (ed. E. Heftmann, Elsevier, Amsterdam) (1992).

Kuksis,A., Marai,L. and Myher,J.J., Reversed-phase liquid chromatography-MS of complex mixtures of natural triacylglycerols with chloride attachment negative chemical ionization. *J. Chromatogr.*, **588**, 73-87 (1992).

Kuksis,A., Myher,J.J., Geher,K., Breckenridge,W.C., Feather,T., McGuire,V. and Little,J.A., GC profiles of plasma total lipids as indicators of dietary history - correlation with carbohydrate and alcohol intake based on 24-h dietary recall. *J. Chromatogr.*, **579**, 13-24 (1992).

Lie Ken Jie,M.S.F., Lam,C.C. and Yan,B.F.Y., Carbon-13 NMR studies on some synthetic saturated glycerol triesters. *J. Chem. Res. (S)*, 12-13 (1992).

Marini,D., HPLC of lipids. In *Food Analysis by HPLC*, (ed. L.M.L. Nollet, Marcel Dekker, New York), pp. 169-240 (1992).

Masterson,C., Fried,B. and Sherma,J., Comparison of mobile phases and detection reagents for the separation of triacylglycerols by silica gel, argentation and reversed-phase TLC. *J. Liqu. Chromatogr.*, **15**, 2967-2980 (1992).

McGill,A.S. and Moffat,C.F., A study of the composition of fish liver and body oil triglycerides. *Lipids*, **27**, 360-370 (1992).

Menguy,L., Christon,R., van Dorsselaer,A. and Leger,C.L., Apparent relative retention of the phosphatidylethanolamine molecular species 18:0-20:5(*n*-3), 16:0-22:6(*n*-3) and the sum 16:0-20:4(*n*-6) + 16:0-20:3(*n*-9) in the liver microsomes of pig on an EFA deficient diet. *Biochim. Biophys. Acta*, **1123**, 41-50 (1992).

Menguy,L., Christon,R., Leger,C.L. and van Dorsselaer,A., Identification by negative FAB

MS of dinitrobenzoyldiradylglycerides - application to the study of the diet-induced changes in the molecular species composition of piglet microvillous intestinal membrane. *Analusis*, **20**, 57-65 (1992).

Mounts,T.L., Abidi,S.L. and Rennick,K.A., HPLC analysis of phospholipids by evaporative laser light-scattering detector. *J. Am. Oil Chem. Soc.*, **69**, 438-442 (1992).

Neff,W.E., Selke,E., Mounts,T.L., Rinsch,W., Frankel,E.N. and Zeitoun,M.A.M., Effect of triacylglycerol composition and structures on oxidative stability of oils from selected soybean germplasm. *J. Am. Oil Chem. Soc.*, **69**, 111-118 (1992).

Nikolova-Damyanova,B., Silver ion chromatography and lipids. In *Advances in Lipid Methodology - One* (ed. W.W. Christie, Oily Press, Ayr), pp. 181-237 (1992).

Olsson,N.U. and Kaufmann,P., Optimized method for the determination of 1,2-diacyl-*sn*-glycero-3-phosphocholine and 1,2-diacyl-*sn*-glycero-3-phosphoethanolamine molecular species by enzymatic hydrolysis and GC. *J. Chromatogr.*, **600**, 257-266 (1992).

Olsson,N.U., Kaufmann,P. and Kroon,C.-G., HPLC separation of molecular species of intact sphingomyelin, utilizing multivariate design and optimization. *Chromatographia*, **34**, 529-534 (1992).

Pasha,M.K. and Ahmad,F., Analysis of triacylglycerols containing cyclopropene fatty acids in *Sterculia foetida* (Linn.) seed lipids. *J. Agric. Food Chem.*, **40**, 626-629 (1992).

Pchelkin,V.P. and Vereshchagin,A.G., Separation of polar lipid classes into their molecular species by planar and column liquid chromatography. *Adv. Chromatogr.*, **32**, 87-129 (1992).

Pchelkin,V.P. and Vereshchagin,A.G., Reversed-phase TLC of diacylglycerols in the presence of silver ions. *J. Chromatogr.*, **603**, 213-222 (1992).

Plank,C. and Lorbeer,E., Quality control of vegetable oil methyl esters used as diesel fuel substitutes - quantitative determination of monoglycerides, diglycerides and triglycerides by capillary GC. *J. High Resolut. Chromatogr.*, **15**, 609-612 (1992).

Prieto,J.A., Ebri,A. and Collar,C., Composition and distribution of individual molecular species of major glycolipids in wheat flour. *J. Am. Oil Chem. Soc.*, **69**, 1019-1022 (1992).

Rezanka,T., Analysis of sterol esters from alga and yeast by HPLC and capillary GC MS with chemical ionization. *J. Chromatogr.*, **598**, 219-226 (1992).

Robinson,B.S., Johnson,D.W. and Poulos,A., Novel molecular species of sphingomyelin containing 2-hydroxylated polyenoic very-long-chain fatty acids in mammalian testes and spermatozoa. *J. Biol. Chem.*, **267**, 1746-1751 (1992).

Sawada,T., Takahashi,K. and Hatano,M., Effect of column temperature on improvement of resolution in separating triglyceride molecular species containing highly unsaturated fatty acids by reversed-phase HPLC. *Nippon Suisan Gakkaishi*, **58**, 1313-1317 (1992).

Sempore,B.G. and Bezard,J.A., Separation of monoacylglycerols by reversed phase HPLC. *J. Chromatogr.*, **596**, 185-196 (1992).

Shansky,R.E. and Kane,R.E., Separation of soy lecithin using gel permeation chromatography. *J. Chromatogr.*, **589**, 165-170 (1992).

Su,C., Curcuruto,O., Catinella,S. and Traldi,P., Identification of phospholipid molecular species containing two fatty acyl chains differing by two daltons by negative-ion fast-atom bombardment with mass-analyzed ion kinetic energy analysis. *Rapid Commun. Mass Spectrom.*, **6**, 454-458 (1992).

Sundin,P., Larsson,P., Wesen,C. and Odham,G., Chlorinated triacylglycerols in fish lipids? Chromatographic and mass spectrometric studies of model compounds. *Biol. Mass Spectrom.*, **21**, 633-641 (1992).

Suzuki,T., Ota,T. and Takagi,T., Temperature effect on enantiomeric separation of diacylglycerol derivatives by HPLC on various chiral columns. *J. Chromatogr. Sci.*, **30**, 315-318 (1992).

Takamatsu,K., Mikami,M., Kiguchi,K., Nozawa,S. and Iwamori,M., Structural characteristics of the ceramides of neutral glycosphingolipids in the human female genital tract - their menstrual cycle-associated changes in the cervical epithelium and uterine endometrium, and their dissociation in the mucosa of the fallopian tube with the menstrual cycle. *Biochim. Biophys. Acta*, **1165**, 177-182 (1992).

Teixidor,P. and Grimalt,J.O., GC determination of isoprenoid alkylglycerol diethers in archaebacterial cultures and environmental samples. *J. Chromatogr.*, **607**, 253-259 (1992).

Turk,J., Bohrer,A., Stump,W.T., Ramanadham,S. and Mangino,M.J., Quantification of distinct molecular species of 2-lyso metabolites of platelet-activating factor by GC-negative ion chemical ionization MS. *J. Chromatogr.*, **575**, 183-196 (1992).

Van der Meeren,P., Vanderdeelen,J. and Baert,L., Phospholipid analysis by HPLC. In *Food Analysis by HPLC*, (ed. L.M.L. Nollet, Marcel Dekker, New York), pp. 241-258 (1992).

Vesterqvist,O., Sargent,C.A., Taylor,S.C., Newburger,J., Tymiak,A.A., Grover,G.J. and Ogletree,M.L., Quantitation of lysophosphatidylcholine molecular species in rat cardiac tissue. *Anal. Biochem.*, **204**, 72-78 (1992).

Wheelan,P., Zirrolli,J.A. and Clay,K.L., Analysis of glycerophosphocholine molecular species as derivatives of 7-[(chlorocarbonyl)-methoxy]-4-methylcoumarin. *J. Lipid Res.*, **33**, 111-121 (1992).

Wiley,M.G., Przetakiewicz,M., Takahashi,M. and Lowenstein,J.M., An extended method for separating and quantitating molecular species of phospholipids. *Lipids*, **27**, 295-301 (1992).

I. STRUCTURAL ANALYSIS OF LIPIDS BY MEANS OF ENZYMATIC HYDROLYSIS

This section corresponds to Chapter 9 in *Lipid Analysis* and relates to simple lipids, phospholipids and glycolipids. Many of the references in the last section are relevant here also and *vice versa*. Some methods for the resolution of chiral lipids are listed here when they deal with methods for determining positional distributions of fatty acids within lipid classes.

Allenmark,S. and Ohlsson,A., Studies of the heterogeneity of a *Candida cylindracea* (Rugosa) lipase: monitoring of esterolytic activity and enantioselectivity by chiral liquid chromatography. *Biocatalysis*, **6**, 211-221 (1992).

Ando,Y., Nishimura,K., Aoyanagi,N. and Takagi,T., Stereospecific analysis of fish oil triacyl-*sn*-glycerols. *J. Am. Oil Chem. Soc.*, **69**, 417-424 (1992).

Armand,M., Borel,P., Ythier,P., Dutot,G., Melin,C., Senft,M., Lafont,M.H. and Lairon,D., Effects of droplet size, triacylglycerol composition, and calcium on the hydrolysis of complex emulsions by pancreatic lipase - an *in vitro* study. *J. Nutr. Biochem.*, **3**, 333-341 (1992).

Christie,W.W., The chromatographic resolution of chiral lipids. In *Advances in Lipid Methodology - One* (ed. W.W. Christie, Oily Press, Ayr), pp. 121-148 (1992).

Christie,W.W., Methods for stereospecific analysis of triacyl-*sn*-glycerols. In *Contemporary Lipid Analysis, 2nd Symposium Proceedings*, pp. 63-71 (ed. N.U. Olsson & B.G. Herslof, LipidTeknik, Stockholm) (1992).

Gunstone,F.D., Structural analysis of lipids by high-resolution ^{13}C NMR spectroscopy. In *Contemporary Lipid Analysis, 2nd Symposium Proceedings*, pp. 7-22 (ed. N.U. Olsson & B.G. Herslof, LipidTeknik, Stockholm) (1992).

Marquez-Ruiz,G., Perez-Camino,M.C. and Dobarganes,M.C., *In vitro* action of pancreatic lipase on complex glycerides from thermally oxidised oils. *Fat Sci. Technol.*, **94**, 307-312 (1992).

Neff,W.E., Zeitoun,M.A.M. and Weisleder,D., Resolution of lipolysis mixtures from soybean oil by a solid-phase extraction procedure. *J. Chromatogr.*, **589**, 353-357 (1992).

Prieto,J.A., Ebri,A. and Collar,C., Composition and distribution of individual molecular species of major glycolipids in wheat flour. *J. Am. Oil Chem. Soc.*, **69**, 1019-1022 (1992).

Sacchi,R., Addeo,F., Giudicianni,I. and Paolillo,L., Analysis of the positional distribution of fatty acids in olive oil triacylglycerols by high resolution ^{13}C-NMR of the carbonyl region. *Ital. J. Food Sci.*, 4, 117-123 (1992).

Santinelli,F., Damiani,P. and Christie,W.W., The triacylglycerol structure of olive oil determined by silver ion HPLC in combination with stereospecific analysis. *J. Am. Oil Chem. Soc.*, **69**, 552-556 (1992).

Tagiri,M., Endo,Y., Fujimoto,K. and Suzuki,T., Hydrolysis of triglycerides of branched-chain fatty acids and castor oil derivatives by pancreatic lipase. *Biosci. Biotech. Biochem.*, **56**, 1490-1491 (1992).

Takagi,T. and Suzuki,T., Effect of temperature on chiral and achiral separation of diacylglycerol derivatives by HPLC on a chiral stationary phase. *J. Chromatogr.*, **625**, 163-168 (1992).

J. THE SEPARATION OF PLASMA LIPOPROTEINS

This section corresponds to Chapter 11 in *Lipid Analysis*, and a only few key papers of particular interest are listed.

Carpenter,A. and Purdy,W.C., Preparation of heparin-glyceryl controlled-pore glass affinity media for the separation of α- and β-lipoproteins. *J. Chromatogr.*, **573**, 132-135 (1992).

Carpenter,A. and Purdy,W.C., A rapid separation of α- and β-lipoproteins by affinity chromatography. *Clin. Biochem.*, **25**, 89-91 (1992).

Gerdes,L.U., Gerdes,C., Klausen,I.C. and Faergeman,O., Generation of analytic plasma lipoprotein profiles using two prepacked Superose 6B columns. *Clin. Chim. Acta*, **205**, 1-9 (1992).

K. SOME MISCELLANEOUS SEPARATIONS

Analyses of lipids such as prostaglandins, acylcarnitines, coenzyme A esters and so forth that do not fit conveniently into other sections are listed here. More complete listings for prostaglandins are available elsewhere (*Prostaglandins, Leukotrienes and Essential Fatty acids*). The decision on whether to list some papers here or in Section E was sometimes arbitrary.

Bhuiyan,A.K.H.M., Jackson,S., Turnbull,D.M., Aynsley-Green,A., Leonard,J.V. and Bartlett,K., The measurement of carnitine and acyl-carnitines - application to the investigation of patients with suspected inherited disorders of mitochondrial fatty acid oxidation. *Clin. Chim. Acta*, **207**, 185-204 (1992).

Bieber,L.L., Quantitation of CoASH and acyl-CoA. *Anal. Biochem.*, **204**, 228-230 (1992).

Bioque,G., Tost,D., Closa,D., Rosello-Catafu, Ramis,I., Cabrer,F., Carcigano,G. and Gelpi,E., Concurrent C_{18}-solid phase extraction of platelet activating factor (PAF) and arachidonic acid metabolites. *J. Liqu. Chromatogr.*, **15**, 1249-1258 (1992).

Capdevila,J.H., Wei,S., Kumar,A., Kobayashi,J., Snapper,J.R., Zeldin,D.C., Bhatt,R.K. and Falck,J.R., Resolution of dihydroxyeicosanoates and of dihydroxyeicosatrienoates by chiral phase chromatography. *Anal. Biochem.*, **207**, 236-240 (1992).

Christie,W.W., The chromatographic resolution of chiral lipids. In *Advances in Lipid Methodology - One* (ed. W.W. Christie, Oily Press, Ayr), pp. 121-148 (1992).

Cole,R.B., Domelsmith,L.N., David,C.M., Laine,R.A. and DeLucca,A.J., Californium-252 plasma desorption MS of lipid A from *Enterobacter agglomerans*. *Rapid Commun. Mass Spectrom.*, **6**, 616-622 (1992).

Deterding,L.J., Curtis,J.F. and Tomer,K.R., Tandem mass spectrometric identification of eicosanoids: leukotrienes and hydroxyeicosatetraenoic acids. *Biol. Mass Spectrom.*, **21**, 597-609 (1992).

Elinder,L.S. and Walldius,G., Simultaneous measurement of serum probucol and lipid-soluble antioxidants. *J. Lipid Res.*, **33**, 131-137 (1992).

Farquarson,J., Jamieson,E.C., Muir,J., Cockburn,F. and Logan,R.W., Direct GC assay of urinary medium-chain fatty acylcarnitines by their thermal decomposition. *Clin. Chim. Acta*, **205**, 233-240 (1992).

Hagemann,J.W. and Rothfus,J.A., Computer modelling: the adjunct microtechnique for lipids. *Food Structure*, **11**, 85-99 (1992).

Hill,E. and Murphy,R.C., Quantitation of 20-hydroxy-5,8,11,14-eicosatetraenoic acid (20-HETE) produced by human polymorphonuclear leukocytes using electron capture ionization GC/MS. *Biol. Mass Spectrom.*, **21**, 249-253 (1992).

Huang,Z.H., Gage,D.A., Bieber,L.L. and Sweeley,C.C., Analysis of acylcarnitines as their *N*-demethylated ester derivatives by GC-chemical ionization-MS: clinical applications. *Prog. Clin. Biol. Res.*, **375**, 363-368 (1992).

Karara,A., Wei,S., Spady,D., Swift,L., Capdevila,J.H. and Falck,J.R., Arachidonic acid epoxygenase: structural characterization and quantification of epoxyeicosatrienostes in plasma. *Biochem. Biophys. Res. Commun.*, **182**, 1320-1325 (1992).

Lehmann,W.D., Stephan,M. and Furstenburger,G., Profiling assay for lipoxygenase products of linoleic and arachidonic acid by GC-MS. *Anal. Biochem.*, **204**, 158-170 (1992).

Lowes,S., Rose,M.E., Mills,G. and Pollitt,R.J., Identification of urinary acylcarnitines using GC-MS: preliminary clinical applications. *J. Chromatogr.*, **577**, 205-214 (1992).

Malik,S., Ahmad,S., Talaat,R.E. and Kenny,M.A., Fast-atom-bombardment (FAB)-MS of new pentafluoropropyl esters of carnitine and its acyl derivatives. *Can. J. Appl. Spectrosc.*, **37**, 142-144 (1992).

Mangino,M.J., Zografikis,J., Murphy,M.K. and Anderson,C.B., Improved and simplified tissue extraction method for quantitating long-chain acyl coenzyme A thioesters with picomolar detection using HPLC. *J. Chromatogr.*, **577**, 157-162 (1992).

Minkler,P.E. and Hoppel,C.L., Determination of free carnitine and 'total' carnitine in human urine: derivatization with 4'-bromophenacyl trifluoromethanesulphonate and HPLC. *Clin. Chim. Acta*, **212**, 55-64 (1992).

O'Neill,C.A. and Schwartz,S.J., Chromatographic analysis of *cis/trans*-carotenoid esters. *J. Chromatogr.*, **624**, 235-252 (1992).

Riutta,A., Mucha,I. and Vapaatalo,H., Solid-phase extraction of urinary 11-dehydrothromboxane B_2 for reliable determination with radioimmunoassay. *Anal. Biochem.*, **202**, 299-305 (1992).

Rose,M.E., Isolation of acylcarnitines from urine - a comparison of methods and application to long-chain acyl-coA dehydrogenase deficiency. *Clin. Chim. Acta*, **211**, 73-81 (1992).

Rosendal,J. and Knudsen,J., A fast and versatile method for extraction and quantitation of long-chain acyl-coA esters from tissue: content of individual long-chain acyl-coA esters in various tissues from fed rat. *Anal. Biochem.*, **207**, 63-67 (1992).

Roughan,G. and Browse,J., A preparation and purification of [1-^{14}C]acetylcarnitine. *Anal. Biochem.*, **207**, 106-108 (1992).

Ruotolo,G., Zhang,H., Bentsianov,V. and Le,N., Protocol for the study of the metabolism of retinyl esters in plasma lipoproteins during postprandial lipemia. *J. Lipid Res.*, **33**, 1541-1549 (1992).

Schmidt-Sommerfeld,E., Penn,D., Duran,M., Rinaldo,P., Bennett,M.J., Sauter,R. and Stanley,C.A., Detection and quantitation of acylcarnitines in plasma and blood spots from patients with inborn errors of fatty acid oxidation. *Prog. Clin. Biol. Res.*, **375**, 355-362 (1992).

Schmidt-Sommerfeld,E., Penn,D., Rinaldo,P., Kossak,B.D., Li,B.U.K., Huang,Z.H. and Gage,D.A., Urinary medium-chain acylcarnitines in medium-chain acyl-CoA dehydrogenase deficiency, medium-chain triglyceride feeding and valproic acid therapy - sensitivity and specificity of the radioisotopic exchange/HPLC method. *Pediat. Res.*, **31**, 545-551 (1992).

Stals,H.K. and Declercq,P.E., A specific and inexpensive assay of radiolabelled long-chain acyl-coenzyme A in isolated hepatocytes. *Anal. Biochem.*, **202**, 117-119 (1992).

Tamvakopoulos,C.S. and Anderson,V.E., Detection of acyl-coenzyme A thioester intermediates of fatty acid β-oxidation as the *N*-acylglycines by negative chemical ionization GC-MS. *Anal. Biochem.*, **200**, 381-387 (1992).

Tardi,P.G., Mukherjee,J.J. and Choy,P.C., The quantitation of long-chain acyl-coA in mammalian tissues. *Lipids*, **27**, 65-67 (1992).

Van Kempen,T.A.T.G. and Odle,J., Quantification of carnitine esters by HPLC - effect of feeding medium-chain triglycerides on the plasma carnitine ester profile. *J. Chromatogr.*, **584**, 157-165 (1992).

VanAlstyne,E.L. and Spaethe,S.M., Rapid method for automated on-line extraction and fractionation of plasma leukotrienes and 12-hydroxy-5,8,10,14-eicosatetraenoic acids by reversed-phase HPLC. *J. Chromatogr.*, **579**, 37-44 (1992).

Yamane,M., Abe,A., Yamane,S. and Ishikawa,F., HPLC-thermospray MS of hydroperoxy polyunsaturated fatty acid acetyl derivatives. *J. Chromatogr.*, **579**, 25-36 (1992).

Acknowledgement

This paper is published as part of a programme funded by the Scottish Office Agriculture and Fisheries Dept.

INDEX

Contents of the Previous Volume

The first volume in this series, *Advances in Lipid Methodology - One*, published in 1992, contains the following chapters